Finance Construction-6, *Tim Asikin, Steve Asikin, Indra Senihardja*

FINANCE

Construction-6

Corporate IFRS-GAAP (B/S-I/S) Engineering

Technologies No. 5,001-5,500 of 111,111 Laws

by

Tim Asikin, Steve Asikin,

Indra Senihardja

`

0

Corporate IFRS-GAAP (B/S-I/S), ISBN 13: **978-1983566332**, ISBN 10: **1983566330**

FINANCE CONSTRUCTION-6
Corporate IFRS-GAAP (B/S-I/S) Engineering
Technology No. 5,001-5,500 of 111,111 Laws
By Tim Asikin, Steve Asikin, Indra Senihardja

Creative Team:Steve Asikin
            & Asikin Business Quant, New York
An Amazon Paperback
First Published in United States 2018
By Amazon Createspace
This 1st edition published in 2018
AMZN, PO. Box 81226, Seattle
WA 98108-1226, United States

ISBN-13: 978-1983566332
ISBN-10: 1983566330

PREFACE

This one is the only Real Finance Engineering Book, written by Engineering & Business Experts with strong Accounting Maths. It comes from a very long world wide experience, analyzing more than 100,000 Corporate Balancesheet and Income Statement reports, in conservative IFRS (International Finance Reporting Standard) and GAAP (Generally Accepted Accounting Principles) of 9,000 days works as Financing Banker, plus Business CEO expert & fresh Engineering views.

This is a manual for healthy Corporate Governance assurance dan practical business management. It is best for Government Authorities, Accounting Regulators, Business Owners & Small Public Investors (all majority & minority share holders as well). This is not just a tool for making debt instrument unrelated to daily Corporate Finance. This only good for sincere Directors & Executive, and truly bad for the lazy and /or incapable one.

The nicest thing about it, is that only requires simple Junior Highschool Algebra. We just need to be consistent with its less number treatment of constellations. Its saddest part is that many professors and doctors are no longer able to und erstand Junior Highschool Algebra, and hiding in Accounting jargons. Please go back to its Introduction for glossary, at anywhere inside the book, so then we can still understand what are those mean with simple Algebraic Rule there. We try to avoid some sentences multiplied, added, eliminated divided, rooted or powered by another sentences, that make many traditional finance book successful (delivering few simple things, seen as difficult and scary), but less useful.

Corporate IFRS-GAAP (B/S-I/S), ISBN 13: **978-1983566332**, ISBN 10: **1983566330**

TABLE OF CONTENTS

Corporate IFRS-GAAP (B/S-I/S), ISBN 13: **978-1983566332**, ISBN 10: **1983566330**

INTRODUCTION:
The Math Fin Law Terminologies

Income Statement   Balance Sheet

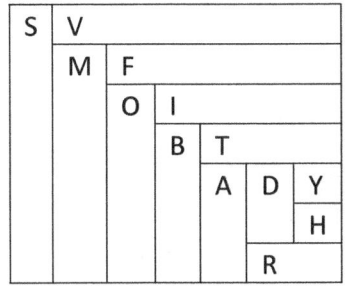

Please carefully look that all of 26 characters (**A** to **Z**) used once, except the **R**, that in Accountancy must be appeared in both diagrams at same identical value, with glossary (of well known abbreviations) in the next page. Only 2 (two) of them are historical as **U=E'** and **S'** means carried forward legal numbers from previous year.

The 13 (thirteen) common measures (**c, d, f, g, i, j, l, p, q, s, t, v,** and **y**), are put up front as desired criteria along the way, so then later analyst could be surprised.

In Balance Sheet, we found that $\mathbf{W= C+N= K+J+P+N= L+E= X+Q+U+R= G+Z+Q+U+R}$ as well, so then we put **W** in the middle of its T-Account balance.

In Income Statement, same applied as $\mathbf{S= V+F+I +T+Y+H+R= V+M= V+F+ O= V+F+I+B= V+F+I+T+A= V+F+I+B= V+F+I+T+D+R}$.

Corporate IFRS-GAAP (B/S-I/S), ISBN 13: **978-1983566332**, ISBN 10: **1983566330**

A= After Tax Income;
B= Before Tax Income,
C= Current Assets, **c**= C/X= Current Ratio,
D= Dividend Paid, **d**= D/A=Dividend Portion or Payout,
E= Equity or Capital, **E'**= Equity of Past Year,
F= Fixed Cost, **f**= F/S=Fixed Portion,
G= Goods or Trade Payables,
    g= 360G/V= Goods Payable Days,
H= Home Taken Dividend,
I= Interest Expense, **i**= I/S= Interest Portion,
J= Jobs or Account Receivable,
    j= 360J/S= Jobs or Trade Account ReceivableDays,
K= Kind of Cash,
L= Liabilities or Debt, **l**= L/E= Leverage or Gearing Ratio,
M= Margin of Contribution,
N= Non Current Assets,
O= Operational Surplus,
P= Procured Inventories,
    p= 360P/V= Procured Inventory Days,
Q= Quoted Longterm Liabilities,
    q= [C-P]/X= Quick or Acid Test Ratio,
R= Retained Earnings,
S= Sales or Revenues, **S'**= Sales of Past Year,
    s= [S/S']-1= Sales Growth,
T= Tax Paid, **t**= T/B= Tax Rate,
U= Utilized or Starting Capital,
V= Variable Cost, **v**= V/S= Variable Portion,
W= Wealth or Total Assets,
X= Xpress or Current Debt,
Y= Yielding Tax of Dividend,
    y= Y/D= Yielding Tax Rate of Dividend,
Z= Zero Trade Current Debt.

Corporate IFRS-GAAP (B/S-I/S), ISBN 13: **978-1983566332**, ISBN 10: **1983566330**

CHAPTER-15:

L= LIABUILITY or Total Debt Optimization

Based on 12/31/15 Pfizer Annual Report
(http://www.nasdaq.com/symbol/pfe/financials?query=income-
statement)

Continuing the Laws No.1-5,000
at Finance Construction 1-5

Law-5001:

If both (d= D/A= Dividend Portion or Payout=
30.00%), (i= I/S= Interest Portion= 1.00%), (s=
[S/S']-1= Sales Growth= 5.00%), (l= L/E= Leverage
or Gearing Ratio= 99.00%), (L= Liabilities or Debt=
69,547 millions USD), (f= F/S= Fixed Portion=
58.00%), (S'= Sales of Past Year= 48,851 millions
USD), (S= Sales or Revenues= 51,294 millions
USD), (v= V/S= Variable Portion= 19.00%), and (t=
T/B= Tax Rate= 30.00%), are known, then it's (U=
Utilized or Starting Capital planned) is:

$$U = L/l - [1-d][1-t]\{S'[1+s][1-i]-S[v+f]\}$$
$$= 69,547/0.9900-[1-0.3000][1-0.3000]$$
$$\{48,851[1+0.0500][1-0.0100]$$
$$-51,294[0.1900+0.5800]\}$$
$$= 64,720 \text{ millions USD}$$

Corporate IFRS-GAAP (B/S-I/S), ISBN 13: **978-1983566332**, ISBN 10: **1983566330**

Law-5002:

If both (**d**= D/A= Dividend Portion or Payout= 30.00%), (**U**= Utilized or Starting Capital= 64,720 millions USD), (**i**= I/S= Interest Portion= 1.00%), (**s**= [S/S']-1= Sales Growth= 5.00%), (**L**= Liabilities or Debt= 69,547 millions USD), (**f**= F/S= Fixed Portion= 58.00%), (**$'**= Sales of Past Year= 48,851 millions USD), (**$**= Sales or Revenues= 51,294 millions USD), (**v**= V/S= Variable Portion= 19.00%), and (**t**= T/B= Tax Rate= 30.00%), are known, then it's (**l** = L/E= Leverage or Gearing Ratio planned) is:

$$l = L/(U+[1-d][1-t]\{\$'[1+s][1-i]-\$[v+f]\})$$
$$= 69,547/(64,720+[1-0.3000]$$
$$[1-0.3000]\{48,851[1+0.0500]$$
$$[1-0.0100]-51,294$$
$$[0.1900+0.5800]\}$$
$$= 99.00\%$$

Corporate IFRS-GAAP (B/S-I/S), ISBN 13: **978-1983566332**, ISBN 10: **1983566330**

<u>Law-5003</u>:

If both (**d**= D/A= Dividend Portion or Payout= 30.00%), (**U**= Utilized or Starting Capital= 64,720 millions USD), (**I**= Interest Expense= 513 millions USD), (**s**= [S/S']-1= Sales Growth= 5.00%), (**I**= L/E= Leverage or Gearing Ratio= 99.00%), (**f**= F/S= Fixed Portion= 58.00%), (**S'**= Sales of Past Year= 48,851 millions USD), (**S**= Sales or Revenues= 51,294 millions USD), (**v**= V/S= Variable Portion= 19.00%), and (**t**= T/B= Tax Rate= 30.00%), are known, then it's (**L**= Liabilities or Debt planned) is:

$$L = I(U+[1-d][1-t]\{S'[1+s]-Sv-S'f[1+s]-I\})$$
$$= I(U+[1-d][1-t]\{S'[1+s][1-f]-Sv-I\})$$
$$= 0.9900/(64,720+[1-0.3000][1-0.3000]\{48,851[1+0.0500][1-0.5800]-51,294*0.1900-513]\})$$
$$= 69,547 \text{ millions USD}$$

Corporate IFRS-GAAP (B/S-I/S), ISBN 13: **978-1983566332**, ISBN 10: **1983566330**

<u>Law-5004</u>:

If both (**d**= D/A= Dividend Portion or Payout=
30.00%), (**U**= Utilized or Starting Capital= <u>64,720</u>
millions USD), (**I**= Interest Expense= <u>513</u> millions
USD), (**s**= [S/S']-1= Sales Growth= <u>5.00</u>%), (**L**=
Liabilities or Debt= <u>69,547</u> millions USD), (**f**= F/S=
Fixed Portion= <u>58.00</u>%), (**l**= L/E= Leverage or
Gearing Ratio= <u>99.00</u>%), (**S**= Sales or Revenues=
<u>51,294</u> millions USD), (**v**= V/S= Variable Portion=
<u>19.00</u>%), and (**t**= T/B= Tax Rate= <u>30.00</u>%), are
known, then it's (**S'**= Sales Past) must be:

$$S' = (Sv+I+L/l-U]\}/[1-d][1-t]\})/\{[1+s][1-f]\}$$
$$= (51,294*0.1900+513+[69,547$$
$$/0.9900-64,720]/\{[1-0.3000]$$
$$[1-0.3000]\})/\{[1+0.0500]$$
$$[1-0.5800]\}$$
$$= \underline{48,851} \text{ millions USD}$$

Corporate IFRS-GAAP (B/S-I/S), ISBN 13: **978-1983566332**, ISBN 10: **1983566330**

<u>Law-5005</u>:

If both (**d**= D/A= Dividend Portion or Payout= <u>30.00</u>%), (**U**= Utilized or Starting Capital= <u>64,720</u> millions USD), (**I**= Interest Expense= <u>513</u> millions USD), (**S'**= Sales of Past Year= <u>48,851</u> millions USD), (**L**= Liabilities or Debt= <u>69,547</u> millions USD), (**f**= F/S= Fixed Portion= <u>58.00</u>%), (**I**= L/E= Leverage or Gearing Ratio= <u>99.00</u>%), (**S**= Sales or Revenues= <u>51,294</u> millions USD), (**v**= V/S= Variable Portion= <u>19.00</u>%), and (**t**= T/B= Tax Rate= <u>30.00</u>%), are known, then it's (**s**= Sales Growth planned) is:

$$s = (\$v+I+L/I-U]\}/[1-d][1-t]\})/\{\$'[1-f]\}-1$$
$$= (51,294*0.1900+513+[69,547$$
$$/0.9900-64,720]/\{[1-0.3000]$$
$$[1-0.3000]\})/\{48,851$$
$$[1-0.5800]\}-1$$
$$= \underline{5.00}\%$$

Corporate IFRS-GAAP (B/S-I/S), ISBN 13: **978-1983566332**, ISBN 10: **1983566330**

Law-5006:

If both (**d**= D/A= Dividend Portion or Payout= 30.00%), (**U**= Utilized or Starting Capital= 64,720 millions USD), (**I**= Interest Expense= 513 millions USD), (**S'**= Sales of Past Year= 48,851 millions USD), (**L**= Liabilities or Debt= 69,547 millions USD), (**s**= [S/S']-1= Sales Growth= 5.00%), (**I**= L/E= Leverage or Gearing Ratio= 99.00%), (**S**= Sales or Revenues= 51,294 millions USD), (**v**= V/S= Variable Portion= 19.00%), and (**t**= T/B= Tax Rate= 30.00%), are known, then it's (**f**= Fixed Portion planned) is:

$$\mathbf{f}= 1\text{-}(\mathbf{S}\mathbf{v}\text{+}\mathbf{I}\text{+}\mathbf{L}/\mathbf{I}\text{-}\mathbf{U}]\}/[1\text{-}\mathbf{d}][1\text{-}\mathbf{t}]\})/\{\mathbf{S'}[1\text{+}\mathbf{s}]\}$$

$$= 1\text{-}(51{,}294*0.1900{+}513{+}[69{,}547$$
$$/0.9900\text{-}64{,}720]/\{[1\text{-}0.3000]$$
$$[1\text{-}0.3000]\})$$
$$/\{48{,}851[1{+}0.0500]\}$$

$$= \underline{58.00\%}$$

Corporate IFRS-GAAP (B/S-I/S), ISBN 13: **978-1983566332**, ISBN 10: **1983566330**

Law-5007:

If both (**d**= D/A= Dividend Portion or Payout= 30.00%), (**U**= Utilized or Starting Capital= 64,720 millions USD), (**I**= Interest Expense= 513 millions USD), (**S'**= Sales of Past Year= 48,851 millions USD), (**L**= Liabilities or Debt= 69,547 millions USD), (**s**= [S/S']-1= Sales Growth= 5.00%), (**I**= L/E= Leverage or Gearing Ratio= 99.00%), (**f**= F/S= Fixed Portion= 58.00%), (**v**= V/S= Variable Portion= 19.00%), and (**t**= T/B= Tax Rate= 30.00%), are known, then it's (**S** = Sales or Revenues planned) is:

$$S = (S'[1+s][1-f]-I-L/I-U]\}/[1-d][1-t]\})/v$$
$$= (48,851[1+0.0500][1-0.5800]-513$$
$$-[69,547/0.9900-64,720]$$
$$/\{[1-0.3000][1-0.3000]\})$$
$$/0.1900$$
$$= \underline{51,294} \text{ millions USD}$$

Corporate IFRS-GAAP (B/S-I/S), ISBN 13: **978-1983566332**, ISBN 10: **1983566330**

<u>Law-5008</u>:

If both (**d**= D/A= Dividend Portion or Payout= 30.00%), (**U**= Utilized or Starting Capital= 64,720 millions USD), (**I**= Interest Expense= 513 millions USD), (**S'**= Sales of Past Year= 48,851 millions USD), (**L**= Liabilities or Debt= 69,547 millions USD), (**s**= [S/S']-1= Sales Growth= 5.00%), (**I**= L/E= Leverage or Gearing Ratio= 99.00%), (**f**= F/S= Fixed Portion= 58.00%), (**S**= Sales or Revenues= 51,294 millions USD), and (**t**= T/B= Tax Rate= 30.00%), are known, then it's (**v**= Variable Portion planned) is:

$$v= (S'[1+s][1-f]-I-L/I-U]\}/[1-d][1-t]\})/S$$
$$= (48,851[1+0.0500][1-0.5800]-513$$
$$-[69,547/0.9900-64,720]$$
$$/\{[1-0.3000][1-0.3000]\})$$
$$/51,294$$
$$= \underline{19.00\%}$$

<u>Law-5009</u>:

If both (**d**= D/A= Dividend Portion or Payout= <u>30.00</u>%), (**U**= Utilized or Starting Capital= <u>64,720</u> millions USD), (**v**= V/S= Variable Portion= <u>19.00</u>%), (**$'**= Sales of Past Year= <u>48,851</u> millions USD), (**L**= Liabilities or Debt= <u>69,547</u> millions USD), (**s**= [S/S']-1= Sales Growth= <u>5.00</u>%), (**/**= L/E= Leverage or Gearing Ratio= <u>99.00</u>%), (**f**= F/S= Fixed Portion= <u>58.00</u>%), (**$**= Sales or Revenues= <u>51,294</u> millions USD), and (**t**= T/B= Tax Rate= <u>30.00</u>%), are known, then it's (**I**= Interest Expense planned) is:

$$I= (\$'[1+s][1-f]-\$v-L/\!/-U]\}/[1-d][1-t]\}$$
$$= (48{,}851[1+0.0500][1-0.5800]$$
$$- 51{,}294*0.1900-[69{,}547$$
$$/0.9900-64{,}720]/\{[1-0.3000]$$
$$[1-0.3000]\}$$
$$= \underline{513} \text{ millions USD}$$

Corporate IFRS-GAAP (B/S-I/S), ISBN 13: **978-1983566332**, ISBN 10: **1983566330**

Law-5010:

If both (**d**= D/A= Dividend Portion or Payout=
30.00%), (**U**= Utilized or Starting Capital= 64,720
millions USD), (**v**= V/S= Variable Portion= 19.00%),
(**S'**= Sales of Past Year= 48,851 millions USD), (**L**=
Liabilities or Debt= 69,547 millions USD), (**s**=
[S/S']-1= Sales Growth= 5.00%), (**l**= Leverage or
Gearing Ratio= 99.00%), (**f**= F/S= Fixed Portion=
58.00%), (**S**= Sales or Revenues= 51,294 millions
USD), and (**I**= Interest Expense= 513 millions USD),
are known, then it's (**t**= Tax Rate planned) is:

$$t= 1-[L/l - U]/([1-d]\{S'[1+s][1-f]-Sv-I\})$$
$$= 1-[69,547/0.9900-64,720]$$
$$/([1-0.3000]\{48,851$$
$$[1+0.0500][1-0.5800]$$
$$-51,294*0.1900-513\})$$
$$= 30.00\%$$

Corporate IFRS-GAAP (B/S-I/S), ISBN 13: **978-1983566332**, ISBN 10: **1983566330**

<u>Law-5011</u>:

If both (**t**= T/B= Tax Rate= <u>30.00%</u>), (**U**= Utilized or Starting Capital= <u>64,720</u> millions USD), (**v**= V/S= Variable Portion= <u>19.00%</u>), (**$'**= Sales of Past Year= <u>48,851</u> millions USD), (**L**= Liabilities or Debt= <u>69,547</u> millions USD), (**s**= [S/S']-1= Sales Growth= <u>5.00%</u>), (**l**= L/E= Leverage or Gearing Ratio= <u>99.00%</u>), (**f**= F/S= Fixed Portion= <u>58.00%</u>), (**$**= Sales or Revenues= <u>51,294</u> millions USD), and (**I**= Interest Expense= <u>513</u> millions USD), are known, then it's (**d**= Dividend Portion or Payout planned) is:

$$d = 1-[\textbf{L}/\textbf{l}-\textbf{U}]/([1-\textbf{t}]\{\textbf{\$'}[1+\textbf{s}][1-\textbf{f}]-\textbf{\$v}-\textbf{I}\})$$
$$= 1-[69,547/0.9900-64,720]/$$
$$([1-0.3000]\{48,851$$
$$[1+0.0500][1-0.5800]$$
$$-51,294*0.1900-513\})$$
$$= \underline{30.00\%}$$

Corporate IFRS-GAAP (B/S-I/S), ISBN 13: **978-1983566332**, ISBN 10: **1983566330**

<u>Law-5012</u>:

If both (**t**= T/B= Tax Rate= <u>30.00</u>%), (**d**= D/A= Dividend Portion or Payout= <u>30.00</u>%), (**v**= V/S= Variable Portion= <u>19.00</u>%), (**$'**= Sales of Past Year= <u>48,851</u> millions USD), (**L**= Liabilities or Debt= <u>69,547</u> millions USD), (**s**= [S/S']-1= Sales Growth= <u>5.00</u>%), (**l**= L/E= Leverage or Gearing Ratio= <u>99.00</u>%), (**f**= F/S= Fixed Portion= <u>58.00</u>%), (**$**= Sales or Revenues= <u>51,294</u> millions USD), and (**I**= Interest Expense= <u>513</u> millions USD), are known, then it's (**U**= Utilized or Starting Capital planned) is:

$$U= L/l -[1-d][1-t]\{\$'[1+s][1-f]-\$v-I\}$$
$$= 69{,}547/0.9900-[1-0.3000][1-0.3000]$$
$$\{48{,}851[1+0.0500][1-0.5800]$$
$$-51{,}294*0.1900-513\}$$
$$= \underline{64{,}420} \text{ millions USD}$$

Corporate IFRS-GAAP (B/S-I/S), ISBN 13: **978-1983566332**, ISBN 10: **1983566330**

Law-5013:

If both (**t**= T/B= Tax Rate= 30.00%), (**d**= D/A= Dividend Portion or Payout= 30.00%), (**v**= V/S= Variable Portion= 19.00%), (**S'**= Sales of Past Year= 48,851 millions USD), (**L**= Liabilities or Debt= 69,547 millions USD), (**s**= [S/S']-1= Sales Growth= 5.00%), (**U**= Utilized or Starting Capital= 64,720 millions USD), (**f**= F/S= Fixed Portion= 58.00%), (**S**= Sales or Revenues= 51,294 millions USD), and (**I**= Interest Expense= 513 millions USD), are known, then it's (**l** = Leverage or Gearing Capital planned) is:

$$l = L/(U+[1-d][1-t]\{S'[1+s][1-f]-Sv-I\})$$
$$= 69,547/(64,720+[1-0.3000]$$
$$[1-0.3000]\{48,851[1+0.0500]$$
$$[1-0.5800]- 51,294*0.1900$$
$$-513\})$$
$$= \underline{99.00}\%$$

Corporate IFRS-GAAP (B/S-I/S), ISBN 13: **978-1983566332**, ISBN 10: **1983566330**

<u>Law-5014</u>:

If both (**t**= T/B= Tax Rate= <u>30.00%</u>), (**d**= D/A= Dividend Portion or Payout= <u>30.00%</u>), (**v**= V/S= Variable Portion= <u>19.00%</u>), (**$'**= Sales of Past Year= <u>48,851</u> millions USD), (**F**= L/E= Leverage or Gearing Ratio= <u>99.00%</u>), (**s**= [S/S']-1= Sales Growth= <u>5.00%</u>), (**U**= Utilized or Starting Capital= <u>64,720</u> millions USD), (**f**= F/S= Fixed Portion= <u>58.00%</u>), (**$**= Sales or Revenues= <u>51,294</u> millions USD), and (**i**= I/S= Interest Portion= <u>1.00%</u>), are known, then it's (**L**= Liabilities or Debt planned) is:

$$\mathbf{L} = \mathbf{\mathit{l}}\,(\mathbf{U}+[1\text{-}\mathbf{d}][1\text{-}\mathbf{t}]\{\mathbf{\$'}[1+\mathbf{s}]\text{-}\mathbf{\$v}\text{-}\mathbf{\$'f}[1+\mathbf{s}]\text{-}\mathbf{\$i}\})$$

$$= \mathbf{\mathit{l}}\,(\mathbf{U}+[1\text{-}\mathbf{d}][1\text{-}\mathbf{t}]\{\mathbf{\$'}[1+\mathbf{s}][1\text{-}\mathbf{f}]\text{-}\mathbf{\$}[\mathbf{v}+\mathbf{i}]\})$$

$$= 0.9900/(64,720+[1\text{-}0.3000]$$
$$[1\text{-}0.3000]\{48,851[1+0.0500]$$
$$[1\text{-}0.5800]\text{-}51,294$$
$$[0.1900+0.0100]\})$$

$$= \underline{69,547} \text{ millions USD}$$

Corporate IFRS-GAAP (B/S-I/S), ISBN 13: **978-1983566332**, ISBN 10: **1983566330**

<u>Law-5015</u>:

If both (**t**= T/B= Tax Rate= <u>30.00</u>%), (**d**= D/A= Dividend Portion or Payout= <u>30.00</u>%), (**v**= V/S= Variable Portion= <u>19.00</u>%), (**L**= Liabilities or Debt= <u>69,547</u> millions USD), (**l**= L/E= Leverage or Gearing Ratio= <u>99.00</u>%), (**s**= [S/S']-1= Sales Growth= <u>5.00</u>%), (**U**= Utilized or Starting Capital= <u>64,720</u> millions USD), (**f**= F/S= Fixed Portion= <u>58.00</u>%), (**\$**= Sales or Revenues= <u>51,294</u> millions USD), and (**i**= I/S= Interest Portion= <u>1.00</u>%), are known, then it's (**\$'**= Sales Past) must be:

$$\mathbf{\$'}= (\mathbf{\$}[\mathbf{v}+\mathbf{i}]+[\mathbf{L}/\mathbf{l} -\mathbf{U}]/\{[1-\mathbf{d}][1-\mathbf{t}]\})/\{[1+\mathbf{s}][1-\mathbf{f}]\}$$

$$= (51,294[0.1900+0.0100]+[69,547$$
$$/0.9900-64,720]/[1-0.3000]$$
$$[1-0.3000]\})/\{[1+0.0500]$$
$$[1-0.5800]\}$$

$$= \underline{48,851} \text{ millions USD}$$

Corporate IFRS-GAAP (B/S-I/S), ISBN 13: **978-1983566332**, ISBN 10: **1983566330**

Law-5016:

If both (**t**= T/B= Tax Rate= 30.00%), (**d**= D/A= Dividend Portion or Payout= 30.00%), (**v**= V/S= Variable Portion= 19.00%), (**L**= Liabilities or Debt= 69,547 millions USD), (**l**= L/E= Leverage or Gearing Ratio= 99.00%), (**$'**= Sales of Past Year= 48,851 millions USD), (**U**= Utilized or Starting Capital= 64,720 millions USD), (**f**= F/S= Fixed Portion= 58.00%), (**$**= Sales or Revenues= 51,294 millions USD), and (**i**= I/S= Interest Portion= 1.00%), are known, then it's (**s**= Sales Growth planned) is:

$$s= (\$[v+i]+[L/l-U]/\{[1-d][1-t]\})/\{\$'[1-f]\}-1$$
$$= (51{,}294[0.1900+0.0100]$$
$$+[69{,}547/0.9900-64{,}720]$$
$$/[1-0.3000][1-0.3000]\})$$
$$/\{[48{,}851[1-0.5800]\}-1$$
$$= \underline{5.00\%}$$

Corporate IFRS-GAAP (B/S-I/S), ISBN 13: **978-1983566332**, ISBN 10: **1983566330**

<u>Law-5017</u>:

If both (**t**= T/B= Tax Rate= <u>30.00</u>%), (**d**= D/A= Dividend Portion or Payout= <u>30.00</u>%), (**v**= V/S= Variable Portion= <u>19.00</u>%), (**L**= Liabilities or Debt= <u>69,547</u> millions USD), (**F**= L/E= Leverage or Gearing Ratio= <u>99.00</u>%), (**S'**= Sales of Past Year= <u>48,851</u> millions USD), (**U**= Utilized or Starting Capital= <u>64,720</u> millions USD), (**s**= [S/S']-1= Sales Growth= <u>5.00</u>%), (**S**= Sales or Revenues= <u>51,294</u> millions USD), and (**i**= I/S= Interest Portion= <u>1.00</u>%), are known, then it's (**f**= Fixed Portion planned) is:

$$f = 1-(\mathbf{S}[\mathbf{v}+\mathbf{i}]+[\mathbf{L}/\mathbf{f}-\mathbf{U}]/\{[1-\mathbf{d}][1-\mathbf{t}]\})/\{\mathbf{S'}[1+\mathbf{s}]\}$$

$$= 1-(51,294[0.1900+0.0100]$$
$$+[69,547/0.9900-64,720]$$
$$/[1-0.3000][1-0.3000]\})$$
$$/\{[48,851[1+0.0500]\}$$

$$= \underline{58.00}\%$$

Corporate IFRS-GAAP (B/S-I/S), ISBN 13: **978-1983566332**, ISBN 10: **1983566330**

Law-5018:

If both (**t**= T/B= Tax Rate= 30.00%), (**d**= D/A= Dividend Portion or Payout= 30.00%), (**v**= V/S= Variable Portion= 19.00%), (**L**= Liabilities or Debt= 69,547 millions USD), (**l**= L/E= Leverage or Gearing Ratio= 99.00%), (**S'**= Sales of Past Year= 48,851 millions USD), (**U**= Utilized or Starting Capital= 64,720 millions USD), (**s**= [S/S']-1= Sales Growth= 5.00%), (**f**= F/S= Fixed Portion= 58.00%), and (**i**= I/S= Interest Portion= 1.00%), are known, then it's (**S**= Sales or Revenues planned) is:

$$S= (S'[1+s][1-f]-[L/l-U]/\{[1-d][1-t]\})/[v+i]$$
$$= (48,851[1+0.0500][1-0.5800]$$
$$-[69,547/0.9900-64,720]$$
$$/[1-0.3000][1-0.3000]\})$$
$$/[0.1900+0.0100]$$
$$= 51,294 \text{ millions USD}$$

Corporate IFRS-GAAP (B/S-I/S), ISBN 13: **978-1983566332**, ISBN 10: **1983566330**

Law-5019:

If both (**t**= T/B= Tax Rate= 30.00%), (**d**= D/A= Dividend Portion or Payout= 30.00%), (**S**= Sales or Revenues= 51,294 millions USD), (**L**= Liabilities or Debt= 69,547 millions USD), (**I**= L/E= Leverage or Gearing Ratio= 99.00%), (**S'**= Sales of Past Year= 48,851 millions USD), (**U**= Utilized or Starting Capital= 64,720 millions USD), (**s**= [S/S']-1= Sales Growth= 5.00%), (**f**= F/S= Fixed Portion= 58.00%), and (**i**= I/S= Interest Portion= 1.00%), are known, then it's (**v**= Variable Portion planned) is:

$$v= (S'[1+s][1-f]-[L/I-U]/\{[1-d][1-t]\})/S-i$$
$$= (48,851[1+0.0500][1-0.5800]$$
$$-[69,547/0.9900-64,720]$$
$$/[1-0.3000][1-0.3000]\})$$
$$/51,294-0.0100$$

$$= 19.00\%$$

Corporate IFRS-GAAP (B/S-I/S), ISBN 13: **978-1983566332**, ISBN 10: **1983566330**

<u>Law-5020</u>:

If both (**t**= T/B= Tax Rate= <u>30.00</u>%), (**d**= D/A= Dividend Portion or Payout= <u>30.00</u>%), (**$**= Sales or Revenues= <u>51,294</u> millions USD), (**L**= Liabilities or Debt= <u>69,547</u> millions USD), (**⌐**= L/E= Leverage or Gearing Ratio= <u>99.00</u>%), (**$'**= Sales of Past Year= <u>48,851</u> millions USD), (**U**= Utilized or Starting Capital= <u>64,720</u> millions USD), (**s**= [S/S']-1= Sales Growth= <u>5.00</u>%), (**f**= F/S= Fixed Portion= <u>58.00</u>%), and (**v**= V/S= Variable Portion= <u>19.00</u>%), are known, then it's (**i** = Interest Portion planned) is:

$$i= (\$'[1+s][1-f]-[L/\!\!\!/ -U]/\{[1-d][1-t]\})/\$-v$$

$$= (48,851[1+0.0500][1-0.5800]$$
$$-[69,547/0.9900-64,720]$$
$$/[1-0.3000][1-0.3000]\})$$
$$/51,294-0.1900$$

$$= \underline{1.00}\%$$

Corporate IFRS-GAAP (B/S-I/S), ISBN 13: **978-1983566332**, ISBN 10: **1983566330**

Law-5021:

If both (**i**= I/S= Interest Portion= 1.00%), (**d**= D/A=
Dividend Portion or Payout= 30.00%), (**$**= Sales or
Revenues= 51,294 millions USD), (**L**= Liabilities or
Debt= 69,547 millions USD), (**⫫**= L/E= Leverage or
Gearing Ratio= 99.00%), (**$'**= Sales of Past Year=
48,851 millions USD), (**U**= Utilized or Starting
Capital= 64,720 millions USD), (**s**= [S/S']-1= Sales
Growth= 5.00%), (**f**= F/S= Fixed Portion= 58.00%),
and (**v**= V/S= Variable Portion= 19.00%), are known,
then it's (**t**= Tax Rate planned) is:

$$t= 1-[\textbf{L}/\textbf{⫫}-\textbf{U}]/([1-\textbf{d}]\{\textbf{\$'}[1+\textbf{s}][1-\textbf{f}]-\textbf{\$}[\textbf{v}+\textbf{i}]\})$$
$$= 1-[69{,}547/0.9900-64{,}720]$$
$$/([1-0.3000]\{48{,}851$$
$$[1+0.0500][1-0.5800]$$
$$-51{,}294[0.1900+0.0100]\})$$

$$= 30.00\%$$

Corporate IFRS-GAAP (B/S-I/S), ISBN 13: **978-1983566332**, ISBN 10: **1983566330**

Law-5022:

If both (**i**= I/S= Interest Portion= 1.00%), (**t**= T/B= Tax Rate= 30.00%), (**S**= Sales or Revenues= 51,294 millions USD), (**L**= Liabilities or Debt= 69,547 millions USD), (**l**= L/E= Leverage or Gearing Ratio= 99.00%), (**S'**= Sales of Past Year= 48,851 millions USD), (**U**= Utilized or Starting Capital= 64,720 millions USD), (**s**= [S/S']-1= Sales Growth= 5.00%), (**f**= F/S= Fixed Portion= 58.00%), and (**v**= V/S= Variable Portion= 19.00%), are known, then it's (**d**= Dividend Portion or Payout planned) is:

$$\mathbf{d} = 1-[\mathbf{L/l}-\mathbf{U}]/([1-\mathbf{t}]\{\mathbf{S'}[1+\mathbf{s}][1-\mathbf{f}]-\mathbf{S}[\mathbf{v}+\mathbf{i}]\})$$

$$= 1-[69,547/0.9900-64,720]$$
$$/([1-0.3000]\{48,851$$
$$[1+0.0500][1-0.5800]$$
$$-51,294[0.1900+0.0100]\})$$

$$= 30.00\%$$

<u>Law-5023</u>:

If both (i= I/S= Interest Portion= <u>1.00%</u>), (t= T/B= Tax Rate= <u>30.00%</u>), (S= Sales or Revenues= <u>51,294</u> millions USD), (**L**= Liabilities or Debt= <u>69,547</u> millions USD), (l= L/E= Leverage or Gearing Ratio= <u>99.00%</u>), (S'= Sales of Past Year= <u>48,851</u> millions USD), (**d**= D/A= Dividend Portion or Payout= <u>30.00%</u>), (s= [S/S']-1= Sales Growth= <u>5.00%</u>), (**f**= F/S= Fixed Portion= <u>58.00%</u>), and (v= V/S= Variable Portion= <u>19.00%</u>), are known, then it's (**U**= Utilized or Starting Capital planned) is:

$$U= L/l -[1-d][1-t]\{S'[1+s][1-f]-S[v+i]\})$$
$$= 69,547/0.9900-[1-0.3000][1-0.3000]$$
$$\{48,851[1+0.0500][1-0.5800]$$
$$-51,294[0.1900+0.0100]\})$$
$$= \underline{64,720} \text{ millions USD}$$

Corporate IFRS-GAAP (B/S-I/S), ISBN 13: **978-1983566332**, ISBN 10: **1983566330**

Law-5024:

If both (**i**= I/S= Interest Portion= 1.00%), (**t**= T/B= Tax Rate= 30.00%), (**$**= Sales or Revenues= 51,294 millions USD), (**L**= Liabilities or Debt= 69,547 millions USD), (**U**= Utilized or Starting Capital= 64,720 millions USD), (**$'**= Sales of Past Year= 48,851 millions USD), (**d**= D/A= Dividend Portion or Payout= 30.00%), (**s**= [S/S']-1= Sales Growth= 5.00%), (**f**= F/S= Fixed Portion= 58.00%), and (**v**= V/S= Variable Portion= 19.00%), are known, then it's (**I** = Leverage or Gearing Ratio planned) is:

$$I = L/(U+[1-d][1-t]\{\$'[1+s][1-f]-\$[v+i]\})$$
$$= 69,547/(64,720+[1-0.3000]$$
$$[1-0.3000]\{48,851[1+0.0500]$$
$$[1-0.5800]-51,294$$
$$[0.1900+0.0100]\})$$
$$= 99.00\%$$

Corporate IFRS-GAAP (B/S-I/S), ISBN 13: 978-1983566332, ISBN 10: 1983566330

<u>Law-5025</u>:

If both (**i**= I/S= Interest Portion= <u>1.00%</u>), (**t**= T/B= Tax Rate= <u>30.00%</u>), (**$**= Sales or Revenues= <u>51,294</u> millions USD), (**f**= L/E= Leverage or Gearing Ratio= <u>99.00%</u>), (**U**= Utilized or Starting Capital= <u>64,720</u> millions USD), (**$'**= Sales of Past Year= <u>48,851</u> millions USD), (**d**= D/A= Dividend Portion or Payout= <u>30.00%</u>), (**s**= [S/S']-1= Sales Growth= <u>5.00%</u>), (**f**= F/S= Fixed Portion= <u>58.00%</u>), and (**v**= V/S= Variable Portion= <u>19.00%</u>), are known, then it's (**L**= Liabilities or Debt planned) is:

$$L = f\,(U+[1-d][1-t]\{\$'[1+s]-\$v-\$'f[1+s] \\ -\$i[1+s]\})$$

$$= f\,(U+[1-d][1-t]\{\$'[1+s][1-f-i]-\$v\})$$

$$= 0.9900\,(64{,}720+[1-0.3000][1-0.3000] \\ \{48{,}851[1+0.0500][1-0.5800 \\ -0.0100]-51{,}294*0.1900\})$$

$$= \underline{69{,}547}\ \text{millions USD}$$

Corporate IFRS-GAAP (B/S-I/S), ISBN 13: **978-1983566332**, ISBN 10: **1983566330**

<u>Law-5026</u>:

If both (**i**= I/S= Interest Portion= <u>1.00%</u>), (**t**= T/B= Tax Rate= <u>30.00%</u>), (**S**= Sales or Revenues= <u>51,294</u> millions USD), (**l**= L/E= Leverage or Gearing Ratio= <u>99.00%</u>), (**U**= Utilized or Starting Capital= <u>64,720</u> millions USD), (**L**= Liabilities or Debt= <u>69,547</u> millions USD), (**d**= D/A= Dividend Portion or Payout= <u>30.00%</u>), (**s**= [S/S']-1= Sales Growth= <u>5.00%</u>), (**f**= F/S= Fixed Portion= <u>58.00%</u>), and (**v**= V/S= Variable Portion= <u>19.00%</u>), are known, then it's (**S'**= Sales Past) must be:

$$\mathbf{S'}= (\mathbf{S}\mathbf{v}+[\mathbf{L}/\mathbf{l}\,\text{-}\mathbf{U}]/[1\text{-}\mathbf{d}][1\text{-}\mathbf{t}])/\{[1+\mathbf{s}][1\text{-}\mathbf{f}\text{-}\mathbf{i}]\}$$

$$= (51,294*0.1900+[69,547/0.9900$$
$$-64,720]/\{[1\text{-}0.3000]$$
$$[1\text{-}0.3000]\})/\{ [1+0.0500]$$
$$[1\text{-}0.5800\text{-}0.0100]\}$$

$$= \underline{48,851} \text{ millions USD}$$

Corporate IFRS-GAAP (B/S-I/S), ISBN 13: **978-1983566332**, ISBN 10: **1983566330**

Law-5027:

If both (**i**= I/S= Interest Portion= 1.00%), (**t**= T/B= Tax Rate= 30.00%), (**$**= Sales or Revenues= 51,294 millions USD), (**⌐**= L/E= Leverage or Gearing Ratio= 99.00%), (**U**= Utilized or Starting Capital= 64,720 millions USD), (**L**= Liabilities or Debt= 69,547 millions USD), (**d**= D/A= Dividend Portion or Payout= 30.00%), (**$'**= Sales of Past Year= 48,851 millions USD), (**f**= F/S=Fixed Portion= 58.00%), and (**v**= V/S= Variable Portion= 19.00%), are known, then it's (**s**= Sales Growth planned) is:

$$s = (\$v+[L/\ell-U]/[1-d][1-t]\})/\{\$'[1-f-i]\}-1$$
$$= (51,294*0.1900+[69,547/0.9900$$
$$-64,720]/\{[1-0.3000]$$
$$[1-0.3000]\})/\{48,851$$
$$[1-0.5800-0.0100]\}-1$$
$$= 5.00\%$$

Corporate IFRS-GAAP (B/S-I/S), ISBN 13: **978-1983566332**, ISBN 10: **1983566330**

Law-5028:

If both (**i**= I/S= Interest Portion= 1.00%), (**t**= T/B= Tax Rate= 30.00%), (**S**= Sales or Revenues= 51,294 millions USD), (**l**= L/E= Leverage or Gearing Ratio= 99.00%), (**U**= Utilized or Starting Capital= 64,720 millions USD), (**L**= Liabilities or Debt= 69,547 millions USD), (**d**= D/A= Dividend Portion or Payout= 30.00%), (**S'**= Sales of Past Year= 48,851 millions USD), (**s**= [S/S']-1= Sales Growth= 5.00%), and (**v**= V/S= Variable Portion= 19.00%), are known, then it's (**f**= Fixed Portion planned) is:

$$f= 1-(Sv+[L/l-U]/[1-d][1-t])/\{S'[1+s]\}-i$$
$$= 1-(51,294*0.1900+[69,547/0.9900$$
$$-64,720]/\{[1-0.3000]$$
$$[1-0.3000]\})/\{48,851$$
$$[1+0.0500]\}-0.0100$$
$$= 58.00\%$$

Corporate IFRS-GAAP (B/S-I/S), ISBN 13: **978-1983566332**, ISBN 10: **1983566330**

<u>Law-5029</u>:

If both (**f**= F/S= Fixed Portion= <u>58.00%</u>), (**t**= T/B= Tax Rate= <u>30.00%</u>), (**$**= Sales or Revenues= <u>51,294</u> millions USD), (**l**= L/E= Leverage or Gearing Ratio= <u>99.00%</u>), (**U**= Utilized or Starting Capital= <u>64,720</u> millions USD), (**L**= Liabilities or Debt= <u>69,547</u> millions USD), (**d**= D/A= Dividend Portion or Payout= <u>30.00%</u>), (**$'**= Sales of Past Year= <u>48,851</u> millions USD), (**s**= [S/S']-1= Sales Growth= <u>5.00%</u>), and (**v**= V/S= Variable Portion= <u>19.00%</u>), are known, then it's (**i**= Interest Portion planned) is:

$$\mathbf{i} = 1 - (\mathbf{\$v} + [\mathbf{L}/\mathbf{l} - \mathbf{U}] / [1 - \mathbf{d}][1 - \mathbf{t}]) / \{\mathbf{\$'}[1 + \mathbf{s}]\} - \mathbf{f}$$

$$= 1 - (51,294*0.1900 + [69,547/0.9900$$
$$-64,720]/\{[1-0.3000]$$
$$[1-0.3000]\})/\{48,851$$
$$[1+0.0500]\}-0.5800$$

$$= \underline{1.00\%}$$

Corporate IFRS-GAAP (B/S-I/S), ISBN 13: **978-1983566332**, ISBN 10: **1983566330**

<u>Law-5030</u>:

If both (f= F/S= Fixed Portion= <u>58.00%</u>), (t= T/B= Tax Rate= <u>30.00%</u>), (i= I/S= Interest Portion= <u>1.00%</u>), (l= L/E= Leverage or Gearing Ratio= <u>99.00%</u>), (U= Utilized or Starting Capital= <u>64,720</u> millions USD), (L= Liabilities or Debt= <u>69,547</u> millions USD), (d= D/A= Dividend Portion or Payout= <u>30.00%</u>), (S'= Sales of Past Year= <u>48,851</u> millions USD), (s= [S/S']-1= Sales Growth= <u>5.00%</u>), and (v= V/S= Variable Portion= <u>19.00%</u>), are known, then it's (S= Sales or Revenues planned) is:

$$S = (S'[1+s][1-f-i]-[L/l-U]/[1-d][1-t]\})/v$$
$$= (48,851[1+0.0500][1-0.5800-0.0100]$$
$$-[69,547/0.9900-64,720]$$
$$/\{[1-0.3000][1-0.3000]\})$$
$$/0.1900$$
$$= \underline{51,294} \text{ millions USD}$$

Corporate IFRS-GAAP (B/S-I/S), ISBN 13: **978-1983566332**, ISBN 10: **1983566330**

Law-5031:

If both (**f**= F/S= Fixed Portion= 58.00%), (**t**= T/B= Tax Rate= 30.00%), (**i**= I/S= Interest Portion= 1.00%), (**l**= L/E= Leverage or Gearing Ratio= 99.00%), (**U**= Utilized or Starting Capital= 64,720 millions USD), (**L**= Liabilities or Debt= 69,547 millions USD), (**d**= D/A= Dividend Portion or Payout= 30.00%), (**$'**= Sales of Past Year= 48,851 millions USD), (**s**= [S/S']-1= Sales Growth= 5.00%), and (**$**= Sales or Revenues= 51,294 millions USD), are known, then it's (**v**= Variable Portion planned) is:

$$v= (\$'[1+s][1-f-i]-[L/l-U]/[1-d][1-t]\})/v$$
$$= (48,851[1+0.0500][1-0.5800-0.0100]$$
$$-[69,547/0.9900-64,720]$$
$$/\{[1-0.3000][1-0.3000]\})$$
$$/51,294$$

$$= \underline{19.00\%}$$

Corporate IFRS-GAAP (B/S-I/S), ISBN 13: **978-1983566332**, ISBN 10: **1983566330**

<u>Law-5032</u>:

If both (**f**= F/S= Fixed Portion= <u>58.00%</u>), (**v**= V/S= Variable Portion= <u>19.00%</u>), (**i**= I/S= Interest Portion= <u>1.00%</u>), (**l**= L/E= Leverage or Gearing Ratio= <u>99.00%</u>), (**U**= Utilized or Starting Capital= <u>64,720</u> millions USD), (**L**= Liabilities or Debt= <u>69,547</u> millions USD), (**d**= D/A= Dividend Portion or Payout= <u>30.00%</u>), (**$'**= Sales of Past Year= <u>48,851</u> millions USD), (**s**= [S/S']-1= Sales Growth= <u>5.00%</u>), and (**$**= Sales or Revenues= <u>51,294</u> millions USD), are known, then it's (**t**= Tax Rate planned) is:

$$t = 1-[\mathbf{L}/\mathbf{l}-\mathbf{U}]/([1-\mathbf{d}]\{\mathbf{\$'}[1+\mathbf{s}][1-\mathbf{f}-\mathbf{i}]-\mathbf{\$v}\})$$

$$= 1-[69,547/0.9900-64,720]$$
$$/([1-0.3000]\{48,851$$
$$[1+0.0500][1-0.5800-0.0100]$$
$$-51,294*0.1900\})$$

$$= \underline{30.00\%}$$

Corporate IFRS-GAAP (B/S-I/S), ISBN 13: **978-1983566332**, ISBN 10: **1983566330**

<u>Law-5033</u>:

If both (f= F/S= Fixed Portion= <u>58.00%</u>), (v= V/S= Variable Portion= <u>19.00%</u>), (i= I/S= Interest Portion= <u>1.00%</u>), (l= L/E= Leverage or Gearing Ratio= <u>99.00%</u>), (U= Utilized or Starting Capital= <u>64,720</u> millions USD), (L= Liabilities or Debt= <u>69,547</u> millions USD), (t= T/B= Tax Rate= <u>30.00%</u>), (S'= Sales of Past Year= <u>48,851</u> millions USD), (s= [S/S']-1= Sales Growth= <u>5.00%</u>), and (S= Sales or Revenues= <u>51,294</u> millions USD), are known, then it's (d= Dividend Portion or Payout planned) is:

$$d = 1-[\mathbf{L}/\mathbf{l} - \mathbf{U}]/([1-\mathbf{t}]\{\mathbf{S'}[1+\mathbf{s}][1-\mathbf{f}-\mathbf{i}]-\mathbf{Sv}\})$$

$$= 1-[69,547/0.9900-64,720]$$
$$/([1-0.3000]\{48,851$$
$$[1+0.0500][1-0.5800-0.0100]$$
$$-51,294*0.1900\})$$

$$= \underline{30.00\%}$$

Corporate IFRS-GAAP (B/S-I/S), ISBN 13: **978-1983566332**, ISBN 10: **1983566330**

<u>Law-5034</u>:

If both (**f**= F/S= Fixed Portion= <u>58.00%</u>), (**v**= V/S= Variable Portion= <u>19.00%</u>), (**i**= I/S= Interest Portion= <u>1.00%</u>), (**l**= L/E= Leverage or Gearing Ratio= <u>99.00%</u>), (**d**= D/A= Dividend Portion or Payout= <u>30.00%</u>), (**L**= Liabilities or Debt= <u>69,547</u> millions USD), (**t**= T/B= Tax Rate= <u>30.00%</u>), (**S'**= Sales of Past Year= <u>48,851</u> millions USD), (**s**= [S/S']-1= Sales Growth= <u>5.00%</u>), and (**S**= Sales or Revenues= <u>51,294</u> millions USD), are known, then it's (**U**= Utilized or Starting Capital planned) is:

$$U= L/l -[1-d][1-t]\{S'[1+s][1-f-i]-Sv\})$$
$$= 69,547/0.9900-[1-0.3000][1-0.3000]$$
$$\{48,851[1+0.0500][1-0.5800$$
$$-0.0100]-51,294*0.1900\})$$
$$= \underline{64,720} \text{ millions USD}$$

Corporate IFRS-GAAP (B/S-I/S), ISBN 13: **978-1983566332**, ISBN 10: **1983566330**

<u>Law-5035</u>:

If both (**f**= F/S= Fixed Portion= <u>58.00%</u>), (**v**= V/S= Variable Portion= <u>19.00%</u>), (**i**= I/S= Interest Portion= <u>1.00%</u>), (**U**= Utilized or Starting Capital= <u>64,720</u> millions USD), (**d**= D/A= Dividend Portion or Payout= <u>30.00%</u>), (**L**= Liabilities or Debt= <u>69,547</u> millions USD), (**t**= T/B= Tax Rate= <u>30.00%</u>), (**$'**= Sales of Past Year= <u>48,851</u> millions USD), (**s**= [S/S']-1= Sales Growth= <u>5.00%</u>), and (**$**= Sales or Revenues= <u>51,294</u> millions USD), are known, then it's (**I** = Leverage or Gearing Ratio planned) is:

$$I = L/(U+[1-d][1-t]\{\$'[1+s][1-f-i]-\$v\})$$
$$= 69,547/(64,720+[1-0.3000]$$
$$[1-0.3000]\{48,851$$
$$[1+0.0500][1-0.5800-0.0100]$$
$$-51,294*0.1900\})$$
$$= \underline{99.00\%}$$

Corporate IFRS-GAAP (B/S-I/S), ISBN 13: **978-1983566332**, ISBN 10: **1983566330**

Law-5036:

If both (**F**= Fixed Cost= 29,740 millions USD), (**v**= V/S= Variable Portion= 19.00%), (**I**= Interest Expense= 513 millions USD), (**l**= L/E= Leverage or Gearing Ratio= 99.00%), (**d**= D/A= Dividend Portion or Payout= 30.00%), (**U**= Utilized or Starting Capital= 64,720 millions USD), (**t**= T/B= Tax Rate= 30.00%), (**s**= [S/S']-1= Sales Growth= 5.00%), and (**S'**= Sales of Past Year= 48,851 millions USD), are known, then it's (**L**= Liabilities or Debt planned) is:

$$L = l(U+[1-d][1-t]\{S'[1+s]-S'v[1+s]-F-I\})$$

$$= l(U+[1-d][1-t]\{S'[1+s][1-v]-F-I\})$$

$$= 0.9900/(64,720+[1-0.3000]$$
$$[1-0.3000]\{48,851$$
$$[1+0.0500][1-0.1900]$$
$$-29,750-513\})$$

$$= 69,547 \text{ millions USD}$$

Corporate IFRS-GAAP (B/S-I/S), ISBN 13: **978-1983566332**, ISBN 10: **1983566330**

Law-5037:

If both (**F**= Fixed Cost= <u>29,740</u> millions USD), (**v**= V/S= Variable Portion= <u>19.00%</u>), (**I**= Interest Expense= <u>513</u> millions USD), (**I**= L/E= Leverage or Gearing Ratio= <u>99.00%</u>), (**d**= D/A= Dividend Portion or Payout= <u>30.00%</u>), (**U**= Utilized or Starting Capital= <u>64,720</u> millions USD), (**t**= T/B= Tax Rate= <u>30.00%</u>), (**s**= [S/S']-1= Sales Growth= <u>5.00%</u>), and (**L**= Liabilities or Debt= <u>69,547</u> millions USD), are known, then it's (**$'** = Sales Past) must be:

$$\$'= (\mathbf{F}+\mathbf{I}+[\mathbf{L}/\mathbf{I}-\mathbf{U}]/\{[1-\mathbf{d}][1-\mathbf{t}]\})/\{[1+\mathbf{s}][1-\mathbf{v}]\}$$

$$= (29,750+513+[69,547/0.9900$$
$$-64,720]/\{[1-0.3000]$$
$$[1-0.3000]\})/\{ [1+0.0500]$$
$$[1-0.1900]\}$$

$$= \underline{48,851} \text{ millions USD}$$

Corporate IFRS-GAAP (B/S-I/S), ISBN 13: **978-1983566332**, ISBN 10: **1983566330**

<u>Law-5038</u>:

If both (**F**= Fixed Cost= <u>29,740</u> millions USD), (**v**= V/S= Variable Portion= <u>19.00%</u>), (**I**= Interest Expense= <u>513</u> millions USD), (**I**= L/E= Leverage or Gearing Ratio= <u>99.00%</u>), (**d**= D/A= Dividend Portion or Payout= <u>30.00%</u>), (**U**= Utilized or Starting Capital= <u>64,720</u> millions USD), (**t**= T/B= Tax Rate= <u>30.00%</u>), (**$'**= Sales of Past Year= <u>48,851</u> millions USD), and (**L**= Liabilities or Debt= <u>69,547</u> millions USD), are known, then it's (**s**= Sales Growth planned) is:

$$s= (F+I+[L/I-U]/\{[1-d][1-t]\})/\{\$'[1-v]\}-1$$
$$= (29,750+513+[69,547/0.9900-64,720]$$
$$/\{[1-0.3000][1-0.3000]\})$$
$$/\{48,851[1-0.1900]\}-1$$
$$= \underline{5.00\%}$$

Corporate IFRS-GAAP (B/S-I/S), ISBN 13: **978-1983566332**, ISBN 10: **1983566330**

Law-5039:

If both (**F**= Fixed Cost= 29,740 millions USD), (**s**= [S/S']-1= Sales Growth= 5.00%), (**I**= Interest Expense= 513 millions USD), (**l**= L/E= Leverage or Gearing Ratio= 99.00%), (**d**= D/A= Dividend Portion or Payout= 30.00%), (**U**= Utilized or Starting Capital= 64,720 millions USD), (**t**= T/B= Tax Rate= 30.00%), (**S'**= Sales of Past Year= 48,851 millions USD), and (**L**= Liabilities or Debt= 69,547 millions USD), are known, then it's (**v**= Variable Portion planned) is:

$$v= 1-(F+I+[L/l-U]/\{[1-d][1-t]\})/\{S'[1+s]\}$$
$$= (29,750+513+[69,547/0.9900$$
$$-64,720]/\{[1-0.3000]$$
$$[1-0.3000]\})/\{48,851$$
$$[1+0.0500]\}$$
$$= 19.00\%$$

Corporate IFRS-GAAP (B/S-I/S), ISBN 13: **978-1983566332**, ISBN 10: **1983566330**

Law-5040:

If both (v= V/S= Variable Portion= 19.00%), (s= [S/S']-1= Sales Growth= 5.00%), (I= Interest Expense= 513 millions USD), (f= L/E= Leverage or Gearing Ratio= 99.00%), (d= D/A= Dividend Portion or Payout= 30.00%), (U= Utilized or Starting Capital= 64,720 millions USD), (t= T/B= Tax Rate= 30.00%), (S'= Sales of Past Year= 48,851 millions USD), and (L= Liabilities or Debt= 69,547 millions USD), are known, then it's (F= Fixed Cost planned) is:

$$F= S'[1+s][1-v]-I-[L/f -U]/\{[1-d][1-t]\}$$
$$= 48,851[1+0.0500][1-0.1900]-513$$
$$-[69,547/0.9900-64,720]$$
$$/\{[1-0.3000][1-0.3000]\}$$
$$= 29,750 \text{ millions USD}$$

Corporate IFRS-GAAP (B/S-I/S), ISBN 13: **978-1983566332**, ISBN 10: **1983566330**

<u>Law-5041</u>:

If both (**v**= V/S= Variable Portion= <u>19.00%</u>), (**s**= [S/S']-1= Sales Growth= <u>5.00%</u>), (**F**= Fixed Cost= <u>29,740</u> millions USD), (**l**= L/E= Leverage or Gearing Ratio= <u>99.00%</u>), (**d**= D/A= Dividend Portion or Payout= <u>30.00%</u>), (**U**= Utilized or Starting Capital= <u>64,720</u> millions USD), (**t**= T/B= Tax Rate= <u>30.00%</u>), (**S'**= Sales of Past Year= <u>48,851</u> millions USD), and (**L**= Liabilities or Debt= <u>69,547</u> millions USD), are known, then it's (**I**= Interest Expense planned) is:

$$I = S'[1+s][1-v]-F-[L/I-U]/\{[1-d][1-t]\}$$
$$= 48,851[1+0.0500][1-0.1900]-29,750$$
$$-[69,547/0.9900-64,720]$$
$$/\{[1-0.3000][1-0.3000]\}$$
$$= \underline{513} \text{ millions USD}$$

Corporate IFRS-GAAP (B/S-I/S), ISBN 13: **978-1983566332**, ISBN 10: **1983566330**

<u>Law-5042</u>:

If both (v= V/S= Variable Portion= <u>19.00%</u>), (s= [S/S']-1= Sales Growth= <u>5.00%</u>), (F= Fixed Cost= <u>29,740</u> millions USD), (I= L/E= Leverage or Gearing Ratio= <u>99.00%</u>), (d= D/A= Dividend Portion or Payout= <u>30.00%</u>), (U= Utilized or Starting Capital= <u>64,720</u> millions USD), (I= Interest Expense= <u>513</u> millions USD), (S'= Sales of Past Year= <u>48,851</u> millions USD), and (L= Liabilities or Debt= <u>69,547</u> millions USD), are known, then it's (t= Tax Rate planned) is:

$$t= 1-[\mathbf{L}/\mathbf{I}-\mathbf{U}]/([1-\mathbf{d}]\{\mathbf{S'}[1+\mathbf{s}][1-\mathbf{v}]-\mathbf{F}-\mathbf{I}\})$$

$$= 1-[69,547/0.9900-64,720]$$
$$/([1-0.3000]\{48,851$$
$$[1+0.0500][1-0.1900]$$
$$-29,750-513\})$$

$$= \underline{30.00\%}$$

Corporate IFRS-GAAP (B/S-I/S), ISBN 13: **978-1983566332**, ISBN 10: **1983566330**

<u>Law-5043</u>:

If both (v= V/S= Variable Portion= <u>19.00%</u>), (s= [S/S']-1= Sales Growth= <u>5.00%</u>), (F= Fixed Cost= <u>29,740</u> millions USD), (l= L/E= Leverage or Gearing Ratio= <u>99.00%</u>), (t= T/B= Tax Rate= <u>30.00%</u>), (U= Utilized or Starting Capital= <u>64,720</u> millions USD), (I= Interest Expense= <u>513</u> millions USD), (S'= Sales of Past Year= <u>48,851</u> millions USD), and (L= Liabilities or Debt= <u>69,547</u> millions USD), are known, then it's (d= Dividend Portion or Payout planned) is:

$$d= 1-[\mathbf{L}/\mathbf{l}-\mathbf{U}]/([1-\mathbf{t}]\{\mathbf{S'}[1+\mathbf{s}][1-\mathbf{v}]-\mathbf{F}-\mathbf{I}\})$$
$$= 1-[69,547/0.9900-64,720]$$
$$/([1-0.3000]\{48,851$$
$$[1+0.0500][1-0.1900]$$
$$-29,750-513\})$$
$$= \underline{30.00\%}$$

Corporate IFRS-GAAP (B/S-I/S), ISBN 13: **978-1983566332**, ISBN 10: **1983566330**

<u>Law-5044</u>:

If both (**v**= V/S= Variable Portion= <u>19.00%</u>), (**s**= [S/S']-1= Sales Growth= <u>5.00%</u>), (**F**= Fixed Cost= <u>29,740</u> millions USD), (**I**= L/E= Leverage or Gearing Ratio= <u>99.00%</u>), (**t**= T/B= Tax Rate= <u>30.00%</u>), (**d**= D/A= Dividend Portion or Payout= <u>30.00%</u>), (**I**= Interest Expense= <u>513</u> millions USD), (**S'**= Sales of Past Year= <u>48,851</u> millions USD), and (**L**= Liabilities or Debt= <u>69,547</u> millions USD), are known, then it's (**U**= Utilized or Starting Capital planned) is:

$$U= L/I - [1-d][1-t]\{S'[1+s][1-v]-F-I\}$$
$$= 69{,}547/0.9900-[1-0.3000][1-0.3000]$$
$$\{48{,}851[1+0.0500][1-0.1900]$$
$$-29{,}750-513\}$$
$$= \underline{64{,}720} \text{ millions USD}$$

Corporate IFRS-GAAP (B/S-I/S), ISBN 13: **978-1983566332**, ISBN 10: **1983566330**

Law-5045:

If both (v= V/S= Variable Portion= 19.00%), (s= [S/S']-1= Sales Growth= 5.00%), (U= Utilized or Starting Capital= 64,720 millions USD), (F= Fixed Cost= 29,740 millions USD), (t= T/B= Tax Rate= 30.00%), (d= D/A= Dividend Portion or Payout= 30.00%), (I= Interest Expense= 513 millions USD), (S'= Sales of Past Year= 48,851 millions USD), and (L= Liabilities or Debt= 69,547 millions USD), are known, then it's (I = Leverage or Gearing Ratio planned) is:

$$I = L/ (U+[1-d][1-t]\{S'[1+s][1-v]-F-I\})$$
$$= 69,547/0.9900-[1-0.3000]$$
$$[1-0.3000]\{48,851$$
$$[1+0.0500][1-0.1900]$$
$$-29,750-513\})$$

$$= 99.00\%$$

Corporate IFRS-GAAP (B/S-I/S), ISBN 13: **978-1983566332**, ISBN 10: **1983566330**

<u>Law-5046</u>:

If both (v= V/S= Variable Portion= <u>19.00%</u>), (s= [S/S']-1= Sales Growth= <u>5.00%</u>), ($\$$= Sales or Revenues= <u>51,294</u> millions USD), (U= Utilized or Starting Capital= <u>64,720</u> millions USD), (F= Fixed Cost= <u>29,740</u> millions USD), (t= T/B= Tax Rate= <u>30.00%</u>), (d= D/A= Dividend Portion or Payout= <u>30.00%</u>), (i= I/S= Interest Portion= <u>1.00%</u>), ($\$'$= Sales of Past Year= <u>48,851</u> millions USD), and (l= L/E= Leverage or Gearing Ratio= <u>99.00%</u>), are known, then it's (L= Liabilities or Debt planned) is:

$$L = l\,(U + [1-d][1-t]\{\$'[1+s][1-v]-F-\$i\})$$
$$= 0.9900/(64,720+[1-0.3000]$$
$$[1-0.3000][1-0.3000]$$
$$\{48,851[1+0.0500]$$
$$[1-0.1900]-29,750-513\})$$
$$= \underline{69,547} \text{ million USD}$$

Corporate IFRS-GAAP (B/S-I/S), ISBN 13: **978-1983566332**, ISBN 10: **1983566330**

<u>Law-5047</u>:

If both (**v**= V/S= Variable Portion= <u>19.00%</u>), (**s**= [S/S']-1= Sales Growth= <u>5.00%</u>), (**S**= Sales or Revenues= <u>51,294</u> millions USD), (**U**= Utilized or Starting Capital= <u>64,720</u> millions USD), (**F**= Fixed Cost= <u>29,740</u> millions USD), (**L**= Liabilities or Debt= <u>69,547</u> millions USD), (**t**= T/B= Tax Rate= <u>30.00%</u>), (**d**= D/A= Dividend Portion or Payout= <u>30.00%</u>), (**i**= I/S= Interest Portion= <u>1.00%</u>),and (**l**= L/E= Leverage or Gearing Ratio= <u>99.00%</u>), are known, then it's (**S'**= Sales Past) must be:

$$S' = (F+Si+[L/l-U]/\{[1-d][1-t]\})/\{[1+s][1-v]\}$$
$$= (29,750+51,294*0.0100+[69,547$$
$$/0.9900-64,720]/\{[1-0.3000]$$
$$[1-0.3000]/\{[1+0.0500]$$
$$[1-0.1900]\}$$
$$= \underline{48,851} \text{ million USD}$$

Corporate IFRS-GAAP (B/S-I/S), ISBN 13: **978-1983566332**, ISBN 10: **1983566330**

Law-5048:

If both (**v**= V/S= Variable Portion= 19.00%), (**S**= Sales or Revenues= 51,294 millions USD), (**U**= Utilized or Starting Capital= 64,720 millions USD), (**F**= Fixed Cost= 29,740 millions USD), (**L**= Liabilities or Debt= 69,547 millions USD), (**S'**= Sales of Past Year= 48,851 millions USD), (**t**= T/B= Tax Rate= 30.00%), (**d**= D/A= Dividend Portion or Payout= 30.00%), (**i**= I/S= Interest Portion= 1.00%), and (**l**= L/E= Leverage or Gearing Ratio= 99.00%), are known, then it's (**s**= Sales Growth planned) is:

$$s= (F+Si+[L/l -U]/\{[1-d][1-t]\})/\{S'[1-v]\}-1$$

$$= (29,750+51,294*0.0100 +[69,547$$
$$/0.9900-64,720]/\{[1-0.3000]$$
$$[1-0.3000]/\{48,851$$
$$[1-0.1900]\}-1$$

$$= \underline{5.00\%}$$

Corporate IFRS-GAAP (B/S-I/S), ISBN 13: **978-1983566332**, ISBN 10: **1983566330**

<u>Law-5049</u>:

If both (**s**= [S/S']-1= Sales Growth= <u>5.00%</u>), (**S**= Sales or Revenues= <u>51,294</u> millions USD), (**U**= Utilized or Starting Capital= <u>64,720</u> millions USD), (**F**= Fixed Cost= <u>29,740</u> millions USD), (**L**= Liabilities or Debt= <u>69,547</u> millions USD), (**S'**= Sales of Past Year= <u>48,851</u> millions USD), (**t**= T/B= Tax Rate= <u>30.00%</u>), (**d**= D/A= Dividend Portion or Payout= <u>30.00%</u>), (**i**= I/S= Interest Portion= <u>1.00%</u>), and (**l**= L/E= Leverage or Gearing Ratio= <u>99.00%</u>), are known, then it's (**v**= Variable Portion planned) is:

$$v= 1-(F+Si+[L/l-U]/\{[1-d][1-t]\})/\{S'[1+s]\}$$
$$= 1-(29,750+51,294*0.0100+[69,547$$
$$/0.9900-64,720]/\{[1-0.3000]$$
$$[1-0.3000]/\{48,851$$
$$[1+0.0500]\}$$
$$= \underline{19.00\%}$$

Corporate IFRS-GAAP (B/S-I/S), ISBN 13: **978-1983566332**, ISBN 10: **1983566330**

<u>Law-5050</u>:

If both (s= [S/S']-1= Sales Growth= <u>5.00%</u>), (S= Sales or Revenues= <u>51,294</u> millions USD), (U= Utilized or Starting Capital= <u>64,720</u> millions USD), (v= V/S= Variable Portion= <u>19.00%</u>), (L= Liabilities or Debt= <u>69,547</u> millions USD), (S'= Sales of Past Year= <u>48,851</u> millions USD), (t= T/B= Tax Rate= <u>30.00%</u>), (d= D/A= Dividend Portion or Payout= <u>30.00%</u>), (i= I/S= Interest Portion= <u>1.00%</u>), and (l= L/E= Leverage or Gearing Ratio= <u>99.00%</u>), are known, then it's (F= Fixed Cost planned) is:

$$F = S'[1+s][1-v]-Si-[L/l-U]/\{[1-d][1-t]\}$$
$$= 48,851[1+0.0500][1-0.1900]$$
$$-51,294*0.0100-[69,547$$
$$/0.9900-64,720]/\{[1-0.3000]$$
$$[1-0.3000]\}$$
$$= \underline{29,750} \text{ millions USD}$$

Corporate IFRS-GAAP (B/S-I/S), ISBN 13: **978-1983566332**, ISBN 10: **1983566330**

Law-5051:

If both (**s**= [S/S']-1= Sales Growth= 5.00%), (**F**= Fixed Cost= 29,740 millions USD), (**U**= Utilized or Starting Capital= 64,720 millions USD), (**v**= V/S= Variable Portion= 19.00%), (**L**= Liabilities or Debt= 69,547 millions USD), (**S'**= Sales of Past Year= 48,851 millions USD), (**t**= T/B= Tax Rate= 30.00%), (**d**= D/A= Dividend Portion or Payout= 30.00%), (**i**= I/S= Interest Portion= 1.00%), and (**I**= L/E= Leverage or Gearing Ratio= 99.00%), are known, then it's (**S**= Sales or Revenues planned) is:

$$S= (S'[1+s][1-v]-F-[L/I-U]/\{[1-d][1-t]\})/I$$
$$= (48,851[1+0.0500][1-0.1900]$$
$$-29,750 -[69,547/0.9900$$
$$-64,720]/\{[1-0.3000]$$
$$[1-0.3000]\})/0.0100$$
$$= 51,294 \text{ millions USD}$$

Corporate IFRS-GAAP (B/S-I/S), ISBN 13: **978-1983566332**, ISBN 10: **1983566330**

<u>Law-5052</u>:

If both (s= [S/S']-1= Sales Growth= <u>5.00%</u>), (**F**= Fixed Cost= <u>29,740</u> millions USD), (**U**= Utilized or Starting Capital= <u>64,720</u> millions USD), (**v**= V/S= Variable Portion= <u>19.00%</u>), (**L**= Liabilities or Debt= <u>69,547</u> millions USD), (**$**= Sales of Past Year= <u>48,851</u> millions USD), (**t**= T/B= Tax Rate= <u>30.00%</u>), (**d**= D/A= Dividend Portion or Payout= <u>30.00%</u>), (**$**= Sales or Revenues= <u>51,294</u> millions USD),and (**l**= L/E= Leverage or Gearing Ratio= <u>99.00%</u>), are known, then it's (**i**= Interest Portion planned) is:

$$i= (\$'[1+s][1-v]-F-[L/l-U]/\{[1-d][1-t]\})/\$$$
$$= (48,851[1+0.0500][1-0.1900]$$
$$-29,750 -[69,547/0.9900$$
$$-64,720]/\{[1-0.3000]$$
$$[1-0.3000]\})/51,294$$
$$= \underline{1.00\%}$$

Corporate IFRS-GAAP (B/S-I/S), ISBN 13: **978-1983566332**, ISBN 10: **1983566330**

Law-5053:

If both (s= [S/S']-1= Sales Growth= 5.00%), (F= Fixed Cost= 29,740 millions USD), (U= Utilized or Starting Capital= 64,720 millions USD), (v= V/S= Variable Portion= 19.00%), (L= Liabilities or Debt= 69,547 millions USD), (S'= Sales of Past Year= 48,851 millions USD), (i= I/S= Interest Portion= 1.00%), (d= D/A= Dividend Portion or Payout= 30.00%), (S= Sales or Revenues= 51,294 millions USD),and (I= L/E= Leverage or Gearing Ratio= 99.00%), are known, then it's (t= Tax Rate planned) is:

$$t= 1-[L/I\text{-}U]/([1\text{-}d]\{S'[1+s][1\text{-}v]\text{-}F\text{-}Si\})$$
$$= 1 -[69,547/0.9900\text{-}64,720]$$
$$/([1\text{-}0.3000]\{48,851$$
$$[1+0.0500][1\text{-}0.1900]$$
$$-29,750\text{-}51,294*0.0100\})$$
$$= 30.00\%$$

Corporate IFRS-GAAP (B/S-I/S), ISBN 13: **978-1983566332**, ISBN 10: **1983566330**

<u>Law-5054</u>:

If both (s= [S/S']-1= Sales Growth= <u>5.00%</u>), (**F**= Fixed Cost= <u>29,740</u> millions USD), (**U**= Utilized or Starting Capital= <u>64,720</u> millions USD), (**v**= V/S= Variable Portion= <u>19.00%</u>), (**L**= Liabilities or Debt= <u>69,547</u> millions USD), (**S'**= Sales of Past Year= <u>48,851</u> millions USD), (**i**= I/S= Interest Portion= <u>1.00%</u>), (**t**= T/B= Tax Rate= <u>30.00%</u>), (**S**= Sales or Revenues= <u>51,294</u> millions USD),and (**l**= L/E= Leverage or Gearing Ratio= <u>99.00%</u>), are known, then it's (**d**= Dividend Portion or Payout planned) is:

$$d = 1-[\mathbf{L}/\mathbf{l}\ \text{-}\mathbf{U}]/([1\text{-}\mathbf{t}]\{\mathbf{S'}[1+\mathbf{s}][1\text{-}\mathbf{v}]\text{-}\mathbf{F}\text{-}\mathbf{S}i\})$$

$$= 1 -[69,547/0.9900-64,720]$$
$$/([1-0.3000]\{48,851$$
$$[1+0.0500][1-0.1900]$$
$$-29,750-51,294*0.0100\})$$

$$= \underline{30.00\%}$$

Corporate IFRS-GAAP (B/S-I/S), ISBN 13: **978-1983566332**, ISBN 10: **1983566330**

Law-5055:

If both (**s**= [S/S']-1= Sales Growth= 5.00%), (**F**= Fixed Cost= 29,740 millions USD), (**d**= D/A= Dividend Portion or Payout= 30.00%), (**v**= V/S= Variable Portion= 19.00%), (**L**= Liabilities or Debt= 69,547 millions USD), (**S'**= Sales of Past Year= 48,851 millions USD), (**i**= I/S= Interest Portion= 1.00%), (**t**= T/B= Tax Rate= 30.00%), (**S**= Sales or Revenues= 51,294 millions USD),and (**l**= L/E= Leverage or Gearing Ratio= 99.00%), are known, then it's (**U**= Utilized or Starting Capital planned) is:

$$U= L/l -[1-d][1-t]\{S'[1+s][1-v]-F-Si\}$$
$$= 69,547/0.9900-[1-0.3000]$$
$$[1-0.3000]\{48,851$$
$$[1+0.0500][1-0.1900]$$
$$-29,750-51,294*0.0100\})$$
$$= 64,720 \text{ millions USD}$$

Corporate IFRS-GAAP (B/S-I/S), ISBN 13: **978-1983566332**, ISBN 10: **1983566330**

<u>Law-5056</u>:

If both (**s**= [S/S']-1= Sales Growth= <u>5.00%</u>), (**F**= Fixed Cost= <u>29,740</u> millions USD), (**d**= D/A= Dividend Portion or Payout= <u>30.00%</u>), (**v**= V/S= Variable Portion= <u>19.00%</u>), (**L**= Liabilities or Debt= <u>69,547</u> millions USD), (**S'**= Sales of Past Year= <u>48,851</u> millions USD), (**i**= I/S= Interest Portion= <u>1.00%</u>), (**t**= T/B= Tax Rate= <u>30.00%</u>), (**S**= Sales or Revenues= <u>51,294</u> millions USD),and (**U**= Utilized or Starting Capital= <u>64,720</u> millions USD),are known, then it's (**I** = Leverage or Gearing Ratio planned) is:

$$I = \text{L}/(\text{U}+[1\text{-}\textbf{d}][1\text{-}\textbf{t}]\{\textbf{S'}[1+\textbf{s}][1\text{-}\textbf{v}]\text{-}\textbf{F}\text{-}\textbf{Si}\})$$

$$= 69,547/(64,720+[1\text{-}0.3000]$$
$$[1\text{-}0.3000]\{48,851$$
$$[1+0.0500][1\text{-}0.1900]$$
$$-29,750\text{-}51,294*0.0100\})$$

$$= \underline{99.00\%}$$

Corporate IFRS-GAAP (B/S-I/S), ISBN 13: **978-1983566332**, ISBN 10: **1983566330**

<u>Law-5057</u>:

If both (s= [S/S']-1= Sales Growth= <u>5.00%</u>), (**F**= Fixed Cost= <u>29,740</u> millions USD), (**d**= D/A= Dividend Portion or Payout= <u>30.00%</u>), (**v**= V/S= Variable Portion= <u>19.00%</u>), (**l**= L/E= Leverage or Gearing Ratio= <u>99.00%</u>), (**S'**= Sales of Past Year= <u>48,851</u> millions USD), (**i**= I/S= Interest Portion= <u>1.00%</u>), (**t**= T/B= Tax Rate= <u>30.00%</u>), and (**U**= Utilized or Starting Capital= <u>64,720</u> millions USD),are known, then it's (**L**= Liabilities or Debt planned) is:

$$\mathbf{L} = \mathit{l}\,(\mathbf{U}+[1\text{-}\mathbf{d}][1\text{-}\mathbf{t}]\{\mathbf{S'}[1+\mathbf{s}]\text{-}\mathbf{S'v}[1+\mathbf{s}]\text{-}\mathbf{F}\text{-}\mathbf{S'i}[1+\mathbf{s}]\})$$

$$= \mathit{l}\,(\mathbf{U}+[1\text{-}\mathbf{d}][1\text{-}\mathbf{t}]\{\mathbf{S'}[1+\mathbf{s}][1\text{-}\mathbf{v}\text{-}\mathbf{i}]\text{-}\mathbf{F}\})$$

$$= 0.9900/(64{,}720+[1\text{-}0.3000][1\text{-}0.3000]\{48{,}851[1+0.0500][1\text{-}0.1900\text{-}0.0100]\text{-}29{,}750\})$$

$$= \underline{69{,}547}\ \text{millions USD}$$

Corporate IFRS-GAAP (B/S-I/S), ISBN 13: **978-1983566332**, ISBN 10: **1983566330**

<u>Law-5058</u>:

If both (s= [S/S']-1= Sales Growth= <u>5.00%</u>), (**F**= Fixed Cost= <u>29,740</u> millions USD), (**d**= D/A = Dividend Portion or Payout= <u>30.00%</u>), (**v**= V/S= Variable Portion= <u>19.00%</u>), (**I**= L/E= Leverage or Gearing Ratio= <u>99.00%</u>), (**L**= Liabilities or Debt= <u>69,547</u> millions USD), (**i**= I/S= Interest Portion= <u>1.00%</u>), (**t**= T/B= Tax Rate= <u>30.00%</u>), and (**U**= Utilized or Starting Capital= <u>64,720</u> millions USD),are known, then it's (**S'**= Sales Past) must be:

$$S'= (F+[L/I-U]/\{[1-d][1-t]\})/\{[1+s][1-v-i]\}$$
$$= (29,750+[69,547/0.9900-64,720]$$
$$/\{[1-0.3000][1-0.3000]\})$$
$$/\{[1+0.0500][1-0.1900$$
$$-0.0100]\}$$
$$= \underline{48,851} \text{ millions USD}$$

Corporate IFRS-GAAP (B/S-I/S), ISBN 13: **978-1983566332**, ISBN 10: **1983566330**

Law-5059:

If both (**$'$**= Sales of Past Year= <u>48,851</u> millions USD), (**F**= Fixed Cost= <u>29,740</u> millions USD), (**d**= D/A= Dividend Portion or Payout= <u>30.00%</u>), (**v**= V/S= Variable Portion= <u>19.00%</u>), (**l**= L/E= Leverage or Gearing Ratio= <u>99.00%</u>), (**L**= Liabilities or Debt= <u>69,547</u> millions USD), (**i**= I/S= Interest Portion= <u>1.00%</u>), (**t**= T/B= Tax Rate= <u>30.00%</u>), and (**U**= Utilized or Starting Capital= <u>64,720</u> millions USD),are known, then it's (**s**= Sales Growth planned) is:

$$s = (F + [L/l - U]/\{[1-d][1-t]\})/\{\$'[1-v-i]\} - 1$$
$$= (29,750 + [69,547/0.9900 - 64,720]$$
$$/\{[1-0.3000][1-0.3000]\})$$
$$/\{48,851[1-0.1900$$
$$-0.0100]\} - 1$$
$$= \underline{5.00\%}$$

Corporate IFRS-GAAP (B/S-I/S), ISBN 13: **978-1983566332**, ISBN 10: **1983566330**

Law-5060:

If both (S'= Sales of Past Year= 48,851 millions USD), (F= Fixed Cost= 29,740 millions USD), (d= D/A= Dividend Portion or Payout= 30.00%), (s= [S/S']-1= Sales Growth= 5.00%), (I= L/E= Leverage or Gearing Ratio= 99.00%), (L= Liabilities or Debt= 69,547 millions USD), (i= I/S= Interest Portion= 1.00%), (t= T/B= Tax Rate= 30.00%), and (U= Utilized or Starting Capital= 64,720 millions USD),are known, then it's (v= Variable Portion planned) is:

$$v = 1-(F+[L/I - U]/\{[1-d][1-t]\})/\{S'[1+s]\}-I$$
$$= 1-(29,750+[69,547/0.9900$$
$$-64,720]/\{[1-0.3000]$$
$$[1-0.3000]\})/\{48,851$$
$$[1+0.0500]\}-0.0100$$
$$= 19.00\%$$

Law-5061:

If both (**S'**= Sales of Past Year= 48,851 millions USD), (**F**= Fixed Cost= 29,740 millions USD), (**d**= D/A= Dividend Portion or Payout= 30.00%), (**s**= [S/S']-1= Sales Growth= 5.00%), (**l**= L/E= Leverage or Gearing Ratio= 99.00%), (**L**= Liabilities or Debt= 69,547 millions USD), (**v**= V/S= Variable Portion= 19.00%), (**t**= T/B= Tax Rate= 30.00%), and (**U**= Utilized or Starting Capital= 64,720 millions USD),are known, then it's (**i**= Interest Portion planned) is:

$$i= 1-(F+[L/l -U]/\{[1-d][1-t]\})/\{S'[1+s]\}-v$$
$$= 1-(29,750+[69,547/0.9900$$
$$-64,720]/\{[1-0.3000]$$
$$[1-0.3000]\})/\{48,851$$
$$[1+0.0500]\}-0.0100$$
$$= 1.00\%$$

Corporate IFRS-GAAP (B/S-I/S), ISBN 13: **978-1983566332**, ISBN 10: **1983566330**

<u>Law-5062</u>:

If both ($\mathbf{\$}$'= Sales of Past Year= <u>48,851</u> millions USD), ($\mathbf{i}$= I/S= Interest Portion= <u>1.00%</u>), ($\mathbf{d}$= D/A= Dividend Portion or Payout= <u>30.00%</u>), ($\mathbf{s}$= [S/S']-1= Sales Growth= <u>5.00%</u>), ($\mathbf{l}$= L/E= Leverage or Gearing Ratio= <u>99.00%</u>), ($\mathbf{L}$= Liabilities or Debt= <u>69,547</u> millions USD), ($\mathbf{v}$= V/S= Variable Portion= <u>19.00%</u>), ($\mathbf{t}$= T/B= Tax Rate= <u>30.00%</u>), and ($\mathbf{U}$= Utilized or Starting Capital= <u>64,720</u> millions USD),are known, then it's ($\mathbf{F}$= Fixed Cost planned) is:

$$\mathbf{F}= \$'[1+\mathbf{s}][1-\mathbf{v}-\mathbf{i}]-[\mathbf{L}/\mathbf{l}-\mathbf{U}]/\{[1-\mathbf{d}][1-\mathbf{t}]\}$$
$$= 48,851[1+0.0500][1-0.1900-0.0100]$$
$$-[69,547/0.9900-64,720]$$
$$/\{[1-0.3000][1-0.3000]\}$$
$$= \underline{29,750} \text{ millions USD}$$

Corporate IFRS-GAAP (B/S-I/S), ISBN 13: **978-1983566332**, ISBN 10: **1983566330**

<u>Law-5063</u>:

If both (**$'**= Sales of Past Year= <u>48,851</u> millions USD), (**i**= I/S= Interest Portion= <u>1.00%</u>), (**d**= D/A= Dividend Portion or Payout= <u>30.00%</u>), (**s**= [S/S']-1= Sales Growth= <u>5.00%</u>), (**l**= L/E= Leverage or Gearing Ratio= <u>99.00%</u>), (**L**= Liabilities or Debt= <u>69,547</u> millions USD), (**v**= V/S= Variable Portion= <u>19.00%</u>), (**F**= Fixed Cost= <u>29,740</u> millions USD), and (**U**= Utilized or Starting Capital= <u>64,720</u> millions USD),are known, then it's (**t**= Tax Rate planned) is:

$$t= 1-[\mathbf{L/l} -\mathbf{U}]/([1-\mathbf{d}]\{\mathbf{\$'}[1+\mathbf{s}][1-\mathbf{v}-\mathbf{i}]-\mathbf{F}\})$$

$$= 1-[69,547/0.9900-64,720]$$
$$/([1-0.3000]\{48,851$$
$$[1+0.0500][1-0.1900$$
$$-0.0100]-29,750\})$$

$$= \underline{30.00\%}$$

Corporate IFRS-GAAP (B/S-I/S), ISBN 13: **978-1983566332**, ISBN 10: **1983566330**

<u>Law-5064</u>:

   If both (**$'**= Sales of Past Year= <u>48,851</u> millions
USD), (**i**= I/S= Interest Portion= <u>1.00%</u>), (**t**= T/B=
Tax Rate= <u>30.00%</u>), (**s**= [S/S']-1= Sales Growth=
<u>5.00%</u>), (**l**= L/E= Leverage or Gearing Ratio=
<u>99.00%</u>), (**L**= Liabilities or Debt= <u>69,547</u> millions
USD), (**v**= V/S= Variable Portion= <u>19.00%</u>), (**F**=
Fixed Cost= <u>29,740</u> millions USD), and (**U**= Utilized
or Starting Capital= <u>64,720</u> millions USD),are
known, then it's (**d**= Dividend Portion planned) is:

$$d= 1-[\mathbf{L}/\mathbf{l}-\mathbf{U}]/([1-\mathbf{t}]\{\mathbf{\$'}[1+\mathbf{s}][1-\mathbf{v}-\mathbf{i}]-\mathbf{F}\})$$
$$= 1-[69,547/0.9900-64,720]$$
$$/([1-0.3000]\{48,851$$
$$[1+0.0500][1-0.1900$$
$$-0.0100]-29,750\})$$
$$= \underline{30.00\%}$$

Corporate IFRS-GAAP (B/S-I/S), ISBN 13: **978-1983566332**, ISBN 10: **1983566330**

Law-5065:

If both (**S'**= Sales of Past Year= 48,851 millions USD), (**i**= I/S= Interest Portion= 1.00%), (**t**= T/B= Tax Rate= 30.00%), (**s**= [S/S']-1= Sales Growth= 5.00%), (**l**= L/E= Leverage or Gearing Ratio= 99.00%), (**L**= Liabilities or Debt= 69,547 millions USD), (**v**= V/S= Variable Portion= 19.00%), (**F**= Fixed Cost= 29,740 millions USD), and (**d**= D/A= Dividend Portion or Payout= 30.00%), then it's (**U**= Utilized or Starting Capital planned) is:

$$U= L/l -[1-d][1-t]\{S'[1+s][1-v-i]-F\})$$
$$= 69,547/0.9900-[1-0.3000]$$
$$[1-0.3000]\{48,851$$
$$[1+0.0500][1-0.1900$$
$$-0.0100]-29,750\}$$
$$= 64,720 \text{ millions USD}$$

Corporate IFRS-GAAP (B/S-I/S), ISBN 13: **978-1983566332**, ISBN 10: **1983566330**

<u>Law-5066</u>:

    If both (**S'**= Sales of Past Year= <u>48,851</u> millions USD), (**i**= I/S= Interest Portion= <u>1.00%</u>), (**t**= T/B= Tax Rate= <u>30.00%</u>), (**s**= [S/S']-1= Sales Growth= <u>5.00%</u>), (**U**= Utilized or Starting Capital= <u>64,720</u> millions USD), (**L**= Liabilities or Debt= <u>69,547</u> millions USD), (**v**= V/S= Variable Portion= <u>19.00%</u>), (**F**= Fixed Cost= <u>29,740</u> millions USD), and (**d**= D/A= Dividend Portion or Payout= <u>30.00%</u>), then it's (**I** = Leverage or Gearing Ratio planned) is:

$$I = L/(U+[1-d][1-t]\{S'[1+s][1-v-i]-F\})$$
$$= 69,547/(64,720+[1-0.3000]$$
$$[1-0.3000]\{48,851$$
$$[1+0.0500][1-0.1900$$
$$-0.0100]-29,750\})$$

$$= \underline{99.00\%}$$

Corporate IFRS-GAAP (B/S-I/S), ISBN 13: **978-1983566332**, ISBN 10: **1983566330**

Law-5067:

If both ($\textbf{S'}$= Sales of Past Year= 48,851 millions USD), ($\textbf{I}$= Interest Expense= 513 millions USD), ($\textbf{t}$= T/B= Tax Rate= 30.00%), ($\textbf{s}$= [S/S']-1= Sales Growth= 5.00%), ($\textbf{U}$= Utilized or Starting Capital= 64,720 millions USD), ($\textbf{l}$= L/E= Leverage or Gearing Ratio= 99.00%), ($\textbf{v}$= V/S= Variable Portion= 19.00%), ($\textbf{S}$= Sales or Revenues= 51,294 millions USD), ($\textbf{f}$= F/S=Fixed Portion= 58.00%), and ($\textbf{d}$= D/A= Dividend Portion or Payout= 30.00%), then it's ($\textbf{L}$= Liabilities or Debt planned) is:

$$L= \textbf{\textit{l}}(U+[1-\textbf{d}][1-\textbf{t}]\{\textbf{S'}[1+\textbf{s}]-\textbf{S'v}[1+\textbf{s}]-\textbf{Sf-I}\})$$

$$= \textbf{\textit{l}}(U+[1-\textbf{d}][1-\textbf{t}]\{\textbf{S'}[1+\textbf{s}][1-\textbf{v}]-\textbf{Sf-I}\})$$

$$= 0.9900/(64,720+[1-0.3000]$$
$$[1-0.3000]\{48,851$$
$$[1+0.0500][1-0.1900]$$
$$-51,294*0.5800-513\})$$

$$= 69,547 \text{ millions USD}$$

Corporate IFRS-GAAP (B/S-I/S), ISBN 13: **978-1983566332**, ISBN 10: **1983566330**

Law-5068:

If both (**L**= Liabilities or Debt= 69,547 millions USD), (**I**= Interest Expense= 513 millions USD), (**t**= T/B= Tax Rate= 30.00%), (**s**= [S/S']-1= Sales Growth= 5.00%), (**U**= Utilized or Starting Capital= 64,720 millions USD), (**I**= L/E= Leverage or Gearing Ratio= 99.00%), (**v**= V/S= Variable Portion= 19.00%), (**S**= Sales or Revenues= 51,294 millions USD), (**f**= F/S= Fixed Portion= 58.00%), and (**d**= D/A= Dividend Portion or Payout= 30.00%), then it's (**S'**= Sales Past) must be:

$$\textbf{S'} = (\textbf{Sf} + \textbf{I} + [\textbf{L}/\textbf{I} - \textbf{U}]/\{[1-\textbf{d}][1-\textbf{t}]\})/\{[1+\textbf{s}][1-\textbf{v}]\}$$

$$= (51,294*0.5800 + 513 + [69,547$$
$$/0.9900 - 64,720]/\{[1-0.3000]$$
$$[1-0.3000]\})/\{[1+0.0500]$$
$$[1-0.1900]\}$$

$$= 48,851 \text{ millions USD}$$

Corporate IFRS-GAAP (B/S-I/S), ISBN 13: **978-1983566332**, ISBN 10: **1983566330**

Law-5069:

If both (**L**= Liabilities or Debt= 69,547 millions USD), (**I**= Interest Expense= 513 millions USD), (**t**= T/B= Tax Rate= 30.00%), (**S'**= Sales of Past Year= 48,851 millions USD), (**U**= Utilized or Starting Capital= 64,720 millions USD), (**I**= L/E= Leverage or Gearing Ratio= 99.00%), (**v**= V/S= Variable Portion= 19.00%), (**S**= Sales or Revenues= 51,294 millions USD), (**f**= F/S= Fixed Portion= 58.00%), and (**d**= D/A= Dividend Portion or Payout= 30.00%), then it's (**s**= Sales Growth planned) is:

$$s = (Sf+I+[L/I-U]/\{[1-d][1-t]\})/\{S'[1-v]\}-1$$
$$= (51{,}294*0.5800+513+[69{,}547$$
$$/0.9900-64{,}720]/\{[1-0.3000]$$
$$[1-0.3000]\})/\{48{,}851$$
$$[1-0.1900]\}-1$$
$$= 5.00\%$$

Corporate IFRS-GAAP (B/S-I/S), ISBN 13: **978-1983566332**, ISBN 10: **1983566330**

Law-5070:

If both (**L**= Liabilities or Debt= 69,547 millions USD), (**I**= Interest Expense= 513 millions USD), (**t**= T/B= Tax Rate= 30.00%), (**S'**= Sales of Past Year= 48,851 millions USD), (**U**= Utilized or Starting Capital= 64,720 millions USD), (**l**= L/E= Leverage or Gearing Ratio= 99.00%), (**s**= [S/S']-1= Sales Growth= 5.00%), (**S**= Sales or Revenues= 51,294 millions USD), (**f**= F/S= Fixed Portion= 58.00%), and (**d**= D/A= Dividend Portion or Payout= 30.00%), then it's (**v**= Variable Portion planned) is:

$$v = 1-(\mathbf{S}\mathbf{f}+\mathbf{I}+[\mathbf{L}/\mathbf{l}-\mathbf{U}]/\{[1-\mathbf{d}][1-\mathbf{t}]\})/\{\mathbf{S'}[1+\mathbf{s}]\}$$
$$= 1-(51,294*0.5800+513+[69,547/0.9900-64,720]/\{[1-0.3000][1-0.3000]\})/\{48,851[1+0.0500]\}$$
$$= 19.00\%$$

Corporate IFRS-GAAP (B/S-I/S), ISBN 13: 978-1983566332, ISBN 10: 1983566330

<u>Law-5071</u>:

If both (**L**= Liabilities or Debt= <u>69,547</u> millions USD), (**I**= Interest Expense= <u>513</u> millions USD), (**t**= T/B= Tax Rate= <u>30.00%</u>), (**$'**= Sales of Past Year= <u>48,851</u> millions USD), (**U**= Utilized or Starting Capital= <u>64,720</u> millions USD), (**I**= L/E= Leverage or Gearing Ratio= <u>99.00%</u>), (**s**= [S/S']-1= Sales Growth= <u>5.00%</u>), (**v**= V/S= Variable Portion= <u>19.00%</u>), (**f**= F/S= Fixed Portion= <u>58.00%</u>), and (**d**= D/A= Dividend Portion or Payout= <u>30.00%</u>), then it's (**$**= Sales or Revenues planned) is:

$$\$= (\$'[1+s][1-v]-I-[L/I-U]/\{[1-d][1-t]\})/f$$
$$= (48,851[1+0.0500][1-0.1900]-513$$
$$-[69,547/0.9900-64,720]$$
$$/\{[1-0.3000][1-0.3000]\})$$
$$/0.5800$$
$$= \underline{51,294} \text{ millions USD}$$

Corporate IFRS-GAAP (B/S-I/S), ISBN 13: **978-1983566332**, ISBN 10: **1983566330**

<u>Law-5072</u>:

If both (**L**= Liabilities or Debt= <u>69,547</u> millions USD), (**I**= Interest Expense= <u>513</u> millions USD), (**t**= T/B= Tax Rate= <u>30.00%</u>), (**$'**= Sales of Past Year= <u>48,851</u> millions USD), (**U**= Utilized or Starting Capital= <u>64,720</u> millions USD), (**I**= L/E= Leverage or Gearing Ratio= <u>99.00%</u>), (**s**= [S/S']-1= Sales Growth= <u>5.00%</u>), (**v**= V/S= Variable Portion= <u>19.00%</u>), (**$**= Sales or Revenues= <u>51,294</u> millions USD),and (**d**= D/A= Dividend Portion or Payout= <u>30.00%</u>), then it's (**f**= Fixed Portion planned) is:

$$\mathbf{f} = (\mathbf{\$'}[1+\mathbf{s}][1-\mathbf{v}]-\mathbf{I}-[\mathbf{L}/\mathbf{I}-\mathbf{U}]/\{[1-\mathbf{d}][1-\mathbf{t}]\})/\mathbf{\$}$$

$$= (48,851[1+0.0500][1-0.1900]-513$$
$$-[69,547/0.9900-64,720]$$
$$/\{[1-0.3000][1-0.3000]\})$$
$$/51,294$$

$$= \underline{58.00\%}$$

Corporate IFRS-GAAP (B/S-I/S), ISBN 13: **978-1983566332**, ISBN 10: **1983566330**

Law-5073:

If both (**L**= Liabilities or Debt= 69,547 millions USD), (**f**= F/S= Fixed Portion= 58.00%), (**t**= T/B= Tax Rate= 30.00%), (**$'**= Sales of Past Year= 48,851 millions USD), (**U**= Utilized or Starting Capital= 64,720 millions USD), (**l**= L/E= Leverage or Gearing Ratio= 99.00%), (**s**= [S/S']-1= Sales Growth= 5.00%), (**v**= V/S= Variable Portion= 19.00%), (**$**= Sales or Revenues= 51,294 millions USD),and (**d**= D/A= Dividend Portion or Payout= 30.00%), then it's (**I**= Interest Expense planned) is:

$$I= S'[1+s][1-v]-Sf-[L/l-U]/\{[1-d][1-t]\}$$
$$= 48,851[1+0.0500][1-0.1900]$$
$$-51,294*0.5800-[69,547$$
$$/0.9900-64,720]/\{[1-0.3000]$$
$$[1-0.3000]\}$$
$$= 513 \text{ millions USD}$$

Corporate IFRS-GAAP (B/S-I/S), ISBN 13: **978-1983566332**, ISBN 10: **1983566330**

Law-5074:

If both (**L**= Liabilities or Debt= 69,547 millions USD), (**f**= F/S= Fixed Portion= 58.00%), (**I**= Interest Expense= 513 millions USD), (**S'**= Sales of Past Year= 48,851 millions USD), (**U**= Utilized or Starting Capital= 64,720 millions USD), (**l**= L/E= Leverage or Gearing Ratio= 99.00%), (**s**= [S/S']-1= Sales Growth= 5.00%), (**v**= V/S= Variable Portion= 19.00%), (**S**= Sales or Revenues= 51,294 millions USD),and (**d**= D/A= Dividend Portion or Payout= 30.00%), then it's (**t**= Tax Rate planned) is:

$$\mathbf{t}= 1-[\mathbf{L}/\mathbf{l}\,-\mathbf{U}]/([1-\mathbf{d}]\{\mathbf{S'}[1+\mathbf{s}][1-\mathbf{v}]-\mathbf{Sf}-\mathbf{I}\})$$

$$= 1-[69{,}547/0.9900-64{,}720]$$
$$/([1-0.3000]\{48{,}851$$
$$[1+0.0500][1-0.1900]$$
$$-51{,}294*0.5800-513\})$$

$$= \underline{30.00\%}$$

Corporate IFRS-GAAP (B/S-I/S), ISBN 13: **978-1983566332**, ISBN 10: **1983566330**

Law-5075:

If both (**L**= Liabilities or Debt= 69,547 millions USD), (**f**= F/S= Fixed Portion= 58.00%), (**I**= Interest Expense= 513 millions USD), (**S'**= Sales of Past Year= 48,851 millions USD), (**U**= Utilized or Starting Capital= 64,720 millions USD), (**I**= L/E= Leverage or Gearing Ratio= 99.00%), (**s**= [S/S']-1= Sales Growth= 5.00%), (**v**= V/S= Variable Portion= 19.00%), (**S**= Sales or Revenues= 51,294 millions USD),and (**t**= T/B= Tax Rate= 30.00%), then it's (**d**= Dividend Portion or Payout planned) is:

$$d= 1-[\mathbf{L}/\mathbf{I}-\mathbf{U}]/([1-\mathbf{t}]\{\mathbf{S'}[1+\mathbf{s}][1-\mathbf{v}]-\mathbf{Sf}-\mathbf{I}\})$$
$$= 1-[69,547/0.9900-64,720]$$
$$/([1-0.3000]\{48,851$$
$$[1+0.0500][1-0.1900]$$
$$-51,294*0.5800-513\})$$
$$= 30.00\%$$

Corporate IFRS-GAAP (B/S-I/S), ISBN 13: **978-1983566332**, ISBN 10: **1983566330**

Law-5076:

If both (**L**= Liabilities or Debt= 69,547 millions USD), (**f**= F/S= Fixed Portion= 58.00%), (**I**= Interest Expense= 513 millions USD), (**$'**= Sales of Past Year= 48,851 millions USD), (**d**= D/A= Dividend Portion or Payout= 30.00%), (**F**= L/E= Leverage or Gearing Ratio= 99.00%), (**s**= [S/S']-1= Sales Growth= 5.00%), (**v**= V/S= Variable Portion= 19.00%), (**$**= Sales or Revenues= 51,294 millions USD),and (**t**= T/B= Tax Rate= 30.00%), then it's (**U**= Utilized or Starting Capital planned) is:

$$\mathbf{U}= \mathbf{L/I} - [1-\mathbf{d}][1-\mathbf{t}]\{\mathbf{\$'}[1+\mathbf{s}][1-\mathbf{v}] - \mathbf{\$f} - \mathbf{I}\})$$

$$= 69,547/0.9900 - [1-0.3000]$$
$$[1-0.3000]\{48,851[1+0.0500]$$
$$[1-0.1900] - 51,294*0.5800$$
$$-513\})$$

$$= 64,720 \text{ millions USD}$$

Corporate IFRS-GAAP (B/S-I/S), ISBN 13: **978-1983566332**, ISBN 10: **1983566330**

<u>Law-5077</u>:

If both (**L**= Liabilities or Debt= <u>69,547</u> millions USD), (**f**= F/S= Fixed Portion= <u>58.00%</u>), (**I**= Interest Expense= <u>513</u> millions USD), (**S'**= Sales of Past Year= <u>48,851</u> millions USD), (**d**= D/A= Dividend Portion or Payout= <u>30.00%</u>), (**U**= Utilized or Starting Capital= <u>64,720</u> millions USD), (**s**= [S/S']-1= Sales Growth= <u>5.00%</u>), (**v**= V/S= Variable Portion= <u>19.00%</u>), (**S**= Sales or Revenues= <u>51,294</u> millions USD),and (**t**= T/B= Tax Rate= <u>30.00%</u>), then it's (**I**= Leverage or Gearing Ratio planned) is:

$$I = L/(U+[1-d][1-t]\{S'[1+s][1-v]-Sf-I\})$$
$$= 69,547/(64,720+[1-0.3000]$$
$$[1-0.3000]\{48,851$$
$$[1+0.0500][1-0.1900]$$
$$-51,294*0.5800-513\})$$
$$= \underline{99.00\%}$$

Corporate IFRS-GAAP (B/S-I/S), ISBN 13: **978-1983566332**, ISBN 10: **1983566330**

Law-5078:

If both (**l**= L/E= Leverage or Gearing Ratio= 99.00%), (**f**= F/S= Fixed Portion= 58.00%), (**i**= I/S= Interest Portion= 1.00%), (**$'**= Sales of Past Year= 48,851 millions USD), (**d**= D/A= Dividend Portion or Payout= 30.00%), (**U**= Utilized or Starting Capital= 64,720 millions USD), (**s**= [S/S']-1= Sales Growth= 5.00%), (**v**= V/S= Variable Portion= 19.00%), (**$**= Sales or Revenues= 51,294 millions USD),and (**t**= T/B= Tax Rate= 30.00%), then it's (**L**= Liabilities Or Debt planned) is:

$$L = l \, (U+[1-d][1-t]\{\$'[1+s]-\$'v[1+s]-\$f-\$i\})$$

$$= l \, (U+[1-d][1-t]\{\$'[1+s][1-v]-\$[f+i]\})$$

$$= 0.9900/(64{,}720+[1-0.3000]$$
$$[1-0.3000]\{48{,}851[1+0.0500]$$
$$[1-0.1900]-51{,}294$$
$$[0.5800+0.0100]\})$$

$$= 69{,}547 \text{ millions USD}$$

Corporate IFRS-GAAP (B/S-I/S), ISBN 13: **978-1983566332**, ISBN 10: **1983566330**

Law-5079:

If both ($\textbf{\textit{I}}$= L/E= Leverage or Gearing Ratio= 99.00%), ($\textbf{f}$= F/S= Fixed Portion= 58.00%), ($\textbf{i}$= I/S= Interest Portion= 1.00%), ($\textbf{L}$= Liabilities or Debt= 69,547 millions USD), ($\textbf{d}$= D/A= Dividend Portion or Payout= 30.00%), ($\textbf{U}$= Utilized or Starting Capital= 64,720 millions USD), ($\textbf{s}$= [S/S']-1= Sales Growth= 5.00%), ($\textbf{v}$= V/S= Variable Portion= 19.00%), ($\textbf{\$}$= Sales or Revenues= 51,294 millions USD),and ($\textbf{t}$= T/B= Tax Rate= 30.00%), then it's ($\textbf{\$'}$= Sales Past) must be:

$$\$' = (\$[\textbf{f}+\textbf{i}]+[\textbf{L}/\textbf{\textit{I}}-\textbf{U}]/[1-\textbf{d}][1-\textbf{t}])/\{[1+\textbf{s}][1-\textbf{v}]\}$$

$$= (51,294[0.5800+0.0100]+[69,547$$
$$/0.9900-64,720]/\{[1-0.3000]$$
$$[1-0.3000]\})/\{[1+0.0500]$$
$$[1-0.1900]\}$$

$$= \underline{48,851} \text{ millions USD}$$

Corporate IFRS-GAAP (B/S-I/S), ISBN 13: **978-1983566332**, ISBN 10: **1983566330**

<u>Law-5080</u>:

If both (**l**= L/E= Leverage or Gearing Ratio= <u>99.00%</u>), (**f**= F/S= Fixed Portion= <u>58.00%</u>), (**i**= I/S= Interest Portion= <u>1.00%</u>), (**L**= Liabilities or Debt= <u>69,547</u> millions USD), (**d**= D/A= Dividend Portion or Payout= <u>30.00%</u>), (**U**= Utilized or Starting Capital= <u>64,720</u> millions USD), (**S'**= Sales of Past Year= <u>48,851</u> millions USD), (**v**= V/S= Variable Portion= <u>19.00%</u>), (**S**= Sales or Revenues= <u>51,294</u> millions USD),and (**t**= T/B= Tax Rate= <u>30.00%</u>), then it's (**s**= Sales Growth planned) is:

$$s = (S[f+i]+[L/l-U]/[1-d][1-t]\})/\{S'[1-v]\}-1$$
$$= (51,294[0.5800+0.0100]+[69,547$$
$$/0.9900-64,720]/\{[1-0.3000]$$
$$[1-0.3000]\})/\{48,851$$
$$[1-0.1900]\}-1$$
$$= \underline{5.00\%}$$

Corporate IFRS-GAAP (B/S-I/S), ISBN 13: **978-1983566332**, ISBN 10: **1983566330**

Law-5081:

If both (I= L/E= Leverage or Gearing Ratio= 99.00%), (f= F/S= Fixed Portion= 58.00%), (i= I/S= Interest Portion= 1.00%), (L= Liabilities or Debt= 69,547 millions USD), (d= D/A= Dividend Portion or Payout= 30.00%), (U= Utilized or Starting Capital= 64,720 millions USD), (S'= Sales of Past Year= 48,851 millions USD), (s= [S/S']-1= Sales Growth= 5.00%), (S= Sales or Revenues= 51,294 millions USD),and (t= T/B= Tax Rate= 30.00%), then it's (v= Variable Portion planned) is:

$$v= 1-(S[f+i]+[L/I-U]/[1-d][1-t]\})/\{S'[1+s]\}$$
$$= 1-(51,294[0.5800+0.0100]+[69,547$$
$$/0.9900-64,720]/\{[1-0.3000]$$
$$[1-0.3000]\})$$
$$/\{48,851[1+0.0500]\}$$
$$= 19.00\%$$

Corporate IFRS-GAAP (B/S-I/S), ISBN 13: **978-1983566332**, ISBN 10: **1983566330**

<u>Law-5082</u>:

If both (l= L/E= Leverage or Gearing Ratio= 99.00%), (f= F/S= Fixed Portion= 58.00%), (i= I/S= Interest Portion= 1.00%), (L= Liabilities or Debt= 69,547 millions USD), (d= D/A= Dividend Portion or Payout= 30.00%), (U= Utilized or Starting Capital= 64,720 millions USD), (S'= Sales of Past Year= 48,851 millions USD), (s= [S/S']-1= Sales Growth= 5.00%), (v= V/S= Variable Portion= 19.00%), and (t= T/B= Tax Rate= 30.00%), then it's (S= Sales or Revenues planned) is:

$$S= (S'[1+s][1-v]-[L/l-U]/[1-d][1-t]\})/[f+i]$$
$$= (48,851[1+0.0500][1-0.1900]$$
$$-[69,547/0.9900-64,720]$$
$$/\{[1-0.3000][1-0.3000]\})$$
$$/[0.5800+0.0100]$$
$$= 51,294 \text{ millions USD}$$

Law-5083:

If both ($\pmb{l}$= L/E= Leverage or Gearing Ratio= 99.00%), ($\pmb{S}$= Sales or Revenues= 51,294 millions USD), ($\pmb{i}$= I/S= Interest Portion= 1.00%), ($\pmb{L}$= Liabilities or Debt= 69,547 millions USD), ($\pmb{d}$= D/A= Dividend Portion or Payout= 30.00%), ($\pmb{U}$= Utilized or Starting Capital= 64,720 millions USD), ($\pmb{S'}$= Sales of Past Year= 48,851 millions USD), ($\pmb{s}$= [S/S']-1= Sales Growth= 5.00%), ($\pmb{v}$= V/S= Variable Portion= 19.00%), and ($\pmb{t}$= T/B= Tax Rate= 30.00%), then it's ($\pmb{f}$= Fixed Portion planned) is:

$$f = (S'[1+s][1-v]-[L/l-U]/[1-d][1-t]\})/S-l$$
$$= (48,851[1+0.0500][1-0.1900]$$
$$-[69,547/0.9900-64,720]$$
$$/\{[1-0.3000][1-0.3000]\})$$
$$/51,294-0.0100$$

$$= 58.00\%$$

Corporate IFRS-GAAP (B/S-I/S), ISBN 13: **978-1983566332**, ISBN 10: **1983566330**

<u>Law-5084</u>:

If both (I= L/E= Leverage or Gearing Ratio= 99.00%), (S= Sales or Revenues= 51,294 millions USD), (f= F/S= Fixed Portion= 58.00%), (L= Liabilities or Debt= 69,547 millions USD), (d= D/A= Dividend Portion or Payout= 30.00%), (U= Utilized or Starting Capital= 64,720 millions USD), (S'= Sales of Past Year= 48,851 millions USD), (s= [S/S']-1= Sales Growth= 5.00%), (v= V/S= Variable Portion= 19.00%), and (t= T/B= Tax Rate= 30.00%), then it's (i= Interest Portion planned) is:

$$i = (S'[1+s][1-v]-[L/I-U]/[1-d][1-t]\})/S-f$$
$$= (48,851[1+0.0500][1-0.1900]$$
$$-[69,547/0.9900-64,720]$$
$$/\{[1-0.3000][1-0.3000]\})$$
$$/51,294-0.5800$$
$$= 58.00\%$$

<u>Law-5085</u>:

If both (**l**= L/E= Leverage or Gearing Ratio= 99.00%), (**$**= Sales or Revenues= 51,294 millions USD), (**f**= F/S= Fixed Portion= 58.00%), (**L**= Liabilities or Debt= 69,547 millions USD), (**d**= D/A= Dividend Portion or Payout= 30.00%), (**U**= Utilized or Starting Capital= 64,720 millions USD), (**$'**= Sales of Past Year= 48,851 millions USD), (**s**= [S/S']-1= Sales Growth= 5.00%), (**v**= V/S= Variable Portion= 19.00%), and (**i**= I/S= Interest Portion= 1.00%), then it's (**t**= Tax Rate planned) is:

$$t = 1-[\mathbf{L}/\mathbf{l}-\mathbf{U}]([1-\mathbf{d}]\{\mathbf{\$'}[1+\mathbf{s}][1-\mathbf{v}]-\mathbf{\$}[\mathbf{f}+\mathbf{i}]\})$$
$$= 1-[69,547/0.9900-64,720]$$
$$/([1-0.3000]\{48,851$$
$$[1+0.0500][1-0.1900]$$
$$-51,294[0.5800+0.0100]\})$$

$$= \underline{30.00\%}$$

Corporate IFRS-GAAP (B/S-I/S), ISBN 13: **978-1983566332**, ISBN 10: **1983566330**

<u>Law-5086</u>:

If both (**I**= L/E= Leverage or Gearing Ratio= <u>99.00%</u>), (**S**= Sales or Revenues= <u>51,294</u> millions USD), (**f**= F/S= Fixed Portion= <u>58.00%</u>), (**L**= Liabilities or Debt= <u>69,547</u> millions USD), (**t**= T/B= Tax Rate= <u>30.00%</u>), (**U**= Utilized or Starting Capital= <u>64,720</u> millions USD), (**S'**= Sales of Past Year= <u>48,851</u> millions USD), (**s**= [S/S']-1= Sales Growth= <u>5.00%</u>), (**v**= V/S= Variable Portion= <u>19.00%</u>), and (**i**= I/S= Interest Portion= <u>1.00%</u>), then it's (**d**= Dividend Portion or Payout planned) is:

$$d = 1-[\textbf{L}/\textbf{I}-\textbf{U}]([1-\textbf{t}]\{\textbf{S}'[1+\textbf{s}][1-\textbf{v}]-\textbf{S}[\textbf{f}+\textbf{i}]\})$$

$$= 1-[69,547/0.9900-64,720]$$
$$/([1-0.3000]\{48,851$$
$$[1+0.0500][1-0.1900]$$
$$-51,294[0.5800+0.0100]\})$$

$$= \underline{30.00\%}$$

Corporate IFRS-GAAP (B/S-I/S), ISBN 13: **978-1983566332**, ISBN 10: **1983566330**

Law-5087:

If both (I= L/E= Leverage or Gearing Ratio= 99.00%), (S= Sales or Revenues= 51,294 millions USD), (f= F/S= Fixed Portion= 58.00%), (L= Liabilities or Debt= 69,547 millions USD), (t= T/B= Tax Rate= 30.00%), (d= D/A= Dividend Portion or Payout= 30.00%), (S'= Sales of Past Year= 48,851 millions USD), (s= [S/S']-1= Sales Growth= 5.00%), (v= V/S= Variable Portion= 19.00%), and (i= I/S= Interest Portion= 1.00%), then it's (U= Utilized or Starting Capital planned) is:

U= L/I -[1-d][1-t]{S'[1+s][1-v]-S[f+i]}
$$= 69,547/0.9900-[1-0.3000]$$
$$[1-0.3000]\{48,851$$
$$[1+0.0500][1-0.1900]$$
$$-51,294[0.5800+0.0100]\})$$

= 64,720 millions USD

Law-5088:

If both (U= Utilized or Starting Capital= 64,720 millions USD), (S= Sales or Revenues= 51,294 millions USD), (f= F/S= Fixed Portion= 58.00%), (L= Liabilities or Debt= 69,547 millions USD), (t= T/B= Tax Rate= 30.00%), (d= D/A= Dividend Portion or Payout= 30.00%), (S'= Sales of Past Year= 48,851 millions USD), (s= [S/S']-1= Sales Growth= 5.00%), (v= V/S= Variable Portion= 19.00%), and (i= I/S= Interest Portion= 1.00%), then it's (l= Leverage or Gearing Ratio planned) is:

$$l = L/(U+[1-d][1-t]\{S'[1+s][1-v]-S[f+i]\}$$
$$= 69,547/(64,720+[1-0.3000]$$
$$[1-0.3000]\{48,851$$
$$[1+0.0500][1-0.1900]$$
$$-51,294[0.5800+0.0100]\})$$
$$= 99.00\%$$

Corporate IFRS-GAAP (B/S-I/S), ISBN 13: **978-1983566332**, ISBN 10: **1983566330**

Law-5089:

If both (**U**= Utilized or Starting Capital= 64,720 millions USD), (**$**= Sales or Revenues= 51,294 millions USD), (**f**= F/S= Fixed Portion= 58.00%), (**I**= L/E= Leverage or Gearing Ratio= 99.00%), (**t**= T/B= Tax Rate= 30.00%), (**d**= D/A= Dividend Portion or Payout= 30.00%), (**$'**= Sales of Past Year= 48,851 millions USD), (**s**= [S/S']-1= Sales Growth= 5.00%), (**v**= V/S= Variable Portion= 19.00%), and (**i**= I/S= Interest Portion= 1.00%), then it's (**L**= Liabilities or Debt planned) is:

$$\mathbf{L} = \mathbf{I}\,(\mathbf{U}+[1-\mathbf{d}][1-\mathbf{t}]\{\mathbf{\$'}[1+\mathbf{s}]-\mathbf{\$'v}[1+\mathbf{s}]$$
$$-\mathbf{\$f}-\mathbf{\$'i}[1+\mathbf{s}]\})$$
$$= \mathbf{I}\,(\mathbf{U}+[1-\mathbf{d}][1-\mathbf{t}]\{\mathbf{\$'}[1+\mathbf{s}][1-\mathbf{v}-\mathbf{i}]-\mathbf{\$f}\})$$
$$= 0.9900/(64,720+[1-0.3000]$$
$$[1-0.3000]\{48,851$$
$$[1+0.0500][1-0.1900$$
$$-0.0100]-51,294*0.5800]\})$$

$= \underline{69,547}$ millions USD

Corporate IFRS-GAAP (B/S-I/S), ISBN 13: **978-1983566332**, ISBN 10: **1983566330**

<u>Law-5090</u>:

If both (**U**= Utilized or Starting Capital= <u>64,720</u> millions USD), (**S**= Sales or Revenues= <u>51,294</u> millions USD), (**f**= F/S= Fixed Portion= <u>58.00%</u>), (**l**= L/E= Leverage or Gearing Ratio= <u>99.00%</u>), (**t**= T/B= Tax Rate= <u>30.00%</u>), (**d**= D/A= Dividend Portion or Payout= <u>30.00%</u>), (**L**= Liabilities or Debt= <u>69,547</u> millions USD), (**s**= [S/S']-1= Sales Growth= <u>5.00%</u>), (**v**= V/S= Variable Portion= <u>19.00%</u>), and (**i**= I/S= Interest Portion= <u>1.00%</u>), then it's (**S'**= Sales Past), must be:

$$S' = (Sf + [L/l - U]/\{[1-d][1-t]\})/\{[1+s][1-v-i]\}$$
$$= (51,294*0.5800 + [69,547/0.9900$$
$$-64,720]/[1-0.3000]$$
$$[1-0.3000]\})/\{[1+0.0500]$$
$$[1-0.1900-0.0100]\}$$
$$= \underline{48,851} \text{ millions USD}$$

Corporate IFRS-GAAP (B/S-I/S), ISBN 13: **978-1983566332**, ISBN 10: **1983566330**

Law-5091:

If both (**U**= Utilized or Starting Capital= 64,720 millions USD), (**S**= Sales or Revenues= 51,294 millions USD), (**f**= F/S= Fixed Portion= 58.00%), (**l**= L/E= Leverage or Gearing Ratio= 99.00%), (**t**= T/B= Tax Rate= 30.00%), (**d**= D/A= Dividend Portion or Payout= 30.00%), (**L**= Liabilities or Debt= 69,547 millions USD), (**S'**= Sales of Past Year= 48,851 millions USD), (**v**= V/S= Variable Portion= 19.00%), and (**i**= I/S= Interest Portion= 1.00%), then it's (**s**= Sales Growth planned), is:

$$s = (Sf + [L/l - U]/\{[1-d][1-t]\})/\{S'[1-v-i]\} - 1$$
$$= (51,294 * 0.5800 + [69,547/0.9900$$
$$-64,720]/[1-0.3000]$$
$$[1-0.3000]\})/\{48,851$$
$$[1-0.1900-0.0100]\} - 1$$
$$= 5.00\%$$

Corporate IFRS-GAAP (B/S-I/S), ISBN 13: **978-1983566332**, ISBN 10: **1983566330**

<u>Law-5092</u>:

If both (**U**= Utilized or Starting Capital= <u>64,720</u> millions USD), (**S**= Sales or Revenues= <u>51,294</u> millions USD), (**f**= F/S= Fixed Portion= <u>58.00%</u>), (**l**= L/E= Leverage or Gearing Ratio= <u>99.00%</u>), (**t**= T/B= Tax Rate= <u>30.00%</u>), (**d**= D/A= Dividend Portion or Payout= <u>30.00%</u>), (**L**= Liabilities or Debt= <u>69,547</u> millions USD), (**S'**= Sales of Past Year= <u>48,851</u> millions USD), (**s**= [S/S']-1= Sales Growth= <u>5.00%</u>), and (**i**= I/S= Interest Portion= <u>1.00%</u>), then it's (**v**= Variable Portion planned), is:

$$\mathbf{v}= 1-(\mathbf{Sf}+[\mathbf{L}/\mathbf{l}-\mathbf{U}]/\{[1-\mathbf{d}][1-\mathbf{t}]\})/\{\mathbf{S'}[1+\mathbf{s}]\}-\mathbf{i}$$
$$= (51,294*0.5800+[69,547/0.9900$$
$$-64,720]/[1-0.3000]$$
$$[1-0.3000]\})/\{48,851$$
$$[1+0.0500]\}-0.0100$$
$$= \underline{19.00\%}$$

Corporate IFRS-GAAP (B/S-I/S), ISBN 13: **978-1983566332**, ISBN 10: **1983566330**

<u>Law-5093</u>:

If both (**U**= Utilized or Starting Capital= <u>64,720</u> millions USD), (**S**= Sales or Revenues= <u>51,294</u> millions USD), (**f**= F/S= Fixed Portion= <u>58.00%</u>), (**l**= L/E= Leverage or Gearing Ratio= <u>99.00%</u>), (**t**= T/B= Tax Rate= <u>30.00%</u>), (**d**= D/A= Dividend Portion or Payout= <u>30.00%</u>), (**L**= Liabilities or Debt= <u>69,547</u> millions USD), (**S'**= Sales of Past Year= <u>48,851</u> millions USD), (**s**= [S/S']-1= Sales Growth= <u>5.00%</u>), and (**v**= V/S= Variable Portion= <u>19.00%</u>), then it's (**i**= Interest Portion planned), is:

$$i = 1-(Sf+[L/l-U]/\{[1-d][1-t]\})/\{S'[1+s]\}-v$$

$$= 1-(51,294*0.5800+[69,547/0.9900$$
$$-64,720]/[1-0.3000]$$
$$[1-0.3000]\})/\{48,851$$
$$[1+0.0500]\}-0.1900$$

$$= \underline{1.00\%}$$

Corporate IFRS-GAAP (B/S-I/S), ISBN 13: **978-1983566332**, ISBN 10: **1983566330**

<u>Law-5094</u>:

If both (**U**= Utilized or Starting Capital= <u>64,720</u> millions USD), (**i**= I/S= Interest Portion= <u>1.00%</u>), (**f**= F/S= Fixed Portion= <u>58.00%</u>), (**/**= L/E= Leverage or Gearing Ratio= <u>99.00%</u>), (**t**= T/B= Tax Rate= <u>30.00%</u>), (**d**= D/A= Dividend Portion or Payout= <u>30.00%</u>), (**L**= Liabilities or Debt= <u>69,547</u> millions USD), (**\$'**= Sales of Past Year= <u>48,851</u> millions USD), (**s**= [S/S']-1= Sales Growth= <u>5.00%</u>), and (**v**= V/S= Variable Portion= <u>19.00%</u>), then it's (**\$**= Sales or Revenues planned), is:

$$\$= (\$'[1+s][1-v-i]-[L//-U]/\{[1-d][1-t]\})/f$$
$$= (48{,}851[1+0.0500][1-0.1900$$
$$-0.0100]-[69{,}547/0.9900$$
$$-64{,}720]/[1-0.3000]$$
$$[1-0.3000]\})/0.5800$$
$$= \underline{51{,}294} \text{ millions USD}$$

Corporate IFRS-GAAP (B/S-I/S), ISBN 13: **978-1983566332**, ISBN 10: **1983566330**

Law-5094:

If both (**U**= Utilized or Starting Capital= 64,720 millions USD), (**i**= I/S= Interest Portion= 1.00%), (**f**= F/S= Fixed Portion= 58.00%), (**l**= L/E= Leverage or Gearing Ratio= 99.00%), (**t**= T/B= Tax Rate= 30.00%), (**d**= D/A= Dividend Portion or Payout= 30.00%), (**L**= Liabilities or Debt= 69,547 millions USD), (**$'**= Sales of Past Year= 48,851 millions USD), (**s**= [S/S']-1= Sales Growth= 5.00%), and (**v**= V/S= Variable Portion= 19.00%), then it's (**$**= Sales or Revenues planned), is:

$$\$ = (\$'[1+s][1-v-i]-[L/l-U]/\{[1-d][1-t]\})/f$$
$$= (48,851[1+0.0500][1-0.1900$$
$$-0.0100]-[69,547/0.9900$$
$$-64,720]/[1-0.3000]$$
$$[1-0.3000]\})/0.5800$$
$$= 51,294 \text{ millions USD}$$

Corporate IFRS-GAAP (B/S-I/S), ISBN 13: **978-1983566332**, ISBN 10: **1983566330**

Law-5095:

If both (U= Utilized or Starting Capital= 64,720 millions USD), (i= I/S= Interest Portion= 1.00%), (S= Sales or Revenues= 51,294 millions USD), (l= L/E= Leverage or Gearing Ratio= 99.00%), (t= T/B= Tax Rate= 30.00%), (d= D/A= Dividend Portion or Payout= 30.00%), (L= Liabilities or Debt= 69,547 millions USD), (S'= Sales of Past Year= 48,851 millions USD), (s= [S/S']-1= Sales Growth= 5.00%), and (v= V/S= Variable Portion= 19.00%), then it's (f= Fixed Portion planned), is:

$$f = (S'[1+s][1-v-i]-[L/l -U]/\{[1-d][1-t]\})/S$$
$$= (48,851[1+0.0500][1-0.1900$$
$$-0.0100]-[69,547/0.9900$$
$$-64,720]/[1-0.3000]$$
$$[1-0.3000]\})/51,294$$
$$= 58.00\%$$

Corporate IFRS-GAAP (B/S-I/S), ISBN 13: **978-1983566332**, ISBN 10: **1983566330**

<u>Law-5096</u>:

If both (**U**= Utilized or Starting Capital= <u>64,720</u> millions USD), (**i**= I/S= Interest Portion= <u>1.00%</u>), (**$**= Sales or Revenues= <u>51,294</u> millions USD), (**l**= L/E= Leverage or Gearing Ratio= <u>99.00%</u>), (**f**= F/S= Fixed Portion= <u>58.00%</u>), (**d**= D/A= Dividend Portion or Payout= <u>30.00%</u>), (**L**= Liabilities or Debt= <u>69,547</u> millions USD), (**$'**= Sales of Past Year= <u>48,851</u> millions USD), (**s**= [S/S']-1= Sales Growth= <u>5.00%</u>), and (**v**= V/S= Variable Portion= <u>19.00%</u>), then it's (**t**= Tax Rate planned), is:

$$t= 1-[\mathbf{L}/\mathbf{l}-\mathbf{U}]/([1-\mathbf{d}]\{\mathbf{\$'}[1+\mathbf{s}][1-\mathbf{v}-\mathbf{i}]-\mathbf{\$f}\})$$

$$= 1-[69,547/0.9900-64,720]$$
$$/([1-0.3000]\{48,851$$
$$[1+0.0500][1-0.1900$$
$$-0.0100]-51,294*0.5800]\})$$

$$= \underline{30.00\%}$$

Corporate IFRS-GAAP (B/S-I/S), ISBN 13: **978-1983566332**, ISBN 10: **1983566330**

<u>Law-5097</u>:

If both (**U**= Utilized or Starting Capital= <u>64,720</u> millions USD), (**i**= I/S= Interest Portion= <u>1.00%</u>), (**S**= Sales or Revenues= <u>51,294</u> millions USD), (**l**= L/E= Leverage or Gearing Ratio= <u>99.00%</u>), (**t**= T/B= Tax Rate= <u>30.00%</u>), (**f**= F/S= Fixed Portion= <u>58.00%</u>), (**L**= Liabilities or Debt= <u>69,547</u> millions USD), (**S'**= Sales of Past Year= <u>48,851</u> millions USD), (**s**= [S/S']-1= Sales Growth= <u>5.00%</u>), and (**v**= V/S= Variable Portion= <u>19.00%</u>), then it's (**d**= Dividend Portion or Payout planned), is:

$$d = 1-[L/l - U]/([1-t]\{S'[1+s][1-v-i]-Sf\})$$
$$= 1-[69,547/0.9900-64,720]$$
$$/([1-0.3000]\{48,851$$
$$[1+0.0500][1-0.1900$$
$$-0.0100]-51,294*0.5800]\})$$
$$= \underline{30.00\%}$$

Corporate IFRS-GAAP (B/S-I/S), ISBN 13: **978-1983566332**, ISBN 10: **1983566330**

<u>Law-5098</u>:

If both (**d**= D/A= Dividend Portion or Payout= <u>30.00%</u>), (**i**= I/S= Interest Portion= <u>1.00%</u>), (**S**= Sales or Revenues= <u>51,294</u> millions USD), (**l**= L/E= Leverage or Gearing Ratio= <u>99.00%</u>), (**t**= T/B= Tax Rate= <u>30.00%</u>), (**f**= F/S= Fixed Portion= <u>58.00%</u>), (**L**= Liabilities or Debt= <u>69,547</u> millions USD), (**S'**= Sales of Past Year= <u>48,851</u> millions USD), (**s**= [S/S']-1= Sales Growth= <u>5.00%</u>), and (**v**= V/S= Variable Portion= <u>19.00%</u>), then it's (**U**= Utilized or Starting Capital planned), is:

$$U= L/l - [1-d][1-t]\{S'[1+s][1-v-i]-Sf\}$$
$$= 69,547/0.9900-[1-0.3000]$$
$$[1-0.3000]\{48,851$$
$$[1+0.0500][1-0.1900$$
$$-0.0100]-51,294*0.5800]\})$$
$$= \underline{64,720} \text{ millions USD}$$

Corporate IFRS-GAAP (B/S-I/S), ISBN 13: **978-1983566332**, ISBN 10: **1983566330**

<u>Law-5099</u>:

If both ($\mathbf{d}$= D/A= Dividend Portion or Payout= 30.00%), ($\mathbf{i}$= I/S= Interest Portion= 1.00%), ($\mathbf{S}$= Sales or Revenues= 51,294 millions USD), ($\mathbf{U}$= Utilized or Starting Capital= 64,720 millions USD), ($\mathbf{t}$= T/B= Tax Rate= 30.00%), ($\mathbf{f}$= F/S= Fixed Portion= 58.00%), ($\mathbf{L}$= Liabilities or Debt= 69,547 millions USD), ($\mathbf{S'}$= Sales of Past Year= 48,851 millions USD), ($\mathbf{s}$= [S/S']-1= Sales Growth= 5.00%), and ($\mathbf{v}$= V/S= Variable Portion= 19.00%), then it's ($\mathbf{I}$= Leverage or Gearing Ratio planned), is:

$$\mathbf{I} = \mathbf{L}/(\mathbf{U}+[1\text{-}\mathbf{d}][1\text{-}\mathbf{t}]\{\mathbf{S'}[1+\mathbf{s}][1\text{-}\mathbf{v}\text{-}\mathbf{i}]\text{-}\mathbf{Sf}\})$$

$$= 69,547/(64,720+[1\text{-}0.3000]$$
$$[1\text{-}0.3000]\{48,851$$
$$[1+0.0500][1\text{-}0.1900$$
$$-0.0100]\text{-}51,294*0.5800]\})$$

$$= \underline{99.00\%}$$

Corporate IFRS-GAAP (B/S-I/S), ISBN 13: **978-1983566332**, ISBN 10: **1983566330**

<u>Law-5100</u>:

If both (**d**= D/A= Dividend Portion or Payout= 30.00%), (**I**= Interest Expense= <u>513</u> millions USD), (**U**= Utilized or Starting Capital= <u>64,720</u> millions USD), (**t**= T/B= Tax Rate= <u>30.00%</u>), (**f**= F/S= Fixed Portion= <u>58.00%</u>), (**I**= L/E= Leverage or Gearing Ratio= <u>99.00%</u>), (**$'**= Sales of Past Year= <u>48,851</u> millions USD), (**s**= [S/S']-1= Sales Growth= <u>5.00%</u>), and (**v**= V/S= Variable Portion= <u>19.00%</u>), then it's (**L**= Liabilities or Debt planned), is:

$$L= I \, (U+[1-d][1-t]\{\$'[1+s]-\$'v[1+s]$$
$$-\$'f[1+s]-I\})$$
$$= I \, (U+[1-d][1-t]\{\$'[1+s][1-v-f]-I\})$$
$$= 0.9900/(64,720+[1-0.3000]$$
$$[1-0.3000]\{48,851$$
$$[1+0.0500][1-0.1900$$
$$-0.5800]-513\})$$
$$= \underline{69,547} \text{ millions USD}$$

Corporate IFRS-GAAP (B/S-I/S), ISBN 13: **978-1983566332**, ISBN 10: **1983566330**

Law-5101:

If both (**d**= D/A= Dividend Portion or Payout= 30.00%), (**I**= Interest Expense= 513 millions USD), (**U**= Utilized or Starting Capital= 64,720 millions USD), (**t**= T/B= Tax Rate= 30.00%), (**f**= F/S= Fixed Portion= 58.00%), (**I**= L/E= Leverage or Gearing Ratio= 99.00%), (**L**= Liabilities or Debt= 69,547 millions USD), (**s**= [S/S']-1= Sales Growth= 5.00%), and (**v**= V/S= Variable Portion= 19.00%), then it's (**S'**= Sales Past), must be:

$$S' = (I+[L/I-U]/\{[1-d][1-t]\})/\{[1+s][1-v-f]\}$$
$$= (513+[69,547/0.9900-64,720]$$
$$/\{[1-0.3000][1-0.3000]\})$$
$$/\{[1+0.0500][1-0.1900$$
$$-0.5800]\}$$
$$= 48,851 \text{ millions USD}$$

Corporate IFRS-GAAP (B/S-I/S), ISBN 13: **978-1983566332**, ISBN 10: **1983566330**

Law-5102:

If both (**d**= D/A=Dividend Portion or Payout= 30.00%), (**I**= Interest Expense= 513 millions USD), (**U**= Utilized or Starting Capital= 64,720 millions USD), (**t**= T/B= Tax Rate= 30.00%), (**f**= F/S= Fixed Portion= 58.00%), (**l**= L/E= Leverage or Gearing Ratio= 99.00%), (**L**= Liabilities or Debt= 69,547 millions USD), (**S'**= Sales of Past Year= 48,851 millions USD), and (**v**= V/S= Variable Portion= 19.00%), then it's (**s**= Sales Growth planned), is:

$$s= (I+[L/l -U]/\{[1-d][1-t]\})/\{S'\{1-v-f]\}-1$$
$$= (513+[69,547/0.9900-64,720]$$
$$/\{[1-0.3000][1-0.3000]\})$$
$$/\{48,851[1-0.1900$$
$$-0.5800]\}-1$$
$$= 5.00\%$$

108

Corporate IFRS-GAAP (B/S-I/S), ISBN 13: **978-1983566332**, ISBN 10: **1983566330**

Law-5103:

If both (**d**= D/A= Dividend Portion or Payout= 30.00%), (**I**= Interest Expense= 513 millions USD), (**U**= Utilized or Starting Capital= 64,720 millions USD), (**t**= T/B= Tax Rate= 30.00%), (**f**= F/S= Fixed Portion= 58.00%), (**I**= L/E= Leverage or Gearing Ratio= 99.00%), (**L**= Liabilities or Debt= 69,547 millions USD), (**S'**= Sales of Past Year= 48,851 millions USD), and (**s**= [S/S']-1= Sales Growth= 5.00%), then it's (**v**= Variable Portion planned), is:

$$v = 1-f-(I+[L/I-U]/\{[1-d][1-t]\})/\{S'\{1+s\}\}$$
$$= 1-0.5800-(513+[69,547/0.9900$$
$$-64,720]/\{[1-0.3000]$$
$$[1-0.3000]\})$$
$$/\{48,851[1+0.0500]\}$$
$$= 19.00\%$$

Corporate IFRS-GAAP (B/S-I/S), ISBN 13: **978-1983566332**, ISBN 10: **1983566330**

Law-5104:

If both (**d**= D/A= Dividend Portion or Payout= 30.00%), (**I**= Interest Expense= 513 millions USD), (**U**= Utilized or Starting Capital= 64,720 millions USD), (**t**= T/B= Tax Rate= 30.00%), (**v**= V/S= Variable Portion= 19.00%), (**l**= L/E= Leverage or Gearing Ratio= 99.00%), (**L**= Liabilities or Debt= 69,547 millions USD), (**S'**= Sales of Past Year= 48,851 millions USD), and (**s**= [S/S']-1= Sales Growth= 5.00%), then it's (**f**= Fixed Portion planned), is:

$$f= 1\text{-}v\text{-}(I+[L/l\text{-}U]/\{[1\text{-}d][1\text{-}t]\})/\{S'\{1+s\}\}$$
$$= 1\text{-}0.1900\text{-}(513+[69,547/0.9900$$
$$-64,720]/\{[1\text{-}0.3000]$$
$$[1\text{-}0.3000]\})$$
$$/\{48,851[1+0.0500]\}$$
$$= 58.00\%$$

Corporate IFRS-GAAP (B/S-I/S), ISBN 13: **978-1983566332**, ISBN 10: **1983566330**

Law-5105:

If both (**d**= D/A= Dividend Portion or Payout= 30.00%), (**f**= F/S= Fixed Portion= 58.00%), (**U**= Utilized or Starting Capital= 64,720 millions USD), (**t**= T/B= Tax Rate= 30.00%), (**v**= V/S= Variable Portion= 19.00%), (**l**= L/E= Leverage or Gearing Ratio= 99.00%), (**L**= Liabilities or Debt= 69,547 millions USD), (**S'**= Sales of Past Year= 48,851 millions USD), and (**s**= [S/S']-1= Sales Growth= 5.00%), then it's (**I**= Interest Expense planned), is:

$$I= S'\{1+s\}[1-v-f]-[L/l-U]/\{[1-d][1-t]\}$$
$$= 48,851[1+0.0500][1-0.1900$$
$$-0.5800]-[69,547/0.9900$$
$$-64,720]/\{[1-0.3000]$$
$$[1-0.3000]\}$$
$$= \underline{513} \text{ millions USD}$$

Corporate IFRS-GAAP (B/S-I/S), ISBN 13: **978-1983566332**, ISBN 10: **1983566330**

Law-5106:

If both (**d**= D/A= Dividend Portion or Payout= 30.00%), (**f**= F/S= Fixed Portion= 58.00%), (**U**= Utilized or Starting Capital= 64,720 millions USD), (**I**= Interest Expense= 513 millions USD), (**v**= V/S= Variable Portion= 19.00%), (**l**= L/E= Leverage or Gearing Ratio= 99.00%), (**L**= Liabilities or Debt= 69,547 millions USD), (**S'**= Sales of Past Year= 48,851 millions USD), and (**s**= [S/S']-1= Sales Growth= 5.00%), then it's (**t**= Tax Rate planned), is:

$$t = 1-[L/l-U]/([1-d]\{S'\{1+s\}[1-v-f]-I\})$$
$$= 1-[69,547/0.9900-64,720]$$
$$/([1-0.3000]\{48,851$$
$$[1+0.0500][1-0.1900$$
$$-0.5800]-513\})$$
$$= 30.00\%$$

Corporate IFRS-GAAP (B/S-I/S), ISBN 13: **978-1983566332**, ISBN 10: **1983566330**

Law-5107:

If both (**t**= T/B= Tax Rate= 30.00%), (**f**= F/S= Fixed Portion= 58.00%), (**U**= Utilized or Starting Capital= 64,720 millions USD), (**I**= Interest Expense= 513 millions USD), (**v**= V/S= Variable Portion= 19.00%), (**l**= L/E= Leverage or Gearing Ratio= 99.00%), (**L**= Liabilities or Debt= 69,547 millions USD), (**S'**= Sales of Past Year= 48,851 millions USD),and (**s**= [S/S']-1= Sales Growth= 5.00%), then it's (**d**= Dividend Portion or Payout planned), is:

$$\mathbf{d}= 1-[\mathbf{L}/\mathbf{l}-\mathbf{U}]/([1-\mathbf{t}]\{\mathbf{S'}\{1+\mathbf{s}\}[1-\mathbf{v}-\mathbf{f}]-\mathbf{I}\})$$

$$= 1-[69,547/0.9900-64,720]$$
$$/([1-0.3000]\{48,851$$
$$[1+0.0500][1-0.1900$$
$$-0.5800]-513\})$$

$$= 30.00\%$$

<u>Law-5108</u>:

If both (**t**= T/B= Tax Rate= <u>30.00%</u>), (**f**= F/S= Fixed Portion= <u>58.00%</u>), (**d**= D/A= Dividend Portion or Payout= <u>30.00%</u>), (**I**= Interest Expense= <u>513</u> millions USD), (**v**= V/S= Variable Portion= <u>19.00%</u>), (**I**= L/E= Leverage or Gearing Ratio= <u>99.00%</u>), (**L**= Liabilities or Debt= <u>69,547</u> millions USD), (**$'**= Sales of Past Year= <u>48,851</u> millions USD),and (**s**= [S/S']-1= Sales Growth= <u>5.00%</u>), then it's (**U**= Utilized or Starting Capital planned), is:

$$U= L/I -[1-d][1-t]\{\$'\{1+s][1-v-f]-I\}$$
$$= 69,547/0.9900-[1-0.3000]$$
$$[1-0.3000]\{48,851$$
$$[1+0.0500][1-0.1900$$
$$-0.5800]-513\})$$
$$= \underline{64,720} \text{ millions USD}$$

Corporate IFRS-GAAP (B/S-I/S), ISBN 13: **978-1983566332**, ISBN 10: **1983566330**

Law-5109:

If both (**t**= T/B= Tax Rate= 30.00%), (**f**= F/S= Fixed
Portion= 58.00%), (**d**= D/A= Dividend Portion or
Payout= 30.00%), (**I**= Interest Expense= 513 millions
USD), (**v**= V/S= Variable Portion= 19.00%), (**U**=
Utilized or Starting Capital= 64,720 millions USD),
(**L**= Liabilities or Debt= 69,547 millions USD), (**$'**=
Sales of Past Year= 48,851 millions USD),and (**s**=
[S/S']-1= Sales Growth= 5.00%), then it's (**I** =
Leverage or Gearing Ratio planned), is:

$$I = L/(U+[1-d][1-t]\{\$'\{1+s\}[1-v-f]-I\})$$
$$= 69,547/(64,720+[1-0.3000]$$
$$[1-0.3000]\{48,851$$
$$[1+0.0500][1-0.1900$$
$$-0.5800]-513\})$$
$$= 99.00\%$$

<u>Law-5110</u>:

If both (**t**= T/B= Tax Rate= <u>30.00%</u>), (**f**= F/S= Fixed Portion= <u>58.00%</u>), (**d**= D/A= Dividend Portion or Payout= <u>30.00%</u>), (**i**= I/S= Interest Portion= <u>1.00%</u>), (**S**= Sales or Revenues= <u>51,294</u> millions USD), (**v**= V/S= Variable Portion= <u>19.00%</u>), (**U**= Utilized or Starting Capital= <u>64,720</u> millions USD), (**l**= L/E= Leverage or Gearing Ratio= <u>99.00%</u>), (**S'**= Sales of Past Year= <u>48,851</u> millions USD),and (**s**= [S/S']-1= Sales Growth= <u>5.00%</u>), then it's (**L**= Liabilities or Debt planned), is:

$$L= l\,(U+[1-d][1-t]\{S'\{1+s\}-S'v[1+s]-S'f[1+s]$$
$$-Si\})$$
$$= l\,(U+[1-d][1-t]\{S'\{1+s\}[1-v-f]-Si\})$$
$$= 0.9900/(64,720+[1-0.3000]$$
$$[1-0.3000]\{48,851$$
$$[1+0.0500][1-0.1900$$
$$-0.5800]-51,294*0.0100\})$$
$$= \underline{69,547}\ \text{millions USD}$$

Corporate IFRS-GAAP (B/S-I/S), ISBN 13: **978-1983566332**, ISBN 10: **1983566330**

Law-5111:

If both (**t**= T/B= Tax Rate= 30.00%), (**f**= F/S= Fixed Portion= 58.00%), (**d**= D/A= Dividend Portion or Payout= 30.00%), (**i**= I/S= Interest Portion= 1.00%), (**S**= Sales or Revenues= 51,294 millions USD), (**v**= V/S= Variable Portion= 19.00%), (**U**= Utilized or Starting Capital= 64,720 millions USD), (**l**= L/E= Leverage or Gearing Ratio= 99.00%), (**L**= Liabilities or Debt= 69,547 millions USD), and (**s**= [S/S']-1= Sales Growth= 5.00%), then it's (**S'**= Sales Past), must be:

$$\mathbf{S'}= (\mathbf{S}\mathbf{i}+[\mathbf{L}/\mathbf{l}-\mathbf{U}]/\{[1-\mathbf{d}][1-\mathbf{t}]\})/\{[1+\mathbf{s}][1-\mathbf{v}-\mathbf{f}]\}$$

$$= (51{,}294*0.0100+[69{,}547/0.9900$$
$$-64{,}720]/\{[1-0.3000]$$
$$[1-0.3000]\})/\{ [1+0.0500]$$
$$[1-0.1900-0.5800]\}$$

$$= \underline{48{,}851} \text{ millions USD}$$

Corporate IFRS-GAAP (B/S-I/S), ISBN 13: **978-1983566332**, ISBN 10: **1983566330**

<u>Law-5112</u>:

If both (**t**= T/B= Tax Rate= <u>30.00%</u>), (**f**= F/S= Fixed Portion= <u>58.00%</u>), (**d**= D/A= Dividend Portion or Payout= <u>30.00%</u>), (**i**= I/S= Interest Portion= <u>1.00%</u>), (**$**= Sales or Revenues= <u>51,294</u> millions USD), (**v**= V/S= Variable Portion= <u>19.00%</u>), (**U**= Utilized or Starting Capital= <u>64,720</u> millions USD), (**/**= L/E= Leverage or Gearing Ratio= <u>99.00%</u>), (**L**= Liabilities or Debt= <u>69,547</u> millions USD), and (**$'**= Sales of Past Year= <u>48,851</u> millions USD), then it's (**s**= Sales Growth planned), is:

$$s = (\$i+[L/\mathit{I} - U]/\{[1-d][1-t]\})/\{\$'[1-v-f]\}-1$$
$$= (51,294*0.0100+[69,547/0.9900$$
$$-64,720]/\{[1-0.3000]$$
$$[1-0.3000]\})/\{48,851$$
$$[1-0.1900-0.5800]\}-1$$
$$= \underline{5.00\%}$$

Corporate IFRS-GAAP (B/S-I/S), ISBN 13: **978-1983566332**, ISBN 10: **1983566330**

<u>Law-5113</u>:

If both (**t**= T/B= Tax Rate= <u>30.00%</u>), (**f**= F/S= Fixed Portion= <u>58.00%</u>), (**d**= D/A= Dividend Portion or Payout= <u>30.00%</u>), (**i**= I/S= Interest Portion= <u>1.00%</u>), (**\$**= Sales or Revenues= <u>51,294</u> millions USD), (**s**= [S/S']-1= Sales Growth= <u>5.00%</u>), (**U**= Utilized or Starting Capital= <u>64,720</u> millions USD), (**l**= L/E= Leverage or Gearing Ratio= <u>99.00%</u>), (**L**= Liabilities or Debt= <u>69,547</u> millions USD), and (**\$'**= Sales of Past Year= <u>48,851</u> millions USD), then it's (**v**= Variable Portion planned), is:

$$\mathbf{v}= 1-(\mathbf{\$i}+[\mathbf{L/l}-\mathbf{U}]/\{[1-\mathbf{d}][1-\mathbf{t}]\})/\{\mathbf{\$'}[1+\mathbf{s}]\}-\mathbf{f}$$

$$= 1-(51,294*0.0100+[69,547/0.9900$$
$$-64,720]/\{[1-0.3000]$$
$$[1-0.3000]\})/\{48,851$$
$$[1+0.0500]\}-0.5800$$

$$= \underline{19.00\%}$$

Corporate IFRS-GAAP (B/S-I/S), ISBN 13: **978-1983566332**, ISBN 10: **1983566330**

Law-5114:

If both (**t**= T/B= Tax Rate= 30.00%), (**v**= V/S= Variable Portion= 19.00%), (**d**= D/A= Dividend Portion or Payout= 30.00%), (**i**= I/S= Interest Portion= 1.00%), (**$**= Sales or Revenues= 51,294 millions USD), (**s**= [S/S']-1= Sales Growth= 5.00%), (**U**= Utilized or Starting Capital= 64,720 millions USD), (**f**= L/E= Leverage or Gearing Ratio= 99.00%), (**L**= Liabilities or Debt= 69,547 millions USD), and (**$'**= Sales of Past Year= 48,851 millions USD), then it's (**f**= Fixed Portion planned), is:

$$f= 1-(\$i+[L/\textbf{\textit{I}}-U]/\{[1-d][1-t]\})/\{\$'[1+s]\}-v$$

$$= 1-(51,294*0.0100+[69,547/0.9900$$
$$-64,720]/\{[1-0.3000]$$
$$[1-0.3000]\})/\{48,851$$
$$[1+0.0500]\}-0.1900$$

$$= \underline{58.00\%}$$

Corporate IFRS-GAAP (B/S-I/S), ISBN 13: **978-1983566332**, ISBN 10: **1983566330**

<u>Law-5115</u>:

If both (**t**= T/B= Tax Rate= <u>30.00%</u>), (**v**= V/S= Variable Portion= <u>19.00%</u>), (**d**= D/A= Dividend Portion or Payout= <u>30.00%</u>), (**i**= I/S= Interest Portion= <u>1.00%</u>), (**f**= F/S= Fixed Portion= <u>58.00%</u>), (**s**= [S/S']-1= Sales Growth= <u>5.00%</u>), (**U**= Utilized or Starting Capital= <u>64,720</u> millions USD), (**l**= L/E= Leverage or Gearing Ratio= <u>99.00%</u>), (**L**= Liabilities or Debt= <u>69,547</u> millions USD), and (**S'**= Sales of Past Year= <u>48,851</u> millions USD), then it's (**S**= Sales or Revenues planned), is:

$$
\begin{aligned}
S &= (S'[1+s][1-v-f]-[L/l-U]/\{[1-d][1-t]\})/i \\
&= (48{,}851[1+0.0500][1-0.1900-0.5800] \\
&\quad -[69{,}547/0.9900-64{,}720] \\
&\quad /\{[1-0.3000][1-0.3000]\}) \\
&\quad /0.0100 \\
&= \underline{51{,}294} \text{ millions USD}
\end{aligned}
$$

Corporate IFRS-GAAP (B/S-I/S), ISBN 13: **978-1983566332**, ISBN 10: **1983566330**

Law-5116:

If both (**t**= T/B= Tax Rate= 30.00%), (**v**= V/S= Variable Portion= 19.00%), (**d**= D/A= Dividend Portion or Payout= 30.00%), (**$**= Sales or Revenues= 51,294 millions USD), (**f**= F/S= Fixed Portion= 58.00%), (**s**= [S/S']-1= Sales Growth= 5.00%), (**U**= Utilized or Starting Capital= 64,720 millions USD), (**⌿**= L/E= Leverage or Gearing Ratio= 99.00%), (**L**= Liabilities or Debt= 69,547 millions USD), and (**$'**= Sales of Past Year= 48,851 millions USD), then it's (**i**= Interest Portion planned), is:

$$i= (\$'[1+s][1-v-f]-[L/⌿-U]/\{[1-d][1-t]\})/\$$$
$$= (48,851[1+0.0500][1-0.1900-$$
$$0.5800]-[69,547/0.9900$$
$$-64,720]/\{[1-0.3000]$$
$$[1-0.3000]\})/51,294$$

$$= \underline{1.00\%}$$

Corporate IFRS-GAAP (B/S-I/S), ISBN 13: **978-1983566332**, ISBN 10: **1983566330**

<u>Law-5117</u>:

If both (**i**= I/S= Interest Portion= <u>1.00%</u>), (**v**= V/S= Variable Portion= <u>19.00%</u>), (**d**= D/A= Dividend Portion or Payout= <u>30.00%</u>), (**S**= Sales or Revenues= <u>51,294</u> millions USD), (**f**= F/S=Fixed Portion= <u>58.00%</u>), (**s**= [S/S']-1= Sales Growth= <u>5.00%</u>), (**U**= Utilized or Starting Capital= <u>64,720</u> millions USD), (**l**= L/E= Leverage or Gearing Ratio= <u>99.00%</u>), (**L**= Liabilities or Debt= <u>69,547</u> millions USD), and (**S'**= Sales of Past Year= <u>48,851</u> millions USD), then it's (**t**= Tax Rate planned), is:

$$t= 1-[\mathbf{L}/\mathbf{l}-\mathbf{U}]/([1-\mathbf{d}]\{\mathbf{S'}[1+\mathbf{s}][1-\mathbf{v}-\mathbf{f}]-\mathbf{Si}\})$$

$$= 1-[69,547/0.9900-64,720]$$
$$/([1-0.3000]\{48,851$$
$$[1+0.0500][1-0.1900$$
$$-0.5800]-51,294*0.0100\})$$

$$= \underline{30.00\%}$$

Corporate IFRS-GAAP (B/S-I/S), ISBN 13: **978-1983566332**, ISBN 10: **1983566330**

Law-5118:

If both (**i**= I/S= Interest Portion= 1.00%), (**v**= V/S= Variable Portion= 19.00%), (**t**= T/B= Tax Rate= 30.00%), (**S**= Sales or Revenues= 51,294 millions USD), (**f**= F/S= Fixed Portion= 58.00%), (**s**= [S/S']-1= Sales Growth= 5.00%), (**U**= Utilized or Starting Capital= 64,720 millions USD), (**l**= L/E= Leverage or Gearing Ratio= 99.00%), (**L**= Liabilities or Debt= 69,547 millions USD), and (**S'**= Sales of Past Year= 48,851 millions USD), then it's (**d**= Dividend Portion or Payout planned), is:

$$d= 1-[\mathbf{L/l}-\mathbf{U}]/([1-\mathbf{t}]\{\mathbf{S'}[1+\mathbf{s}][1-\mathbf{v}-\mathbf{f}]-\mathbf{Si}\})$$
$$= 1-[69,547/0.9900-64,720]$$
$$/([1-0.3000]\{48,851$$
$$[1+0.0500][1-0.1900$$
$$-0.5800]-51,294*0.0100\})$$
$$= 30.00\%$$

Corporate IFRS-GAAP (B/S-I/S), ISBN 13: **978-1983566332**, ISBN 10: **1983566330**

<u>Law-5119</u>:

If both (i= I/S= Interest Portion= <u>1.00%</u>), (v= V/S= Variable Portion= <u>19.00%</u>), (t= T/B= Tax Rate= <u>30.00%</u>), (S= Sales or Revenues= <u>51,294</u> millions USD), (f= F/S= Fixed Portion= <u>58.00%</u>), (s= [S/S']-1= Sales Growth= <u>5.00%</u>), (d= D/A= Dividend Portion or Payout= <u>30.00%</u>), (l= L/E= Leverage or Gearing Ratio= <u>99.00%</u>), (L= Liabilities or Debt= <u>69,547</u> millions USD), and (S'= Sales of Past Year= <u>48,851</u> millions USD), then it's (U= Utilized or Starting Capital planned), is:

$$U= L/l -[1-d][1-t]\{S'[1+s][1-v-f]-Si\}$$
$$= 69,547/0.9900-[1-0.3000]$$
$$[1-0.3000]\{48,851$$
$$[1+0.0500][1-0.1900$$
$$-0.5800]-51,294*0.0100\})$$
$$= \underline{64,720} \text{ millions USD}$$

Corporate IFRS-GAAP (B/S-I/S), ISBN 13: **978-1983566332**, ISBN 10: **1983566330**

<u>Law-5120</u>:

If both (**i**= I/S= Interest Portion= <u>1.00%</u>), (**v**= V/S= Variable Portion= <u>19.00%</u>), (**t**= T/B= Tax Rate= <u>30.00%</u>), (**$**= Sales or Revenues= <u>51,294</u> millions USD), (**f**= F/S= Fixed Portion= <u>58.00%</u>), (**s**= [S/S']-1= Sales Growth= <u>5.00%</u>), (**d**= D/A= Dividend Portion or Payout= <u>30.00%</u>), (**U**= Utilized or Starting Capital= <u>64,720</u> millions USD), (**L**= Liabilities or Debt= <u>69,547</u> millions USD), and (**$'**= Sales of Past Year= <u>48,851</u> millions USD), then it's (**l** = Leverage or Gearing Ratio planned), is:

$$l = L/(U+[1-d][1-t]\{\$'[1+s][1-v-f]-\$i\}$$
$$= 69,547/(64,720+[1-0.3000]$$
$$[1-0.3000]\{48,851$$
$$[1+0.0500][1-0.1900$$
$$-0.5800]-51,294*0.0100\})$$
$$= \underline{99.00\%}$$

Corporate IFRS-GAAP (B/S-I/S), ISBN 13: **978-1983566332**, ISBN 10: **1983566330**

<u>Law-5121</u>:

If both (**i**= I/S= Interest Portion= <u>1.00%</u>), (**v**= V/S= Variable Portion= <u>19.00%</u>), (**t**= T/B= Tax Rate= <u>30.00%</u>), (**$**= Sales or Revenues= <u>51,294</u> millions USD), (**f**= F/S= Fixed Portion= <u>58.00%</u>), (**s**= [S/S']-1= Sales Growth= <u>5.00%</u>), (**d**= D/A= Dividend Portion or Payout= <u>30.00%</u>), (**U**= Utilized or Starting Capital= <u>64,720</u> millions USD), (**l**= L/E= Leverage or Gearing Ratio= <u>99.00%</u>), and (**$'**= Sales of Past Year= <u>48,851</u> millions USD), then it's (**L**= Liabilities or Debt planned), is:

$$L= l\,(U+[1\text{-}d][1\text{-}t]\{\$'[1+s]\text{-}\$'v[1+s]$$
$$-\$'f[1+s]]\text{-}\$'i[1+s]\})$$
$$= l\,(U+[1\text{-}d][1\text{-}t]\{\$'[1+s][1\text{-}v\text{-}f\text{-}i]\}$$
$$= 0.9900/(64,720]+[1\text{-}0.3000]$$
$$[1\text{-}0.3000]\{48,851$$
$$[1+0.0500][1\text{-}0.1900\text{-}0.5800$$
$$-0.0100]\})$$
$$= \underline{69,547}\ \text{millions USD}$$

Corporate IFRS-GAAP (B/S-I/S), ISBN 13: **978-1983566332**, ISBN 10: **1983566330**

Law-5122:

If both (**i**= I/S= Interest Portion= <u>1.00%</u>), (**v**= V/S= Variable Portion= <u>19.00%</u>), (**t**= T/B= Tax Rate= <u>30.00%</u>), (**S**= Sales or Revenues= <u>51,294</u> millions USD), (**f**= F/S= Fixed Portion= <u>58.00%</u>), (**s**= [S/S']- 1= Sales Growth= <u>5.00%</u>), (**d**= D/A= Dividend Portion or Payout= <u>30.00%</u>), (**U**= Utilized or Starting Capital= <u>64,720</u> millions USD), (**l**= L/E= Leverage or Gearing Ratio= <u>99.00%</u>), and (**L**= Liabilities or Debt= <u>69,547</u> millions USD), then it's (**S'**= Sales Past), must be:

$$S' = [L/l - U]/\{[1-d][1-t]\})/\{[1+s][1-v-f-i]\}$$
$$= [69,547/0.9900 - 64,720]$$
$$/\{[1-0.3000][1-0.3000]\})$$
$$/\{[1+0.0500][1-0.1900$$
$$-0.5800-0.0100]\}$$
$$= \underline{69,547} \text{ millions USD}$$

Corporate IFRS-GAAP (B/S-I/S), ISBN 13: **978-1983566332**, ISBN 10: **1983566330**

<u>Law-5123</u>:

If both (**i**= I/S= Interest Portion= <u>1.00%</u>), (**v**= V/S= Variable Portion= <u>19.00%</u>), (**t**= T/B= Tax Rate= <u>30.00%</u>), (**$**= Sales or Revenues= <u>51,294</u> millions USD), (**f**= F/S= Fixed Portion= <u>58.00%</u>), (**$'**= Sales of Past Year= <u>48,851</u> millions USD), (**d**= D/A= Dividend Portion or Payout= <u>30.00%</u>), (**U**= Utilized or Starting Capital= <u>64,720</u> millions USD), (**l**= L/E= Leverage or Gearing Ratio= <u>99.00%</u>), and (**L**= Liabilities or Debt= <u>69,547</u> millions USD), then it's (**s**= Sales Growth planned), is:

$$s = [\mathbf{L}/\mathbf{l} - \mathbf{U}]/\{[1-\mathbf{d}][1-\mathbf{t}]\})/\{\mathbf{\$'}[1-\mathbf{v}-\mathbf{f}-\mathbf{i}]\} - 1$$

$$= [69,547/0.9900 - 64,720]$$
$$/\{[1-0.3000][1-0.3000]\})$$
$$/\{48,851[1-0.1900-0.5800$$
$$-0.0100]\} - 1$$

$$= \underline{5.00\%}$$

Corporate IFRS-GAAP (B/S-I/S), ISBN 13: **978-1983566332**, ISBN 10: **1983566330**

<u>Law-5124</u>:

If both (**i**= I/S= Interest Portion= <u>1.00%</u>), (**s**= [S/S']-1= Sales Growth= <u>5.00%</u>), (**t**= T/B= Tax Rate= <u>30.00%</u>), (**$**= Sales or Revenues= <u>51,294</u> millions USD), (**f**= F/S= Fixed Portion= <u>58.00%</u>), (**$'**= Sales of Past Year= <u>48,851</u> millions USD), (**d**= D/A= Dividend Portion or Payout= <u>30.00%</u>), (**U**= Utilized or Starting Capital= <u>64,720</u> millions USD), (**l**= L/E= Leverage or Gearing Ratio= <u>99.00%</u>), and (**L**= Liabilities or Debt= <u>69,547</u> millions USD), then it's (**v**= Variable Portion planned), is:

$$\mathbf{v}= 1\text{-}\mathbf{f}\text{-}\mathbf{i}\text{-}[\mathbf{L}/\mathbf{l}\text{-}\mathbf{U}]/\{[1\text{-}\mathbf{d}][1\text{-}\mathbf{t}]\})/\{\mathbf{\$'}[1+\mathbf{s}]\}$$

$$= 1\text{-}0.5800\text{-}0.0100\text{-}[69,547/0.9900$$
$$-64,720]/\{[1\text{-}0.3000]$$
$$[1\text{-}0.3000]\})/\{48,851$$
$$[1+0.0500]\}$$

$$= \underline{19.00\%}$$

Corporate IFRS-GAAP (B/S-I/S), ISBN 13: **978-1983566332**, ISBN 10: **1983566330**

Law-5125:

If both (**i**= I/S= Interest Portion= 1.00%), (**s**= [S/S']-1= Sales Growth= 5.00%), (**t**= T/B= Tax Rate= 30.00%), (**S**= Sales or Revenues= 51,294 millions USD), (**v**= V/S= Variable Portion= 19.00%), (**S'**= Sales of Past Year= 48,851 millions USD), (**d**= D/A= Dividend Portion or Payout= 30.00%), (**U**= Utilized or Starting Capital= 64,720 millions USD), (**I**= L/E= Leverage or Gearing Ratio= 99.00%), and (**L**= Liabilities or Debt= 69,547 millions USD), then it's (**f**= Fixed Portion planned), is:

$$\mathbf{f}= 1-\mathbf{v}-\mathbf{i}-[\mathbf{L/I}-\mathbf{U}]/\{[1-\mathbf{d}][1-\mathbf{t}]\})/\{\mathbf{S'}[1+\mathbf{s}]\}$$

$$= 1-0.1900-0.0100-[69,547/0.9900$$
$$-64,720]/\{[1-0.3000]$$
$$[1-0.3000]\})/$$
$$\{48,851[1+0.0500]\}$$

$$= 58.00\%$$

Corporate IFRS-GAAP (B/S-I/S), ISBN 13: **978-1983566332**, ISBN 10: **1983566330**

Law-5126:

If both (**f**= F/S= Fixed Portion= 58.00%), (**s**= [S/S']-1= Sales Growth= 5.00%), (**t**= T/B= Tax Rate= 30.00%), (**S**= Sales or Revenues= 51,294 millions USD), (**v**= V/S= Variable Portion= 19.00%), (**S'**= Sales of Past Year= 48,851 millions USD), (**d**= D/A= Dividend Portion or Payout= 30.00%), (**U**= Utilized or Starting Capital= 64,720 millions USD), (**l**= L/E= Leverage or Gearing Ratio= 99.00%), and (**L**= Liabilities or Debt= 69,547 millions USD), then it's (**i**= Interest Portion planned), is:

$$i = 1-v-f-[L/l-U]/\{[1-d][1-t]\})/\{S'[1+s]\}$$
$$= 1-0.1900-0.5800-[69,547/0.9900$$
$$-64,720]/\{[1-0.3000]$$
$$[1-0.3000]\})/\{48,851$$
$$[1+0.0500]\}$$
$$= 1.00\%$$

Corporate IFRS-GAAP (B/S-I/S), ISBN 13: **978-1983566332**, ISBN 10: **1983566330**

<u>Law-5127</u>:

If both (**f**= F/S= Fixed Portion= <u>58.00%</u>), (**s**= [S/S']-1= Sales Growth= <u>5.00%</u>), (**i**= I/S= Interest Portion= <u>1.00%</u>), (**$**= Sales or Revenues= <u>51,294</u> millions USD), (**v**= V/S= Variable Portion= <u>19.00%</u>), (**$'**= Sales of Past Year= <u>48,851</u> millions USD), (**d**= D/A= Dividend Portion or Payout= <u>30.00%</u>), (**U**= Utilized or Starting Capital= <u>64,720</u> millions USD), (**l**= L/E= Leverage or Gearing Ratio= <u>99.00%</u>), and (**L**= Liabilities or Debt= <u>69,547</u> millions USD), then it's (**t**= Tax Rate planned), is:

$$t= 1-[\mathbf{L}/\mathbf{l}-\mathbf{U}]/([1-\mathbf{d}]\{\mathbf{\$'}[1+\mathbf{s}][1-\mathbf{v}-\mathbf{f}-\mathbf{i}]\})$$

$$= 1-[69,547/0.9900-64,720]$$
$$/([1-0.3000]\{48,851$$
$$[1+0.0500][1-0.1900$$
$$-0.5800-0.0100]\})$$

$$= \underline{30.00\%}$$

Corporate IFRS-GAAP (B/S-I/S), ISBN 13: **978-1983566332**, ISBN 10: **1983566330**

<u>Law-5128</u>:

If both (**f**= F/S= Fixed Portion= <u>58.00%</u>), (**s**= [S/S']-1= Sales Growth= <u>5.00%</u>), (**i**= I/S= Interest Portion= <u>1.00%</u>), (**S**= Sales or Revenues= <u>51,294</u> millions USD), (**v**= V/S= Variable Portion= <u>19.00%</u>), (**S'**= Sales of Past Year= <u>48,851</u> millions USD), (**t**= T/B= Tax Rate= <u>30.00%</u>), (**U**= Utilized or Starting Capital= <u>64,720</u> millions USD), (**l**= L/E= Leverage or Gearing Ratio= <u>99.00%</u>), and (**L**= Liabilities or Debt= <u>69,547</u> millions USD), then it's (**d**= Dividend Portion or Payout planned), is:

$$d= 1-[\textbf{L}/\textbf{l}-\textbf{U}]/([1-\textbf{t}]\{\textbf{S'}[1+\textbf{s}][1-\textbf{v}-\textbf{f}-\textbf{i}]\})$$

$$= 1-[69,547/0.9900-64,720]$$
$$/([1-0.3000]\{48,851$$
$$[1+0.0500][1-0.1900$$
$$-0.5800-0.0100]\})$$

$$= \underline{30.00\%}$$

Corporate IFRS-GAAP (B/S-I/S), ISBN 13: **978-1983566332**, ISBN 10: **1983566330**

<u>Law-5129</u>:

If both (f= F/S= Fixed Portion= <u>58.00%</u>), (s= [S/S']-1= Sales Growth= <u>5.00%</u>), (i= I/S= Interest Portion= <u>1.00%</u>), ($\$$= Sales or Revenues= <u>51,294</u> millions USD), (v= V/S= Variable Portion= <u>19.00%</u>), ($\$'$= Sales of Past Year= <u>48,851</u> millions USD), (t= T/B= Tax Rate= <u>30.00%</u>), (d= D/A= Dividend Portion or Payout= <u>30.00%</u>), (l= L/E= Leverage or Gearing Ratio= <u>99.00%</u>), and (L= Liabilities or Debt= <u>69,547</u> millions USD), then it's (U= Utilized or Starting Capital planned), is:

$$U = L/l - [1-d][1-t]\{\$'[1+s][1-v-f-i]\}$$
$$= 69,547/0.9900 - [1-0.3000]$$
$$[1-0.3000]\{48,851$$
$$[1+0.0500][1-0.1900$$
$$-0.5800-0.0100]\}$$
$$= \underline{64,720} \text{ millions USD}$$

Corporate IFRS-GAAP (B/S-I/S), ISBN 13: **978-1983566332**, ISBN 10: **1983566330**

<u>Law-5130</u>:

If both (**f**= F/S= Fixed Portion= <u>58.00%</u>), (**s**= [S/S']-1= Sales Growth= <u>5.00%</u>), (**i**= I/S= Interest Portion= <u>1.00%</u>), (**S**= Sales or Revenues= <u>51,294</u> millions USD), (**v**= V/S= Variable Portion= <u>19.00%</u>), (**S'**= Sales of Past Year= <u>48,851</u> millions USD), (**t**= T/B= Tax Rate= <u>30.00%</u>), (**d**= D/A= Dividend Portion or Payout= <u>30.00%</u>), (**U**= Utilized or Starting Capital= <u>64,720</u> millions USD),and (**L**= Liabilities or Debt= <u>69,547</u> millions USD), then it's (**l** = Leverage or Gearing Ratio planned), is:

$$l = L/(U+[1-d][1-t]\{S'[1+s][1-v-f-i]\}$$
$$= 69,547/(64,720+[1-0.3000]$$
$$[1-0.3000]\{48,851$$
$$[1+0.0500][1-0.1900$$
$$-0.5800-0.0100]\}$$
$$= \underline{99.00\%}$$

Corporate IFRS-GAAP (B/S-I/S), ISBN 13: **978-1983566332**, ISBN 10: **1983566330**

CHAPTER-16:
L= LIABUILITY or Total Debt Optimization

Based on 12/31/15 Pfizer Annual Report
(http://www.nasdaq.com/symbol/pfe/financials?query=income-statement)

Law-5131:
 If both (**E**= Equity or Capital= 70,249 millions USD),
 and (**L**= Liabilities or Debt= 69,547 millions USD),
 then it's (**W**= Wealth or Total Assets planned), is:
 W= **E**+**L**
$$= 70,249+69,547$$
$$= 139,796 \text{ millions USD}$$

Law-5132:
If both (**W**= Wealth or Total Assets= 139,796 millions
USD),and (**L**= Liabilities or Debt= 69,547 millions
USD), then it's (**E**= Equity or Capital planned), is:
 $E = W-L$

$$= 139,796-69,547$$
$$= 70,249 \text{ millions USD}$$

Corporate IFRS-GAAP (B/S-I/S), ISBN 13: **978-1983566332**, ISBN 10: **1983566330**

Law-5133:

   If both (**W**= Wealth or Total Assets= <u>139,796</u> millions USD),and (**E**= Equity or Capital= <u>70,249</u> millions USD), then it's (**L**= Liabilities or Debt planned), is:

   L= W-E
   $$= 139,796\text{-}70,249$$
   $$= \underline{69,547} \text{ millions USD}$$

Law-5134:

   If both (**I**= L/E= Leverage or Gearing Ratio= <u>99.00%</u>), and (**E**= Equity or Capital= <u>70,249</u> millions USD), then it's (**W**= Wealth or Total Assets planned), is:

   W= E+L= E[1+**I**]
   $$= 70,249[1+0.9900]$$
   $$= \underline{139,796} \text{ millions USD}$$

Corporate IFRS-GAAP (B/S-I/S), ISBN 13: **978-1983566332**, ISBN 10: **1983566330**

Law-5135:

If both (**I**= L/E= Leverage or Gearing Ratio= 99.00%), and (**W**= Wealth or Total Assets= 139,796 millions USD), then it's (**E**= Equity or Capital planned), is:

$$\mathbf{E} = \mathbf{W}/[1+\mathbf{I}]$$
$$= 139,796/[1+0.9900]$$
$$= \underline{70,249} \text{ millions USD}$$

Law-5136:

If both (**E**= Equity or Capital= 70,249 millions USD), and (**W**= Wealth or Total Assets= 139,796 millions USD), then it's (**I** = Leverage or Gearing Ratio planned), is:

$$\mathbf{I} = [\mathbf{W}/\mathbf{E}]-1$$
$$= [139,796/70,249]-1$$
$$= \underline{99.00}\%$$

Corporate IFRS-GAAP (B/S-I/S), ISBN 13: **978-1983566332**, ISBN 10: **1983566330**

Law-5137:

If both (**U**= Utilized or Starting Capital= 64,720 millions USD), (**R**= Retained Earnings= 5,529 millions USD),and (**I**= L/E= Leverage or Gearing Ratio= 99.00%), then it's (**W**= Wealth or Total Assets planned), is:

$$\textbf{W}= \textbf{E}[1+\textbf{I}]= [1+\textbf{I}][\textbf{U}+\textbf{R}]$$
$$= [1+0.9900][64,720+5,529]$$
$$= 139,796 \text{ millions USD}$$

Law-5138:

If both (**U**= Utilized or Starting Capital= 64,720 millions USD), (**W**= Wealth or Total Assets= 139,796 millions USD),and (**I**= L/E= Leverage or Gearing Ratio= 99.00%), then it's (**R**= Retained Earnings planned), is:

$$\textbf{R}= \{\textbf{W}/[1+\textbf{I}]\}\textbf{-U}$$
$$= \{139,796/[1+0.9900]\}-64,720$$
$$= 5,529 \text{ millions USD}$$

Corporate IFRS-GAAP (B/S-I/S), ISBN 13: **978-1983566332**, ISBN 10: **1983566330**

<u>Law-5139</u>:

If both (**R**= Retained Earnings= <u>5,529</u> millions USD), (**W**= Wealth or Total Assets= <u>139,796</u> millions USD),and (**I**= L/E= Leverage or Gearing Ratio= <u>99.00%</u>), then it's (**U**= Utilized or Starting Capital planned), is:

$$\mathbf{U} = \{\mathbf{W}/[1+\mathbf{I}]\} - \mathbf{R}$$
$$= \{139,796/[1+0.9900]\} - 5,529$$
$$= \underline{64,720} \text{ millions USD}$$

<u>Law-5140</u>:

If both (**R**= Retained Earnings= <u>5,529</u> millions USD), (**W**= Wealth or Total Assets= <u>139,796</u> millions USD),and (**U**= Utilized or Starting Capital= <u>64,720</u> millions USD),then it's (**I** = Leverage or Gearing Ratio planned), is:

$$\mathbf{I} = \{\mathbf{W}/[\mathbf{U}+\mathbf{R}]\} - 1$$
$$= \{139,796/[64,720+5,529]\} - 1$$
$$= \underline{99.00\%}$$

Corporate IFRS-GAAP (B/S-I/S), ISBN 13: **978-1983566332**, ISBN 10: **1983566330**

Law-5141:

If both (**D**= Dividend Paid= 2,370 millions USD), (**A**= After Tax Income= 7,899 millions USD), (**I**= L/E= Leverage or Gearing Ratio= 99.00%), and (**U**= Utilized or Starting Capital= 64,720 millions USD),then it's (**W**= Wealth or Total Assets planned), is:

$$\mathbf{W}= \mathbf{E}[1+\mathbf{I}]= [1+\mathbf{I}][\mathbf{U}+\mathbf{A}\text{-}\mathbf{D}]$$
$$= [1+0.9900][64,720+7,899\text{-}2,370]\}$$
$$= 139,796 \text{ millions USD}$$

Law-5142:

If both (**D**= Dividend Paid= 2,370 millions USD), (**W**= Wealth or Total Assets= 139,796 millions USD), (**I**= L/E= Leverage or Gearing Ratio= 99.00%), and (**U**= Utilized or Starting Capital= 64,720 millions USD),then it's (**A**= After Tax Income planned), is:

$$\mathbf{A}= \{\mathbf{W}[1+\mathbf{I}]\}+\mathbf{D}\text{-}\mathbf{U}$$
$$= \{139,796/[1+0.9900]\}+2,370$$
$$-64,720$$
$$= 7,899 \text{ millions USD}$$

Corporate IFRS-GAAP (B/S-I/S), ISBN 13: **978-1983566332**, ISBN 10: **1983566330**

Law-5143:

If both (**A**= After Tax Income= 7,899 millions USD), (**W**= Wealth or Total Assets= 139,796 millions USD), (**I**= L/E= Leverage or Gearing Ratio= 99.00%), and (**U**= Utilized or Starting Capital= 64,720 millions USD),then it's (**D**= Dividend Paid planned), is:

$$\mathbf{D}= \mathbf{U}+\mathbf{A}-\{\mathbf{W}[1+\mathbf{I}]\}$$
$$= 64,720+7,899-\{139,796$$
$$/[1+0.9900]\}$$
$$= 2,370 \text{ millions USD}$$

Law-5144:

If both (**A**= After Tax Income= 7,899 millions USD), (**W**= Wealth or Total Assets= 139,796 millions USD), (**I**= L/E= Leverage or Gearing Ratio= 99.00%),and (**D**= Dividend Paid= 2,370 millions USD),then it's (**U**= Utilized or Starting Capital planned), is:

$$\mathbf{U}= \{\mathbf{W}[1+\mathbf{I}]\}+\mathbf{D}-\mathbf{A}$$
$$= \{139,796/[1+0.9900]\}+2,370$$
$$-7,899$$
$$= 64,720 \text{ millions USD}$$

Corporate IFRS-GAAP (B/S-I/S), ISBN 13: **978-1983566332**, ISBN 10: **1983566330**

Law-5145:

  If both (**A**= After Tax Income= 7,899 millions USD),
  (**W**= Wealth or Total Assets= 139,796 millions
  USD), (**U**= Utilized or Starting Capital= 64,720
  millions USD), and (**D**= Dividend Paid= 2,370
  millions USD),then it's (**I** = Leverage or Gearing
  Ratio planned), is:

$$I = \{W/[U+A-D]\}-1$$
$$= \{139,796/[64,720+7,899$$
$$-2,370]\}-1$$
$$= 99.00\%$$

Law-5146:

  If both (**A**= After Tax Income= 7,899 millions USD),
  (**I**= L/E= Leverage or Gearing Ratio= 99.00%), (**U**=
  Utilized or Starting Capital= 64,720 millions USD),
  and (**d**= D/A= Dividend Portion or Payout= 30.00%),
  then it's (**W**= Wealth or Total Assets planned), is:

$$W = [1+I][U+A-Ad] = [1+I]\{U+A[1-d]\}$$
$$= [1+0.9900]\{64,720+7,899$$
$$[1-0.3000]\}$$
$$= 139,796 \text{ millions USD}$$

Corporate IFRS-GAAP (B/S-I/S), ISBN 13: **978-1983566332**, ISBN 10: **1983566330**

<u>Law-5147</u>:

If both (**W**= Wealth or Total Assets= <u>139,796</u> millions USD), (**I**= L/E= Leverage or Gearing Ratio= <u>99.00%</u>), (**U**= Utilized or Starting Capital= <u>64,720</u> millions USD), and (**d**= D/A= Dividend Portion or Payout= <u>30.00%</u>), then it's (**A**= After Tax Income planned), is:

$$\mathbf{A} = \{\mathbf{W}/[1+\mathbf{I}]-\mathbf{U}\}/[1-\mathbf{d}]$$
$$= \{139{,}796/[1+0.9900]-64{,}720\}$$
$$/[1-0.3000]$$
$$= \underline{7{,}899} \text{ millions USD}$$

<u>Law-5148</u>:

If both (**W**= Wealth or Total Assets= <u>139,796</u> millions USD), (**I**= L/E= Leverage or Gearing Ratio= <u>99.00%</u>), (**U**= Utilized or Starting Capital= <u>64,720</u> millions USD), and (**A**= After Tax Income= <u>7,899</u> millions USD), then it's (**d**= Dividend Portion or Payout planned), is:

$$\mathbf{d} = 1 - \{\mathbf{W}/[1+\mathbf{I}]-\mathbf{U}\}/\mathbf{A}$$
$$= 1 - \{139{,}796/[1+0.9900]-64{,}720\}$$
$$/7{,}899$$
$$= \underline{30.00\%}$$

Corporate IFRS-GAAP (B/S-I/S), ISBN 13: **978-1983566332**, ISBN 10: **1983566330**

Law-5149:

If both (**W**= Wealth or Total Assets= 139,796 millions USD), (**I**= L/E= Leverage or Gearing Ratio= 99.00%), (**d**= D/A= Dividend Portion or Payout= 30.00%), and (**A**= After Tax Income= 7,899 millions USD), then it's (**U**= Utilized or Starting Capital planned), is:

$$\mathbf{U}= \{\mathbf{W}/[1+\mathbf{I}]-\mathbf{A}[1-\mathbf{d}]$$
$$= \{139,796/[1+0.9900]$$
$$-7,899[1-0.3000]$$
$$= 64,720 \text{ millions USD}$$

Law-5150:

If both (**W**= Wealth or Total Assets= 139,796 millions USD), (**U**= Utilized or Starting Capital= 64,720 millions USD), (**d**= D/A= Dividend Portion or Payout= 30.00%), and (**A**= After Tax Income= 7,899 millions USD), then it's (**I** = Leverage or Gearing Ratio planned), is:

$$\mathbf{I} = \mathbf{W}/\{\mathbf{U}+\mathbf{A}[1-\mathbf{d}]\}-1$$
$$= 139,796/\{64,720+7,899$$
$$[1-0.3000]\}-1$$
$$= 99.00\%$$

Corporate IFRS-GAAP (B/S-I/S), ISBN 13: **978-1983566332**, ISBN 10: **1983566330**

<u>Law-5151</u>:

If both (**T**= Tax Paid= <u>3,385</u> millions USD), (**I**= L/E= Leverage or Gearing Ratio= <u>99.00%</u>), (**U**= Utilized or Starting Capital= <u>64,720</u> millions USD), (**d**= D/A= Dividend Portion or Payout= <u>30.00%</u>), and (**B**= Before Tax Income = <u>11,285</u> millions USD), then it's (**W**= Wealth or Total Assets planned), is:

$$W= [1+I]\{U+[1-d][B-T]$$
$$= [1+0.9900]\{64,720+[1-0.3000]$$
$$[11,285-3,385]$$
$$= \underline{139,796} \text{ millions USD}$$

Corporate IFRS-GAAP (B/S-I/S), ISBN 13: **978-1983566332**, ISBN 10: **1983566330**

Law-5152:

If both (**T**= Tax Paid= 3,385 millions USD), (**I**= L/E= Leverage or Gearing Ratio= 99.00%), (**U**= Utilized or Starting Capital= 64,720 millions USD), (**d**= D/A= Dividend Portion or Payout= 30.00%), and (**W**= Wealth or Total Assets= 139,796 millions USD), then it's (**B**= Before Tax Income planned), is:

$$\mathbf{B}= \mathbf{T}+\{\mathbf{W}/[1+\mathbf{I}]-\mathbf{U}\}/[1-\mathbf{d}]$$
$$= 3,385+\{139,796/[1+0.9900]$$
$$-64,720\}/[1-0.3000]$$
$$= 11,285 \text{ millions USD}$$

Corporate IFRS-GAAP (B/S-I/S), ISBN 13: **978-1983566332**, ISBN 10: **1983566330**

<u>Law-5153</u>:

If both (**B**= Before Tax Income = <u>11,285</u> millions USD), (**I**= L/E= Leverage or Gearing Ratio= <u>99.00%</u>), (**U**= Utilized or Starting Capital= <u>64,720</u> millions USD), (**d**= D/A= Dividend Portion or Payout= <u>30.00%</u>), and (**W**= Wealth or Total Assets= <u>139,796</u> millions USD), then it's (**T**= Tax Paid planned), is:

$$\mathbf{T}= \mathbf{B}-\{\mathbf{W}/[1+\mathbf{I}]-\mathbf{U}\}/[1-\mathbf{d}]$$
$$= 11{,}285-\{139{,}796/[1+0.9900]$$
$$-64{,}720\}/[1-0.3000]$$
$$= \underline{3{,}385}\ \text{millions USD}$$

Corporate IFRS-GAAP (B/S-I/S), ISBN 13: **978-1983566332**, ISBN 10: **1983566330**

Law-5154:

If both (**B**= Before Tax Income = 11,285 millions USD), (**I**= L/E= Leverage or Gearing Ratio= 99.00%), (**U**= Utilized or Starting Capital= 64,720 millions USD), (**T**= Tax Paid= 3,385 millions USD),and (**W**= Wealth or Total Assets= 139,796 millions USD), then it's (**d**= Dividend Portion or Payout planned), is:

$$\mathbf{d}= 1-\{\mathbf{W}/[1+\mathbf{I}]-\mathbf{U}\}/[\mathbf{B}-\mathbf{T}]$$
$$= 1-\{139,796/[1+0.9900]-64,720\}/[11,285-3,385]$$
$$= 30.00\%$$

Corporate IFRS-GAAP (B/S-I/S), ISBN 13: **978-1983566332**, ISBN 10: **1983566330**

<u>Law-5155</u>:

If both (**B**= Before Tax Income = <u>11,285</u> millions USD), (**I**= L/E= Leverage or Gearing Ratio= <u>99.00%</u>), (**d**= D/A= Dividend Portion or Payout= <u>30.00%</u>), (**T**= Tax Paid= <u>3,385</u> millions USD),and (**W**= Wealth or Total Assets= <u>139,796</u> millions USD), then it's (**U**= Utilized or Starting Capital planned), is:

$$U= W/[1+I]-[1-d][B-T]$$
$$= 139,796/[1+0.9900]-[1-0.3000]$$
$$[11,285-3,385]$$
$$= \underline{64,720} \text{ millions USD}$$

Corporate IFRS-GAAP (B/S-I/S), ISBN 13: **978-1983566332**, ISBN 10: **1983566330**

Law-5156:

If both (**B**= Before Tax Income = 11,285 millions USD), (**U**= Utilized or Starting Capital= 64,720 millions USD), (**d**= D/A= Dividend Portion or Payout= 30.00%), (**T**= Tax Paid= 3,385 millions USD),and (**W**= Wealth or Total Assets= 139,796 millions USD), then it's (**I** = Leverage or Gearing Ratio planned), is:

$$I = W/\{U+[1-d][B-T]\}-1$$
$$= 139,796/\{64,720+[1-0.3000]$$
$$[11,285-3,385]\}-1$$
$$= 99.00\%$$

Corporate IFRS-GAAP (B/S-I/S), ISBN 13: **978-1983566332**, ISBN 10: **1983566330**

Law-5157:

    If both (**B**= Before Tax Income = 11,285 millions USD), (**U**= Utilized or Starting Capital= 64,720 millions USD), (**d**= D/A= Dividend Portion or Payout= 30.00%), (**t**= T/B= Tax Rate= 30.00%), and (**l**= L/E= Leverage or Gearing Ratio= 99.00%), then it's (**W** = Wealth or Total Assets planned), is:

$$\mathbf{W}= [1+\mathbf{l}]\{\mathbf{U}+[1-\mathbf{d}][\mathbf{B}-\mathbf{Bt}]\}$$
$$= [1+\mathbf{l}]\{\mathbf{U}+\mathbf{B}[1-\mathbf{d}][1-\mathbf{t}]\}$$
$$= [1+0.9900]\{64,720+11,285$$
$$[1-0.3000][1-0.3000]\}$$
$$= 139,796 \text{ millions USD}$$

Corporate IFRS-GAAP (B/S-I/S), ISBN 13: **978-1983566332**, ISBN 10: **1983566330**

Law-5158:

If both (**W**= Wealth or Total Assets= 139,796 millions USD), (**U**= Utilized or Starting Capital= 64,720 millions USD), (**d**= D/A= Dividend Portion or Payout= 30.00%), (**t**= T/B= Tax Rate= 30.00%), and (**I**= L/E= Leverage or Gearing Ratio= 99.00%), then it's (**B**= Before Tax Income planned), is:

$$\mathbf{B} = (\{\mathbf{W}/[1+\mathbf{I}]-\mathbf{U}\})/\{[1-\mathbf{d}][1-\mathbf{t}]\}$$
$$= (\{139,796/[1+0.9900]-64,720\})$$
$$/\{[1-0.3000][1-0.3000]\}$$
$$= 1,285 \text{ millions USD}$$

Corporate IFRS-GAAP (B/S-I/S), ISBN 13: **978-1983566332**, ISBN 10: **1983566330**

Law-5159:

If both (**W**= Wealth or Total Assets= 139,796 millions USD), (**U**= Utilized or Starting Capital= 64,720 millions USD), (**d**= D/A=Dividend Portion or Payout= 30.00%), (**B**= Before Tax Income = 11,285 millions USD), and (**I**= L/E= Leverage or Gearing Ratio= 99.00%), then it's (**t**= Tax Rate planned), is:

$$t= 1-(\{W/[1+I]-U\})/\{B[1-d]\}$$
$$= 1-(\{139,796/[1+0.9900]-64,720\}$$
$$/\{11,285[1-0.3000]\}\})$$
$$= 30.00\%$$

Corporate IFRS-GAAP (B/S-I/S), ISBN 13: **978-1983566332**, ISBN 10: **1983566330**

Law-5160:

If both (**W**= Wealth or Total Assets= 139,796 millions USD), (**U**= Utilized or Starting Capital= 64,720 millions USD), (**t**= T/B= Tax Rate= 30.00%), (**B**= Before Tax Income = 11,285 millions USD), and (**I**= L/E= Leverage or Gearing Ratio= 99.00%), then it's (**d**= Dividend Portion or Payout planned), is:

$$\mathbf{d}= 1-(\{\mathbf{W}/[1+\mathbf{I}]-\mathbf{U}\})/\{\mathbf{B}[1-\mathbf{t}]\}$$
$$= 1-(\{139{,}796/[1+0.9900]-64{,}720\})/\{11{,}285[1-0.3000]\}]\}$$
$$= 30.00\%$$

Corporate IFRS-GAAP (B/S-I/S), ISBN 13: **978-1983566332**, ISBN 10: **1983566330**

Law-5161:

If both (**W**= Wealth or Total Assets= 139,796 millions USD), (**d**= D/A= Dividend Portion or Payout= 30.00%), (**t**= T/B= Tax Rate= 30.00%), (**B**= Before Tax Income = 11,285 millions USD), and (**I**= L/E= Leverage or Gearing Ratio= 99.00%), then it's (**U**= Utilized or Starting Capital planned), is:

$$U= W/[1+I]-B[1-d][1-t]$$
$$= 139,796/[1+0.9900]-11,285$$
$$[1-0.3000][1-0.3000]\}$$
$$= 64,720 \text{ millions USD}$$

Corporate IFRS-GAAP (B/S-I/S), ISBN 13: **978-1983566332**, ISBN 10: **1983566330**

Law-5162:

If both (**W**= Wealth or Total Assets= 139,796 millions USD), (**d**= D/A= Dividend Portion or Payout= 30.00%), (**t**= T/B= Tax Rate= 30.00%), (**B**= Before Tax Income = 11,285 millions USD), and (**U**= Utilized or Starting Capital= 64,720 millions USD), then it's (**I** = Leverage or Starting Capital planned), is:

$$I = W/\{U+B[1-d][1-t]\}-1$$
$$= 139,796/\{64,720+11,285$$
$$[1-0.3000][1-0.3000]\}-1$$
$$= 99.00\%$$

Corporate IFRS-GAAP (B/S-I/S), ISBN 13: **978-1983566332**, ISBN 10: **1983566330**

Law-5163:

If both (I= L/E= Leverage or Gearing Ratio= 99.00%), (d= D/A= Dividend Portion or Payout= 30.00%), (t= T/B= Tax Rate= 30.00%), (O= Operational Surplus= 11,798 millions USD), (I= Interest Expense= 513 millions USD), and (U= Utilized or Starting Capital= 64,720 millions USD), then it's (W= Wealth or Total Assets planned), is:

$$W= [1+I]\{U+[1-d][1-t]\}[O-I]$$
$$= [1+0.9900]\{64,720+[1-0.3000]$$
$$[1-0.3000]\}[11,798-513]$$
$$= 139,796 \text{ millions USD}$$

Corporate IFRS-GAAP (B/S-I/S), ISBN 13: **978-1983566332**, ISBN 10: **1983566330**

Law-5164:

If both (I= L/E= Leverage or Gearing Ratio= 99.00%), (d= D/A= Dividend Portion or Payout= 30.00%), (W= Wealth or Total Assets= 139,796 millions USD), (O= Operational Surplus= 11,798 millions USD), (I= Interest Expense= 513 millions USD), and (U= Utilized or Starting Capital= 64,720 millions USD), then it's (t= Tax Rate planned), is:

$$t = (1-\{W/[1+I]-U\})/\{[1-d][O-I]\}$$
$$= 1-\{139,796/[1+0.9900]-64,720\}$$
$$/\{[1-0.3000][11,798-513]\}$$
$$= 30.00\%$$

Corporate IFRS-GAAP (B/S-I/S), ISBN 13: **978-1983566332**, ISBN 10: **1983566330**

Law-5165:

If both ($\textbf{\textit{I}}$= L/E= Leverage or Gearing Ratio= 99.00%), ($\textbf{t}$= T/B= Tax Rate= 30.00%), ($\textbf{W}$= Wealth or Total Assets= 139,796 millions USD), ($\textbf{O}$= Operational Surplus= 11,798 millions USD), ($\textbf{I}$= Interest Expense= 513 millions USD), and ($\textbf{U}$= Utilized or Starting Capital= 64,720 millions USD), then it's ($\textbf{d}$= Dividend Portion or Payout planned), is:

$$\textbf{d}= (1-\{\textbf{W}/[1+\textbf{\textit{I}}]-\textbf{U}\})/\{[1-\textbf{t}][\textbf{O}-\textbf{I}]\}$$
$$= 1-\{139,796/[1+0.9900]-64,720\}$$
$$/\{[1-0.3000][11,798-513]\}$$
$$= 30.00\%$$

Corporate IFRS-GAAP (B/S-I/S), ISBN 13: **978-1983566332**, ISBN 10: **1983566330**

<u>Law-5166</u>:

If both (**I**= L/E= Leverage or Gearing Ratio= 99.00%), (**t**= T/B= Tax Rate= 30.00%), (**W**= Wealth or Total Assets= 139,796 millions USD), (**d**= D/A= Dividend Portion or Payout= 30.00%), (**I**= Interest Expense= 513 millions USD), and (**U**= Utilized or Starting Capital= 64,720 millions USD), then it's (**O**= Operational Surplus planned), is:

$$\mathbf{O}= \mathbf{I}+\{\mathbf{W}/[1+\mathbf{I}]-\mathbf{U}\}/\{[1-\mathbf{d}][1-\mathbf{t}]\}$$
$$= 513+\{139,796/[1+0.9900]-64,720\}$$
$$/\{[1-0.3000][1-0.3000]\}$$
$$= 11,798 \text{ millions USD}$$

Corporate IFRS-GAAP (B/S-I/S), ISBN 13: **978-1983566332**, ISBN 10: **1983566330**

Law-5167:

If both (I= L/E= Leverage or Gearing Ratio=
99.00%), (t= T/B= Tax Rate= 30.00%), (W= Wealth
or Total Assets= 139,796 millions USD), (d= D/A=
Dividend Portion or Payout= 30.00%), (O=
Operational Surplus= 11,798 millions USD), and (U=
Utilized or Starting Capital= 64,720 millions USD),
then it's (I= Interest Expense planned), is:

$$I = O - \{W/[1+I] - U\}/\{[1-d][1-t]\}$$
$$= 11,798 - \{139,796/[1+0.9900]$$
$$-64,720\}/\{[1-0.3000]$$
$$[1-0.3000]\}$$
$$= \underline{513} \text{ millions USD}$$

Corporate IFRS-GAAP (B/S-I/S), ISBN 13: **978-1983566332**, ISBN 10: **1983566330**

Law-5168:

If both (I= L/E= Leverage or Gearing Ratio= 99.00%), (t= T/B= Tax Rate= 30.00%), (W= Wealth or Total Assets= 139,796 millions USD), (d= D/A= Dividend Portion or Payout= 30.00%), (O= Operational Surplus= 11,798 millions USD), and (I= Interest Expense= 513 millions USD), then it's (U= Utilized or Starting Capital planned), is:

$$U = \{W/[1+I]\} - [1-d][1-t][O-I]$$
$$= \{139,796/[1+0.9900]\} - [1-0.3000]$$
$$[1-0.3000][11,798-513]$$
$$= 64,720 \text{ millions USD}$$

Corporate IFRS-GAAP (B/S-I/S), ISBN 13: **978-1983566332**, ISBN 10: **1983566330**

Law-5169:

If both (**U**= Utilized or Starting Capital= 64,720 millions USD), (**t**= T/B= Tax Rate= 30.00%), (**W**= Wealth or Total Assets= 139,796 millions USD), (**d**= D/A= Dividend Portion or Payout= 30.00%), (**O**= Operational Surplus= 11,798 millions USD), and (**l**= Interest Expense= 513 millions USD), then it's (***l*** = Leverage or Gearing Ratio planned), is:

$$\textbf{\textit{l}} = \textbf{W}/\{\textbf{U}+[1\text{-}\textbf{d}][1\text{-}\textbf{t}][\textbf{O-l}]\}\text{-}1$$
$$= 139,796/\{64,720+[1\text{-}0.3000]$$
$$[1\text{-}0.3000][11,798\text{-}513]\}\text{-}1$$
$$= \underline{99.00\%}$$

Corporate IFRS-GAAP (B/S-I/S), ISBN 13: **978-1983566332**, ISBN 10: **1983566330**

<u>Law-5170</u>:

If both (I= L/E= Leverage or Gearing Ratio= <u>99.00%</u>), (i= I/S= Interest Portion= <u>1.00%</u>), (t= T/B= Tax Rate= <u>30.00%</u>), (U= Utilized or Starting Capital= <u>64,720</u> millions USD), (d= D/A= Dividend Portion or Payout= <u>30.00%</u>), (O= Operational Surplus= <u>11,798</u> millions USD), and (S= Sales or Revenues= <u>51,294</u> millions USD), then it's (W= Wealth or Total Assets planned), is:

$$W= [1+I]\{U+[1-d][1-t][O-Si]\}$$
$$= [1+0.9900]\{64,720+[1-0.3000]$$
$$[1-0.3000][11,798-51,294$$
$$*0.0100]\}$$
$$= \underline{139,796} \text{ millions USD}$$

Corporate IFRS-GAAP (B/S-I/S), ISBN 13: **978-1983566332**, ISBN 10: **1983566330**

Law-5171:

If both (I= L/E= Leverage or Gearing Ratio= 99.00%), (i= I/S= Interest Portion= 1.00%), (W= Wealth or Total Assets= 139,796 millions USD), (U= Utilized or Starting Capital= 64,720 millions USD), (d= D/A= Dividend Portion or Payout= 30.00%), (O= Operational Surplus= 11,798 millions USD), and (S= Sales or Revenues= 51,294 millions USD), then it's (t= Tax Rate planned), is:

$$t= 1-\{W/[1+I]-U\}/\{[1-d][O-Si]\}$$
$$= 1-\{139,796/[1+0.9900]-64,720\}$$
$$/\{[1-0.3000][11,798-51,294$$
$$*0.0100]\}$$
$$= 30.00\%$$

Corporate IFRS-GAAP (B/S-I/S), ISBN 13: **978-1983566332**, ISBN 10: **1983566330**

Law-5172:

If both (I= L/E= Leverage or Gearing Ratio= 99.00%), (i= I/S= Interest Portion= 1.00%), (W= Wealth or Total Assets= 139,796 millions USD), (U= Utilized or Starting Capital= 64,720 millions USD), (t= T/B= Tax Rate= 30.00%), (O= Operational Surplus= 11,798 millions USD), and (S= Sales or Revenues= 51,294 millions USD), then it's (d= Dividend Portion or Payout planned), is:

$$d = 1-\{W/[1+I]-U\}/\{[1-t][O-Si]\}$$
$$= 1-\{139,796/[1+0.9900]-64,720\}/\{[1-0.3000][11,798-51,294*0.0100]\}$$
$$= 30.00\%$$

Corporate IFRS-GAAP (B/S-I/S), ISBN 13: **978-1983566332**, ISBN 10: **1983566330**

<u>Law-5173</u>:

If both (**⌐**= L/E= Leverage or Gearing Ratio=
99.00%), (**i**= I/S= Interest Portion= 1.00%), (**W**=
Wealth or Total Assets= 139,796 millions USD), (**U**=
Utilized or Starting Capital= 64,720 millions USD),
(**t**= T/B= Tax Rate= 30.00%), (**d**= D/A= Dividend
Portion or Payout= 30.00%), and (**$**= Sales or
Revenues= 51,294 millions USD), then it's (**O**=
Operational Surplus planned), is:

$$O = \$i + \{W/[1+I]-U\}/\{[1-d][1-t]\}$$
$$= 51{,}294*0.0100 + \{139{,}796$$
$$/[1+0.9900]-64{,}720\}$$
$$/\{[1-0.3000][1-0.3000]\}$$
$$= \underline{11{,}798}\ \text{millions USD}$$

Corporate IFRS-GAAP (B/S-I/S), ISBN 13: **978-1983566332**, ISBN 10: **1983566330**

Law-5174:

If both (I= L/E= Leverage or Gearing Ratio= 99.00%), (i= I/S= Interest Portion= 1.00%), (W= Wealth or Total Assets= 139,796 millions USD), (U= Utilized or Starting Capital= 64,720 millions USD), (t= T/B= Tax Rate= 30.00%), (d= D/A= Dividend Portion or Payout= 30.00%), and (O= Operational Surplus= 11,798 millions USD), then it's (S= Sales or Revenues planned), is:

$$S= (O-\{W/[1+I]-U\}/\{[1-d][1-t]\})/I$$
$$= (11,798-\{139,796/[1+0.9900]$$
$$-64,720\}/\{[1-0.3000]$$
$$[1-0.3000]\})/0.0100$$
$$= 51,294 \text{ millions USD}$$

Corporate IFRS-GAAP (B/S-I/S), ISBN 13: **978-1983566332**, ISBN 10: **1983566330**

Law-5175:

If both (I= L/E= Leverage or Gearing Ratio= 99.00%), (S= Sales or Revenues= 51,294 millions USD), (W= Wealth or Total Assets= 139,796 millions USD), (U= Utilized or Starting Capital= 64,720 millions USD), (t= T/B= Tax Rate= 30.00%), (d= D/A= Dividend Portion or Payout= 30.00%), and (O= Operational Surplus= 11,798 millions USD), then it's (i= Interest Portion planned), is:

$$i= (O-\{W/[1+I]-U\}/\{[1-d][1-t]\})/S$$
$$= (11,798-\{139,796/[1+0.9900]$$
$$-64,720\}/\{[1-0.3000]$$
$$[1-0.3000]\})/51,294$$
$$= \underline{1.00\%}$$

Corporate IFRS-GAAP (B/S-I/S), ISBN 13: **978-1983566332**, ISBN 10: **1983566330**

Law-5176:

If both (**I**= L/E= Leverage or Gearing Ratio= 99.00%), (**S**= Sales or Revenues= 51,294 millions USD), (**W**= Wealth or Total Assets= 139,796 millions USD), (**i**= I/S= Interest Portion= 1.00%), (**t**= T/B= Tax Rate= 30.00%), (**d**= D/A= Dividend Portion or Payout= 30.00%), and (**O**= Operational Surplus= 11,798 millions USD), then it's (**U**= Utilized or Starting Capital planned), is:

$$U= W/[1+I]-[O-Si][1-d][1-t]\}$$
$$= 139{,}796/[1+0.9900]-[11{,}798$$
$$-51{,}294*0.0100][1-0.3000]$$
$$[1-0.3000]\}$$
$$= \underline{64{,}720} \text{ millions USD}$$

Corporate IFRS-GAAP (B/S-I/S), ISBN 13: **978-1983566332**, ISBN 10: **1983566330**

<u>Law-5177</u>:

If both (I= L/E= Leverage or Gearing Ratio= 99.00%), (S= Sales or Revenues= <u>51,294</u> millions USD), (W= Wealth or Total Assets= <u>139,796</u> millions USD), (i= I/S= Interest Portion= <u>1.00%</u>), (t= T/B= Tax Rate= <u>30.00%</u>), (d= D/A= Dividend Portion or Payout= <u>30.00%</u>), and (O= Operational Surplus= <u>11,798</u> millions USD), then it's (I= Leverage or Gearing Ratio planned), is:

$$I = W/\{[U+[1-d][1-t][O-Si]\}-1$$
$$= 139,796/\{[64,720+[1-0.3000]$$
$$[1-0.3000][11,798-51,294$$
$$*0.0100]\}-1$$
$$= \underline{99.00\%}$$

Corporate IFRS-GAAP (B/S-I/S), ISBN 13: **978-1983566332**, ISBN 10: **1983566330**

Law-5178:

If both (I= L/E= Leverage or Gearing Ratio= 99.00%), (S'= Sales of Past Year= 48,851 millions USD), (s= [S/S']-1= Sales Growth= 5.00%), (U= Utilized or Starting Capital= 64,720 millions USD), (i= I/S= Interest Portion= 1.00%), (t= T/B= Tax Rate= 30.00%), (d= D/A= Dividend Portion or Payout= 30.00%), and (O= Operational Surplus= 11,798 millions USD), then it's (W= Wealth or Total Assets planned), is:

$$W = [1+I](U+[1-d][1-t]\{O-S'i[1+s]\})$$
$$= [1+0.9900](64,720+[1-0.3000]$$
$$[1-0.3000]\{11,798-48,851$$
$$*0.0100[1+0.0500]\})$$
$$= 139,796 \text{ millions USD}$$

Corporate IFRS-GAAP (B/S-I/S), ISBN 13: **978-1983566332**, ISBN 10: **1983566330**

Law-5179:

If both (**I**= L/E= Leverage or Gearing Ratio= 99.00%), (**S'**= Sales of Past Year= 48,851 millions USD), (**s**= [S/S']-1= Sales Growth= 5.00%), (**U**= Utilized or Starting Capital= 64,720 millions USD), (**i**= I/S= Interest Portion= 1.00%), (**t**= T/B= Tax Rate= 30.00%), (**d**= D/A= Dividend Portion or Payout= 30.00%), and (**W**= Wealth or Total Assets= 139,796 millions USD), then it's (**O**= Operational Surplus planned), is:

$$O = S'i[1+s] + W/\{[1+I]-U\}/\{[1-d][1-t]\}$$
$$= 48,851*0.0100[1+0.0500]$$
$$+\{139,796/[1+0.9900]$$
$$-64,720\}/\{[1-0.3000]$$
$$[1-0.3000]\}$$
$$= 11,798 \text{ millions USD}$$

Corporate IFRS-GAAP (B/S-I/S), ISBN 13: **978-1983566332**, ISBN 10: **1983566330**

Law-5180:

If both (I= L/E= Leverage or Gearing Ratio= 99.00%), (O= Operational Surplus= 11,798 millions USD), (s= [S/S']-1= Sales Growth= 5.00%), (U= Utilized or Starting Capital= 64,720 millions USD), (i= I/S= Interest Portion= 1.00%), (t= T/B= Tax Rate= 30.00%), (d= D/A= Dividend Portion or Payout= 30.00%), and (W= Wealth or Total Assets= 139,796 millions USD), then it's (S'= Sales Past), must be:

$$S' = (O-W/\{[1+I]-U\}/\{[1-d][1-t]\})/\{i[1+s]\}$$
$$= (11,798-\{139,796/[1+0.9900]$$
$$-64,720\}/\{[1-0.3000]$$
$$[1-0.3000]\})/\{0.0100$$
$$[1+0.0500]\}$$
$$= \underline{48,851} \text{ millions USD}$$

Corporate IFRS-GAAP (B/S-I/S), ISBN 13: **978-1983566332**, ISBN 10: **1983566330**

<u>Law-5181</u>:

If both (**I**= L/E= Leverage or Gearing Ratio= <u>99.00%</u>), (**O**= Operational Surplus= <u>11,798</u> millions USD), (**s**= [S/S']-1= Sales Growth= <u>5.00%</u>), (**U**= Utilized or Starting Capital= <u>64,720</u> millions USD), (**S'**= Sales of Past Year= <u>48,851</u> millions USD), (**t**= T/B= Tax Rate= <u>30.00%</u>), (**d**= D/A= Dividend Portion or Payout= <u>30.00%</u>), and (**W**= Wealth or Total Assets= <u>139,796</u> millions USD), then it's (**i**= Interest Portion planned), is:

$$i= (O-W/\{[1+I]-U\}/\{[1-d][1-t]\})/\{S'[1+s]\}$$
$$= (11,798-\{139,796/[1+0.9900]$$
$$-64,720\}/\{[1-0.3000]$$
$$[1-0.3000]\})/\{48,851$$
$$[1+0.0500]\}$$

$$= \underline{1.00\%}$$

Corporate IFRS-GAAP (B/S-I/S), ISBN 13: **978-1983566332**, ISBN 10: **1983566330**

Law-5182:

If both (**I**= L/E= Leverage or Gearing Ratio= 99.00%), (**O**= Operational Surplus= 11,798 millions USD), (**i**= I/S= Interest Portion= 1.00%), (**U**= Utilized or Starting Capital= 64,720 millions USD), (**S'**= Sales of Past Year= 48,851 millions USD), (**t**= T/B= Tax Rate= 30.00%), (**d**= D/A= Dividend Portion or Payout= 30.00%), and (**W**= Wealth or Total Assets= 139,796 millions USD), then it's (**s**= Sales Growth planned), is:

$$s= (O\text{-}W/\{[1+I]\text{-}U\}/\{[1\text{-}d][1\text{-}t]\})/[S'i]\text{-}1$$
$$= (11,798\text{-}\{139,796/[1+0.9900]$$
$$-64,720\}/\{[1\text{-}0.3000]$$
$$[1\text{-}0.3000]\})/[48,851$$
$$*0.0100]\text{-}1$$
$$= 5.00\%$$

Corporate IFRS-GAAP (B/S-I/S), ISBN 13: **978-1983566332**, ISBN 10: **1983566330**

<u>Law-5183</u>:

If both (I= L/E= Leverage or Gearing Ratio= <u>99.00%</u>), (O= Operational Surplus= <u>11,798</u> millions USD), (i= I/S= Interest Portion= <u>1.00%</u>), (U= Utilized or Starting Capital= <u>64,720</u> millions USD), (S'= Sales of Past Year= <u>48,851</u> millions USD), (s= [S/S']-1= Sales Growth= <u>5.00%</u>), (d= D/A= Dividend Portion or Payout= <u>30.00%</u>), and (W= Wealth or Total Assets= <u>139,796</u> millions USD), then it's (t= Tax Rate planned), is:

$$t = 1 - \{W/[1+I]-U\}/([1-d]\{O-S'i[1+s]\})$$
$$= 1 - \{139,796/[1+0.9900]-64,720\}$$
$$/([1-0.3000]\{11,798$$
$$-48,851*0.0100$$
$$[1+0.0500]\})$$
$$= \underline{30.00\%}$$

Corporate IFRS-GAAP (B/S-I/S), ISBN 13: **978-1983566332**, ISBN 10: **1983566330**

Law-5184:

If both (I= L/E= Leverage or Gearing Ratio= 99.00%), (O= Operational Surplus= 11,798 millions USD), (i= I/S= Interest Portion= 1.00%), (U= Utilized or Starting Capital= 64,720 millions USD), (S'= Sales of Past Year= 48,851 millions USD), (s= [S/S']-1= Sales Growth= 5.00%), (t= T/B= Tax Rate= 30.00%), and (W= Wealth or Total Assets= 139,796 millions USD), then it's (d= Dividend Portion or Payout planned), is:

$$d = 1 - \{W/[1+I]-U\}/([1-t]\{O-S'i[1+s]\})$$
$$= 1 - \{139,796/[1+0.9900]-64,720\}$$
$$/([1-0.3000]\{11,798$$
$$-48,851*0.0100$$
$$[1+0.0500]\})$$
$$= 30.00\%$$

Corporate IFRS-GAAP (B/S-I/S), ISBN 13: **978-1983566332**, ISBN 10: **1983566330**

<u>Law-5185</u>:

If both (**/**= L/E= Leverage or Gearing Ratio= <u>99.00%</u>), (**O**= Operational Surplus= <u>11,798</u> millions USD), (**i**= I/S= Interest Portion= <u>1.00%</u>), (**d**= D/A= Dividend Portion or Payout= <u>30.00%</u>), (**$'**= Sales of Past Year= <u>48,851</u> millions USD), (**s**= [S/S']-1= Sales Growth= <u>5.00%</u>), (**t**= T/B= Tax Rate= <u>30.00%</u>), and (**W**= Wealth or Total Assets= <u>139,796</u> millions USD), then it's (**U**= Utilized or Starting Capital planned), is:

$$U = \{W/[1+/]-[1-d][1-t]\{O-\$'i[1+s]\}$$
$$= \{139,796/[1+0.9900]-[1-0.3000]$$
$$[1-0.3000]\{11,798-48,851$$
$$*0.0100[1+0.0500]\}$$
$$= \underline{64,720} \text{ millions USD}$$

Corporate IFRS-GAAP (B/S-I/S), ISBN 13: **978-1983566332**, ISBN 10: **1983566330**

<u>Law-5186</u>:

If both (**U**= Utilized or Starting Capital= <u>64,720</u> millions USD), (**O**= Operational Surplus= <u>11,798</u> millions USD), (**i**= I/S= Interest Portion= <u>1.00%</u>), (**d**= D/A= Dividend Portion or Payout= <u>30.00%</u>), (**S'**= Sales of Past Year= <u>48,851</u> millions USD), (**s**= [S/S']-1= Sales Growth= <u>5.00%</u>), (**t**= T/B= Tax Rate= <u>30.00%</u>), and (**W**= Wealth or Total Assets= <u>139,796</u> millions USD), then it's (**I** = Leverage or Gearing Ratio planned), is:

$$I = W/(U+[1-d][1-t]\{O-S'i[1+s]\})-1$$
$$= 139,796/(64,720+[1-0.3000]$$
$$[1-0.3000]\{11,798-48,851$$
$$*0.0100[1+0.0500]\})-1$$

$$= \underline{99.00\%}$$

Corporate IFRS-GAAP (B/S-I/S), ISBN 13: **978-1983566332**, ISBN 10: **1983566330**

Law-5187:

If both (**U**= Utilized or Starting Capital= 64,720 millions USD), (**M**= Margin of Contribution= 41,548 millions USD), (**F**= Fixed Cost= 29,740 millions USD), (**I**= Interest Expense= 513 millions USD), (**d**= D/A= Dividend Portion or Payout= 30.00%), (**t**= T/B= Tax Rate= 30.00%), and (**I**= L/E= Leverage or Gearing Ratio= 99.00%), then it's (**W**= Wealth or Total Assests planned), is:

$$W= [1+I]\{U+[1-d][1-t][M-F-I]\}$$
$$= [1+0.9900]\{64,720+[1-0.3000]$$
$$[1-0.3000][41,548-29,750$$
$$-513]\}$$
$$= \underline{139,796} \text{ millions USD}$$

Corporate IFRS-GAAP (B/S-I/S), ISBN 13: **978-1983566332**, ISBN 10: **1983566330**

Law-5188:

If both (**U**= Utilized or Starting Capital= 64,720 millions USD), (**W**= Wealth or Total Assets= 139,796 millions USD), (**F**= Fixed Cost= 29,740 millions USD), (**I**= Interest Expense= 513 millions USD), (**d**= D/A= Dividend Portion or Payout= 30.00%), (**t**= T/B= Tax Rate= 30.00%), and (**I**= L/E= Leverage or Gearing Ratio= 99.00%), then it's (**M**= Margin of Contribution planned), is:

$$M= F+I+\{W/[1+I]-U\}/\{[1-d][1-t]\}$$
$$= 29,750+513+\{139,796$$
$$/[1+0.9900-64,720\}$$
$$/\{[1-0.3000][1-0.3000]\}$$
$$= 41,548 \text{ millions USD}$$

Corporate IFRS-GAAP (B/S-I/S), ISBN 13: **978-1983566332**, ISBN 10: **1983566330**

Law-5189:

If both (**U**= Utilized or Starting Capital= 64,720 millions USD), (**W**= Wealth or Total Assets= 139,796 millions USD), (**M**= Margin of Contribution= 41,548 millions USD), (**I**= Interest Expense= 513 millions USD), (**d**= D/A= Dividend Portion or Payout= 30.00%), (**t**= T/B= Tax Rate= 30.00%), and (**I**= L/E= Leverage or Gearing Ratio= 99.00%), then it's (**F**= Fixed Cost planned), is:

$$F= M-I-\{W/[1+I]-U\}/\{[1-d][1-t]\}$$
$$= 41,548-513-\{139,796$$
$$/[1+0.9900-64,720\}$$
$$/\{[1-0.3000][1-0.3000]\}$$
$$= 29,750 \text{ millions USD}$$

Corporate IFRS-GAAP (B/S-I/S), ISBN 13: **978-1983566332**, ISBN 10: **1983566330**

Law-5190:

If both (**U**= Utilized or Starting Capital= 64,720 millions USD), (**W**= Wealth or Total Assets= 139,796 millions USD), (**M**= Margin of Contribution= 41,548 millions USD), (**F**= Fixed Cost= 29,740 millions USD), (**d**= D/A= Dividend Portion or Payout= 30.00%), (**t**= T/B= Tax Rate= 30.00%), and (**I**= L/E= Leverage or Gearing Ratio= 99.00%), then it's (**I**= Interest Expense planned), is:

$$I = M-F-\{W/[1+I]-U\}/\{[1-d][1-t]\}$$
$$= 41,548-29,750-\{139,796$$
$$/[1+0.9900-64,720\}$$
$$/\{[1-0.3000][1-0.3000]\}$$
$$= \underline{513} \text{ millions USD}$$

Corporate IFRS-GAAP (B/S-I/S), ISBN 13: **978-1983566332**, ISBN 10: **1983566330**

Law-5191:

If both (**U**= Utilized or Starting Capital= 64,720 millions USD), (**W**= Wealth or Total Assets= 139,796 millions USD), (**M**= Margin of Contribution= 41,548 millions USD), (**F**= Fixed Cost= 29,740 millions USD), (**d**= D/A= Dividend Portion or Payout= 30.00%), (**I**= Interest Expense= 513 millions USD), and (**I**= L/E= Leverage or Gearing Ratio= 99.00%), then it's (**t**= Tax Rate planned), is:

$$t= 1-\{W/[1+I]-U\}/\{[1-d][M-F-I]\}$$
$$= 1-\{139,796/[1+0.9900]-64,720\}$$
$$/\{[1-0.3000][41,548$$
$$-29,750-513]\}$$
$$= 30.00\%$$

Corporate IFRS-GAAP (B/S-I/S), ISBN 13: **978-1983566332**, ISBN 10: **1983566330**

Law-5192:

If both (**U**= Utilized or Starting Capital= 64,720 millions USD), (**W**= Wealth or Total Assets= 139,796 millions USD), (**M**= Margin of Contribution= 41,548 millions USD), (**F**= Fixed Cost= 29,740 millions USD), (**t**= T/B= Tax Rate= 30.00%), (**I**= Interest Expense= 513 millions USD), and (**I**= L/E= Leverage or Gearing Ratio= 99.00%), then it's (**d**= Dividend Portion or Payout planned), is:

$$d = 1 - \{W/[1+I] - U\}/\{[1-t][M-F-I]\}$$
$$= 1 - \{139,796/[1+0.9900] - 64,720\}$$
$$/\{[1-0.3000][41,548$$
$$-29,750 - 513]\}$$
$$= 30.00\%$$

Law-5193:

If both (**d**= D/A= Dividend Portion or Payout= 30.00%), (**W**= Wealth or Total Assets= 139,796 millions USD), (**M**= Margin of Contribution= 41,548 millions USD), (**F**= Fixed Cost= 29,740 millions USD), (**t**= T/B= Tax Rate= 30.00%), (**I**= Interest Expense= 513 millions USD), and (**I**= L/E= Leverage or Gearing Ratio= 99.00%), then it's (**U**= Utilized or Starting Capital planned), is:

$$U= W/[1+I]-[1-d][1-t][M-F-I]$$
$$= 139,796/[1+0.9900]-[1-0.3000]$$
$$[1-0.3000][41,548-29,750$$
$$-513]$$
$$= 64,720 \text{ millions USD}$$

Corporate IFRS-GAAP (B/S-I/S), ISBN 13: **978-1983566332**, ISBN 10: **1983566330**

<u>Law-5194</u>:

If both (**d**= D/A= Dividend Portion or Payout= 30.00%), (**W**= Wealth or Total Assets= <u>139,796</u> millions USD), (**M**= Margin of Contribution= <u>41,548</u> millions USD), (**F**= Fixed Cost= <u>29,740</u> millions USD), (**t**= T/B= Tax Rate= <u>30.00%</u>), (**I**= Interest Expense= <u>513</u> millions USD), and (**U**= Utilized or Starting Capital= <u>64,720</u> millions USD), then it's (**I** = Leverage or Gearing Ratio planned), is:

$$I = W/\{U+[1-d][1-t][M-F-I]\}-1$$
$$= 139,796/\{64,720+[1-0.3000]$$
$$[1-0.3000][41,548-29,750$$
$$-513]\}-1$$
$$= \underline{99.00\%}$$

Corporate IFRS-GAAP (B/S-I/S), ISBN 13: **978-1983566332**, ISBN 10: **1983566330**

Law-5195:

If both (**d**= D/A= Dividend Portion or Payout= 30.00%), (**I**= L/E= Leverage or Gearing Ratio= 99.00%), (**M**= Margin of Contribution= 41,548 millions USD), (**F**= Fixed Cost= 29,740 millions USD), (**t**= T/B= Tax Rate= 30.00%), (**S**= Sales or Revenues= 51,294 millions USD), (**i**= I/S= Interest Portion= 1.00%), and (**U**= Utilized or Starting Capital= 64,720 millions USD), then it's (**W**= Wealth or Total Assets planned), is:

$$\mathbf{W}= [1+\mathbf{I}]\{\mathbf{U}+[1-\mathbf{d}][1-\mathbf{t}][\mathbf{M}-\mathbf{F}-\mathbf{Si}]\}$$
$$= [1+0.9900]\{64,720+[1-0.3000]$$
$$[1-0.3000][41,548$$
$$-29,750-51,294*0.0100]\}$$
$$= \underline{139,796} \text{ millions USD}$$

Corporate IFRS-GAAP (B/S-I/S), ISBN 13: **978-1983566332**, ISBN 10: **1983566330**

Law-5196:

If both (**d**= D/A= Dividend Portion or Payout= 30.00%), (**⌐**= L/E= Leverage or Gearing Ratio= 99.00%), (**W**= Wealth or Total Assets= 139,796 millions USD), (**F**= Fixed Cost= 29,740 millions USD), (**t**= T/B= Tax Rate= 30.00%), (**$**= Sales or Revenues= 51,294 millions USD), (**i**= I/S= Interest Portion= 1.00%), and (**U**= Utilized or Starting Capital= 64,720 millions USD), then it's (**M**= Margin of Contribution planned), is:

$$M= F+\$i+\{W/[1+\text{⌐}]-U\}/\{[1-d][1-t]\}$$
$$= 29,750+51,294*0.0100$$
$$+\{139,796/[1+0.9900]$$
$$-64,720\}/\{[1-0.3000]$$
$$[1-0.3000]\}$$
$$= 41,548 \text{ millions USD}$$

Corporate IFRS-GAAP (B/S-I/S), ISBN 13: **978-1983566332**, ISBN 10: **1983566330**

Law-5197:

If both (**d**= D/A= Dividend Portion or Payout= 30.00%), (**I**= L/E= Leverage or Gearing Ratio= 99.00%), (**W**= Wealth or Total Assets= 139,796 millions USD), (**M**= Margin of Contribution= 41,548 millions USD), (**t**= T/B= Tax Rate= 30.00%), (**$**= Sales or Revenues= 51,294 millions USD), (**i**= I/S= Interest Portion= 1.00%), and (**U**= Utilized or Starting Capital= 64,720 millions USD), then it's (**F**= Fixed Cost planned), is:

$$F= M\text{-}\$i\text{-}\{W/[1+I]\text{-}U\}/\{[1\text{-}d][1\text{-}t]\}$$
$$= 41,548\text{-}51,294*0.0100\text{-}\{139,796$$
$$/[1+0.9900]\text{-}64,720\}$$
$$/\{[1\text{-}0.3000][1\text{-}0.3000]\}$$
$$= 29,750 \text{ millions USD}$$

Corporate IFRS-GAAP (B/S-I/S), ISBN 13: **978-1983566332**, ISBN 10: **1983566330**

<u>Law-5198</u>:

If both (**d**= D/A= Dividend Portion or Payout= 30.00%), (**l**= L/E= Leverage or Gearing Ratio= 99.00%), (**W**= Wealth or Total Assets= <u>139,796</u> millions USD), (**F**= Fixed Cost= <u>29,740</u> millions USD), (**M**= Margin of Contribution= <u>41,548</u> millions USD), (**t**= T/B= Tax Rate= <u>30.00%</u>), (**i**= I/S= Interest Portion= <u>1.00%</u>), and (**U**= Utilized or Starting Capital= <u>64,720</u> millions USD), then it's (**S**= Sales or Revenues planned), is:

$$S= M-F-\{W/[1+l]-U\}/\{[1-d][1-t]\})/i$$

$$
\begin{aligned}
&= (41,548-29,750-\{139,796 \\
&\quad /[1+0.9900]-64,720\} \\
&\quad /\{[1-0.3000][1-0.3000]\}) \\
&\quad /0.0100
\end{aligned}
$$

= <u>51,294</u> millions USD

Corporate IFRS-GAAP (B/S-I/S), ISBN 13: **978-1983566332**, ISBN 10: **1983566330**

Law-5199:

If both (**d**= D/A= Dividend Portion or Payout= 30.00%), (**I**= L/E= Leverage or Gearing Ratio= 99.00%), (**W**= Wealth or Total Assets= 139,796 millions USD), (**M**= Margin of Contribution= 41,548 millions USD), (**t**= T/B= Tax Rate= 30.00%), (**S**= Sales or Revenues= 51,294 millions USD), (**F**= Fixed Cost= 29,740 millions USD), and (**U**= Utilized or Starting Capital= 64,720 millions USD), then it's (**i**= Interest Portion planned), is:

$$\mathbf{i}= \mathbf{M}\text{-}\mathbf{F}\text{-}\{\mathbf{W}/[1+\mathbf{I}]\text{-}\mathbf{U}\}/\{[1\text{-}\mathbf{d}][1\text{-}\mathbf{t}]\})/\mathbf{S}$$

$$= (41,548\text{-}29,750\text{-}\{139,796 /[1+0.9900]\text{-}64,720\} /\{[1\text{-}0.3000][1\text{-}0.3000]\}) /51,294$$

$$= \underline{1.00\%}$$

Corporate IFRS-GAAP (B/S-I/S), ISBN 13: **978-1983566332**, ISBN 10: **1983566330**

Law-5200:

If both (**d**= D/A= Dividend Portion or Payout= 30.00%), (**I**= L/E= Leverage or Gearing Ratio= 99.00%), (**W**= Wealth or Total Assets= 139,796 millions USD), (**M**= Margin of Contribution= 41,548 millions USD), (**i**= I/S= Interest Portion= 1.00%), (**S**= Sales or Revenues= 51,294 millions USD), (**F**= Fixed Cost= 29,740 millions USD), and (**U**= Utilized or Starting Capital= 64,720 millions USD), then it's (**t**= Tax Rate planned), is:

$$\mathbf{t}= 1-\{\mathbf{W}/[1+\mathbf{I}]-\mathbf{U}\}/\{[1-\mathbf{d}][\mathbf{M}-\mathbf{F}-\mathbf{Si}]\}$$
$$= 1-\{139,796/[1+0.9900]-64,720\}$$
$$/\{[1-0.3000]\{41,548-29,750$$
$$-51,294*0.0100]\}$$
$$= 30.00\%$$

Corporate IFRS-GAAP (B/S-I/S), ISBN 13: **978-1983566332**, ISBN 10: **1983566330**

Law-5201:

If both (**t**= T/B= Tax Rate= 30.00%), (**l**= L/E= Leverage or Gearing Ratio= 99.00%), (**W**= Wealth or Total Assets= 139,796 millions USD), (**M**= Margin of Contribution= 41,548 millions USD), (**i**= I/S= Interest Portion= 1.00%), (**S**= Sales or Revenues= 51,294 millions USD), (**F**= Fixed Cost= 29,740 millions USD), and (**U**= Utilized or Starting Capital= 64,720 millions USD), then it's (**d**= Dividend Portion or Payout planned), is:

$$d= 1-\{W/[1+l]-U\}/\{[1-t][M-F-Si]\}$$
$$= 1-\{139,796/[1+0.9900]-64,720\}$$
$$/\{[1-0.3000]\{41,548-29,750$$
$$-51,294*0.0100]\}$$
$$= 30.00\%$$

Corporate IFRS-GAAP (B/S-I/S), ISBN 13: **978-1983566332**, ISBN 10: **1983566330**

Law-5202:

If both (**t**= T/B= Tax Rate= 30.00%), (**I**= L/E= Leverage or Gearing Ratio= 99.00%), (**W**= Wealth or Total Assets= 139,796 millions USD), (**M**= Margin of Contribution= 41,548 millions USD), (**i**= I/S= Interest Portion= 1.00%), (**S**= Sales or Revenues= 51,294 millions USD), (**F**= Fixed Cost= 29,740 millions USD), and (**d**= D/A= Dividend Portion or Payout= 30.00%), then it's (**U**= Utilized or Starting Capital planned), is:

$$U= W/[1+I]-[1-d][1-t][M-F-Si]$$
$$= 139,796/[1+0.9900]-[1-0.3000]$$
$$[1-0.3000][41,548-29,750$$
$$-51,294*0.0100]$$
$$= 64,720 \text{ millions USD}$$

Corporate IFRS-GAAP (B/S-I/S), ISBN 13: **978-1983566332**, ISBN 10: **1983566330**

Law-5203:

If both (**t**= T/B= Tax Rate= 30.00%), (**U**= Utilized or Starting Capital= 64,720 millions USD), (**W**= Wealth or Total Assets= 139,796 millions USD), (**M**= Margin of Contribution= 41,548 millions USD), (**i**= I/S= Interest Portion= 1.00%), (**S**= Sales or Revenues= 51,294 millions USD), (**F**= Fixed Cost= 29,740 millions USD), and (**d**= D/A= Dividend Portion or Payout= 30.00%), then it's (**I**= Leverage or Gearing Ratio planned), is:

$$I = W/\{U+[1-d][1-t][M-F-Si]\}-1$$
$$= 139,796/\{64,720+[1-0.3000]$$
$$[1-0.3000][41,548-29,750$$
$$-51,294*0.0100]\}-1$$
$$= 99.00\%$$

Corporate IFRS-GAAP (B/S-I/S), ISBN 13: **978-1983566332**, ISBN 10: **1983566330**

<u>Law-5204</u>:

If both (**t**= T/B= Tax Rate= <u>30.00%</u>), (**U**= Utilized or Starting Capital= <u>64,720</u> millions USD), (**I**= L/E= Leverage or Gearing Ratio= <u>99.00%</u>), (**M**= Margin of Contribution= <u>41,548</u> millions USD), (**i**= I/S= Interest Portion= <u>1.00%</u>), (**S'**= Sales of Past Year= <u>48,851</u> millions USD), (**s**= [S/S']-1= Sales Growth= <u>5.00%</u>), (**F**= Fixed Cost= <u>29,740</u> millions USD), and (**d**= D/A= Dividend Portion or Payout= <u>30.00%</u>), then it's (**W**= Wealth or Total Assets planned), is:

$$W= [1+I](U+[1-d][1-t]\{M-F-S'i[1+s]\}$$
$$= [1+0.9900](64,720+[1-0.3000]$$
$$[1-0.3000]\{41,548-29,750$$
$$-48,851*0.0100][1+0.0500]\}$$
$$= \underline{139,796} \text{ millions USD}$$

Corporate IFRS-GAAP (B/S-I/S), ISBN 13: **978-1983566332**, ISBN 10: **1983566330**

<u>Law-5205</u>:

If both (**t**= T/B= Tax Rate= <u>30.00%</u>), (**U**= Utilized or Starting Capital= <u>64,720</u> millions USD), (**I**= L/E= Leverage or Gearing Ratio= <u>99.00%</u>), (**W**= Wealth or Total Assets= <u>139,796</u> millions USD), (**i**= I/S= Interest Portion= <u>1.00%</u>), (**S'**= Sales of Past Year= <u>48,851</u> millions USD), (**s**= [S/S']-1= Sales Growth= <u>5.00%</u>), (**F**= Fixed Cost= <u>29,740</u> millions USD), and (**d**= D/A= Dividend Portion or Payout= <u>30.00%</u>), then it's (**M**= Margin of Contribution planned), is:

$$M= F+S'i[1+s]+\{W/[1+I]-U\}/\{[1-d][1-t]\}$$
$$= 29,750+48,851*0.0100[1+0.0500]$$
$$+\{139,796/ [1+0.9900]$$
$$-64,720\}/\{[1-0.3000]$$
$$[1-0.3000]\}$$
$$= \underline{41,548} \text{ millions USD}$$

Corporate IFRS-GAAP (B/S-I/S), ISBN 13: **978-1983566332**, ISBN 10: **1983566330**

<u>Law-5206</u>:

If both (**t**= T/B= Tax Rate= <u>30.00%</u>), (**U**= Utilized or Starting Capital= <u>64,720</u> millions USD), (**⫫**= L/E= Leverage or Gearing Ratio= <u>99.00%</u>), (**W**= Wealth or Total Assets= <u>139,796</u> millions USD), (**i**= I/S= Interest Portion= <u>1.00%</u>), (**$'**= Sales of Past Year= <u>48,851</u> millions USD), (**s**= [S/S']-1= Sales Growth= <u>5.00%</u>), (**M**= Margin of Contribution= <u>41,548</u> millions USD), and (**d**= D/A= Dividend Portion or Payout= <u>30.00%</u>), then it's (**F**= Fixed Cost planned), is:

$$F= M\text{-}\$'i[1+s]\text{-}\{W/[1+\textit{⫫}]\text{-}U\}/\{[1\text{-}d][1\text{-}t]\}$$
$$= 41,548\text{-}48,851*0.0100[1+0.0500]$$
$$-\{139,796/\ [1+0.9900]$$
$$-64,720\}/\{[1\text{-}0.3000]$$
$$[1\text{-}0.3000]\}$$
$$= \underline{29,750}\ \text{millions USD}$$

Corporate IFRS-GAAP (B/S-I/S), ISBN 13: **978-1983566332**, ISBN 10: **1983566330**

Law-5207:

If both (**t**= T/B= Tax Rate= 30.00%), (**U**= Utilized or Starting Capital= 64,720 millions USD), (**I**= L/E= Leverage or Gearing Ratio= 99.00%), (**W**= Wealth or Total Assets= 139,796 millions USD), (**i**= I/S= Interest Portion= 1.00%), (**F**= Fixed Cost= 29,740 millions USD), (**s**= [S/S']-1= Sales Growth= 5.00%), (**M**= Margin of Contribution= 41,548 millions USD), and (**d**= D/A= Dividend Portion or Payout= 30.00%), then it's (**S'**= Sales Past), must be:

$$S' = (M-F-\{W/[1+I]-U\}/\{[1-d][1-t]\})/\{i[1+s]\}$$
$$= (41,548-29,750-\{139,796$$
$$/[1+0.9900]-64,720\}$$
$$/\{[1-0.3000][1-0.3000]\})$$
$$/\{0.0100[1+0.0500]\}$$
$$= 48,851 \text{ millions USD}$$

Corporate IFRS-GAAP (B/S-I/S), ISBN 13: **978-1983566332**, ISBN 10: **1983566330**

Law-5208:

If both (**t**= T/B= Tax Rate= 30.00%), (**U**= Utilized or Starting Capital= 64,720 millions USD), (**I**= L/E= Leverage or Gearing Ratio= 99.00%), (**W**= Wealth or Total Assets= 139,796 millions USD), (**S'**= Sales of Past Year= 48,851 millions USD), (**F**= Fixed Cost= 29,740 millions USD), (**s**= [S/S']-1= Sales Growth= 5.00%), (**M**= Margin of Contribution= 41,548 millions USD), and (**d**= D/A= Dividend Portion or Payout= 30.00%), then it's (**i**= Interest Portion planned), is:

$$\mathbf{i} = (\mathbf{M}\text{-}\mathbf{F}\text{-}\{\mathbf{W}/[1+\mathbf{I}]\text{-}\mathbf{U}\}/\{[1\text{-}\mathbf{d}][1\text{-}\mathbf{t}]\})/\{\mathbf{S'}[1+\mathbf{s}]\}$$

$$= (41{,}548\text{-}29{,}750\text{-}\{139{,}796$$
$$/\,[1+0.9900]\text{-}64{,}720\}$$
$$/\{[1\text{-}0.3000][1\text{-}0.3000]\})$$
$$/\{48{,}851[1+0.0500]\}$$

$$= \underline{1.00\%}$$

Corporate IFRS-GAAP (B/S-I/S), ISBN 13: **978-1983566332**, ISBN 10: **1983566330**

<u>Law-5209</u>:

If both (**t**= T/B= Tax Rate= <u>30.00%</u>), (**U**= Utilized or Starting Capital= <u>64,720</u> millions USD), (**I**= L/E= Leverage or Gearing Ratio= <u>99.00%</u>), (**W**= Wealth or Total Assets= <u>139,796</u> millions USD), (**S'**= Sales of Past Year= <u>48,851</u> millions USD), (**F**= Fixed Cost= <u>29,740</u> millions USD), (**i**= I/S= Interest Portion= <u>1.00%</u>), (**M**= Margin of Contribution= <u>41,548</u> millions USD), and (**d**= D/A= Dividend Portion or Payout= <u>30.00%</u>), then it's (**s**= Sales Growth planned), is:

$$s= (\mathbf{M}\text{-}\mathbf{F}\text{-}\{\mathbf{W}/[1+\mathbf{I}]\text{-}\mathbf{U}\}/\{[1\text{-}\mathbf{d}][1\text{-}\mathbf{t}]\})/[\mathbf{S'i}]\text{-}1$$

$$= (41{,}548\text{-}29{,}750\text{-}\{139{,}796$$
$$/[1+0.9900]\text{-}64{,}720\}$$
$$/\{[1\text{-}0.3000][1\text{-}0.3000]\})$$
$$/[48{,}851{*}0.0100]\text{-}1$$

$$= \underline{5.00\%}$$

Corporate IFRS-GAAP (B/S-I/S), ISBN 13: **978-1983566332**, ISBN 10: **1983566330**

Law-5210:

If both (**s**= [S/S']-1= Sales Growth= 5.00%), (**U**= Utilized or Starting Capital= 64,720 millions USD), (**f**= L/E= Leverage or Gearing Ratio= 99.00%), (**W**= Wealth or Total Assets= 139,796 millions USD), (**S'**= Sales of Past Year= 48,851 millions USD), (**F**= Fixed Cost= 29,740 millions USD), (**i**= I/S= Interest Portion= 1.00%), (**M**= Margin of Contribution= 41,548 millions USD), and (**d**= D/A= Dividend Portion or Payout= 30.00%), then it's (**t**= Tax Rate planned), is:

$$\mathbf{t}= 1-\{\mathbf{W}/[1+\mathbf{f}]-\mathbf{U}\}/([1-\mathbf{d}]\{\mathbf{M}-\mathbf{F}-\mathbf{S'}i[1+\mathbf{s}]\})$$
$$= 1-\{139,796/[1+0.9900]-64,720\}$$
$$/([1-0.3000]\{41,548-29,750$$
$$-48,851*0.0100[1+0.0500]\})$$
$$= 30.00\%$$

Corporate IFRS-GAAP (B/S-I/S), ISBN 13: **978-1983566332**, ISBN 10: **1983566330**

<u>Law-5211</u>:

If both (**s**= [S/S']-1= Sales Growth= <u>5.00%</u>), (**U**= Utilized or Starting Capital= <u>64,720</u> millions USD), (**I**= L/E= Leverage or Gearing Ratio= <u>99.00%</u>), (**W**= Wealth or Total Assets= <u>139,796</u> millions USD), (**S'**= Sales of Past Year= <u>48,851</u> millions USD), (**F**= Fixed Cost= <u>29,740</u> millions USD), (**i**= I/S= Interest Portion= <u>1.00%</u>), (**M**= Margin of Contribution= <u>41,548</u> millions USD), and (**t**= T/B= Tax Rate= <u>30.00%</u>), then it's (**d**= Dividend Portion or Payout planned), is:

$$\mathbf{d}= 1-\{\mathbf{W}/[1+\mathbf{I}]-\mathbf{U}\}/([1-\mathbf{t}]\{\mathbf{M}-\mathbf{F}-\mathbf{S'i}[1+\mathbf{s}]\})$$
$$= 1-\{139,796/[1+0.9900]-64,720\}$$
$$/([1-0.3000]\{41,548-29,750$$
$$-48,851*0.0100[1+0.0500]\})$$
$$= \underline{30.00\%}$$

Corporate IFRS-GAAP (B/S-I/S), ISBN 13: **978-1983566332**, ISBN 10: **1983566330**

Law-5212:

If both (s= [S/S']-1= Sales Growth= 5.00%), (d= D/A= Dividend Portion or Payout= 30.00%), (l= L/E= Leverage or Gearing Ratio= 99.00%), (W= Wealth or Total Assets= 139,796 millions USD), (S'= Sales of Past Year= 48,851 millions USD), (F= Fixed Cost= 29,740 millions USD), (i= I/S= Interest Portion= 1.00%), (M= Margin of Contribution= 41,548 millions USD), and (t= T/B= Tax Rate= 30.00%), then it's (U= Utilized or Starting Capital planned), is:

$$U = W/[1+l]-[1-d][1-t]\{M-F-S'i[1+s]\}$$
$$= 139,796/[1+0.9900]-[1-0.3000]$$
$$[1-0.3000]\{41,548-29,750$$
$$-48,851*0.0100[1+0.0500]\}$$
$$= 64,720 \text{ millions USD}$$

Corporate IFRS-GAAP (B/S-I/S), ISBN 13: **978-1983566332**, ISBN 10: **1983566330**

<u>Law-5213</u>:

If both (s= [S/S']-1= Sales Growth= <u>5.00%</u>), (d= D/A= Dividend Portion or Payout= <u>30.00%</u>), (U= Utilized or Starting Capital= <u>64,720</u> millions USD), (W= Wealth or Total Assets= <u>139,796</u> millions USD), (S'= Sales of Past Year= <u>48,851</u> millions USD), (F= Fixed Cost= <u>29,740</u> millions USD), (i= I/S= Interest Portion= <u>1.00%</u>), (M= Margin of Contribution= <u>41,548</u> millions USD), and (t= T/B= Tax Rate= <u>30.00%</u>), then it's (I= Leverage or Gearing Ratio planned), is:

$$I = W/(U+[1-d][1-t]\{M-F-S'i[1+s]\}$$
$$= 139,796/(64,720+[1-0.3000]$$
$$[1-0.3000]\{41,548-29,750$$
$$-48,851*0.0100[1+0.0500]\})$$
$$= \underline{99.00\%}$$

Corporate IFRS-GAAP (B/S-I/S), ISBN 13: **978-1983566332**, ISBN 10: **1983566330**

Law-5214:

If both (**$**= Sales or Revenues= 51,294 millions USD), (**d**= D/A= Dividend Portion or Payout= 30.00%), (**U**= Utilized or Starting Capital= 64,720 millions USD), (**I**= L/E= Leverage or Gearing Ratio= 99.00%), (**f**= F/S= Fixed Portion= 58.00%), (**I**= Interest Expense= 513 millions USD), (**M**= Margin of Contribution= 41,548 millions USD), and (**t**= T/B= Tax Rate= 30.00%), then it's (**W**= Wealth or Total Assets planned), is:

$$\mathbf{W} = [1+\mathbf{I}]\{\mathbf{U}+[1-\mathbf{d}][1-\mathbf{t}][\mathbf{M}\text{-}\mathbf{Sf}\text{-}\mathbf{I}]$$
$$= [1+0.9900]\{64,720+[1-0.3000]$$
$$[1-0.3000][41,548\text{-}51,294$$
$$*0.5800\text{-}513]$$
$$= 139,796 \text{ millions USD}$$

Corporate IFRS-GAAP (B/S-I/S), ISBN 13: **978-1983566332**, ISBN 10: **1983566330**

<u>Law-5215</u>:

If both (**$**= Sales or Revenues= <u>51,294</u> millions USD), (**d**= D/A= Dividend Portion or Payout= <u>30.00%</u>), (**U**= Utilized or Starting Capital= <u>64,720</u> millions USD), (**/**= L/E= Leverage or Gearing Ratio= <u>99.00%</u>), (**f**= F/S= Fixed Portion= <u>58.00%</u>), (**I**= Interest Expense= <u>513</u> millions USD), (**W**= Wealth or Total Assets= <u>139,796</u> millions USD), and (**t**= T/B= Tax Rate= <u>30.00%</u>), then it's (**M**= Margin of Contribution planned), is:

$$M = \$f + I + \{W/[1+I] - U\} / \{[1-d][1-t]\}$$
$$= 51{,}294 * 0.5800 + 513 + \{139{,}796$$
$$/[1+0.9900] - 64{,}720\}$$
$$/\{[1-0.3000][1-0.3000]\}$$
$$= \underline{41{,}548} \text{ millions USD}$$

Corporate IFRS-GAAP (B/S-I/S), ISBN 13: **978-1983566332**, ISBN 10: **1983566330**

Law-5216:

If both (**M**= Margin of Contribution= 41,548 millions USD), (**d**= D/A= Dividend Portion or Payout= 30.00%), (**U**= Utilized or Starting Capital= 64,720 millions USD), (**l**= L/E= Leverage or Gearing Ratio= 99.00%), (**f**= F/S= Fixed Portion= 58.00%), (**I**= Interest Expense= 513 millions USD), (**W**= Wealth or Total Assets= 139,796 millions USD), and (**t**= T/B= Tax Rate= 30.00%), then it's (**$**= Sales or Revenues planned), is:

$$\$= (M\text{-}I\text{-}\{W/[1+l]\text{-}U\}/\{[1\text{-}d][1\text{-}t]\})/f$$
$$= (41,548\text{-}513\text{-}\{139,796/[1+0.9900]$$
$$-64,720\}/\{[1\text{-}0.3000]$$
$$[1\text{-}0.3000]\})/0.5800$$
$$= 51,294 \text{ millions USD}$$

Corporate IFRS-GAAP (B/S-I/S), ISBN 13: **978-1983566332**, ISBN 10: **1983566330**

Law-5217:

If both (**M**= Margin of Contribution= 41,548 millions USD), (**d**= D/A= Dividend Portion or Payout= 30.00%), (**$**= Sales or Revenues= 51,294 millions USD), (**U**= Utilized or Starting Capital= 64,720 millions USD), (**I**= L/E= Leverage or Gearing Ratio= 99.00%), (**I**= Interest Expense= 513 millions USD), (**W**= Wealth or Total Assets= 139,796 millions USD), and (**t**= T/B= Tax Rate= 30.00%), then it's (**f**= Fixed Portion planned), is:

$$\mathbf{f} = (\mathbf{M}\text{-}\mathbf{I}\text{-}\{\mathbf{W}/[1+\mathbf{I}]\text{-}\mathbf{U}\}/\{[1\text{-}\mathbf{d}][1\text{-}\mathbf{t}]\})/\mathbf{\$}$$

$$= (41{,}548\text{-}513\text{-}\{139{,}796/[1+0.9900]$$
$$-64{,}720\}/\{[1\text{-}0.3000]$$
$$[1\text{-}0.3000]\})/51{,}294$$

$$= \underline{58.00\%}$$

Corporate IFRS-GAAP (B/S-I/S), ISBN 13: **978-1983566332**, ISBN 10: **1983566330**

<u>Law-5218</u>:

If both (**M**= Margin of Contribution= <u>41,548</u> millions USD), (**d**= D/A= Dividend Portion or Payout= <u>30.00%</u>), (**S**= Sales or Revenues= <u>51,294</u> millions USD), (**U**= Utilized or Starting Capital= <u>64,720</u> millions USD), (**F**= L/E= Leverage or Gearing Ratio= <u>99.00%</u>), (**I**= Interest Expense= <u>513</u> millions USD), (**W**= Wealth or Total Assets= <u>139,796</u> millions USD), and (**t**= T/B= Tax Rate= <u>30.00%</u>), then it's (**f**= Fixed Portion planned), is:

$$f= (M-I-\{W/[1+F]-U\}/\{[1-d][1-t]\})/S$$
$$= (41{,}548-513-\{139{,}796/[1+0.9900]$$
$$-64{,}720\}/\{[1-0.3000]$$
$$[1-0.3000]\})/51{,}294$$
$$= \underline{58.00\%}$$

Corporate IFRS-GAAP (B/S-I/S), ISBN 13: **978-1983566332**, ISBN 10: **1983566330**

Law-5218:

If both (**M**= Margin of Contribution= 41,548 millions USD), (**I**= L/E= Leverage or Gearing Ratio= 99.00%), (**d**= D/A= Dividend Portion or Payout= 30.00%), (**S**= Sales or Revenues= 51,294 millions USD), (**U**= Utilized or Starting Capital= 64,720 millions USD), (**W**= Wealth or Total Assets= 139,796 millions USD), (**f**= F/S=Fixed Portion= 58.00%), and (**t**= T/B= Tax Rate= 30.00%), then it's (**I**= Interest Expense planned), is:

$$\mathbf{I = M\text{-}Sf\text{-}\{W/[1+\mathit{I}]\text{-}U\}/\{[1\text{-}d][1\text{-}t]\}}$$
$$= 41,548\text{-}51,294*0.5800\text{-}\{139,796$$
$$/[1+0.9900]\text{-}64,720\}$$
$$/\{[1\text{-}0.3000][1\text{-}0.3000]\}$$
$$= \underline{513} \text{ millions USD}$$

Corporate IFRS-GAAP (B/S-I/S), ISBN 13: **978-1983566332**, ISBN 10: **1983566330**

Law-5219:

If both (**M**= Margin of Contribution= 41,548 millions USD), (**I**= L/E= Leverage or Gearing Ratio= 99.00%), (**d**= D/A= Dividend Portion or Payout= 30.00%), (**S**= Sales or Revenues= 51,294 millions USD), (**U**= Utilized or Starting Capital= 64,720 millions USD), (**W**= Wealth or Total Assets= 139,796 millions USD), (**f**= F/S= Fixed Portion= 58.00%), and (**I**= Interest Expense= 513 millions USD), then it's (**t**= Tax rate planned), is:

$$t = 1 - \{W/[1+I]-U\}/\{[1-d][M-Sf-I]\}$$
$$= 1 - \{139,796/[1+0.9900]-64,720\}$$
$$/\{[1-0.3000][41,548$$
$$-51,294*0.5800-513]\}$$
$$= 30.00\%$$

Corporate IFRS-GAAP (B/S-I/S), ISBN 13: **978-1983566332**, ISBN 10: **1983566330**

<u>Law-5220</u>:

If both (**M**= Margin of Contribution= <u>41,548</u> millions USD), (**I**= L/E= Leverage or Gearing Ratio= <u>99.00%</u>), (**t**= T/B= Tax Rate= <u>30.00%</u>), (**S**= Sales or Revenues= <u>51,294</u> millions USD), (**U**= Utilized or Starting Capital= <u>64,720</u> millions USD), (**W**= Wealth or Total Assets= <u>139,796</u> millions USD), (**f**= F/S= Fixed Portion= <u>58.00%</u>), and (**I**= Interest Expense= <u>513</u> millions USD), then it's (**d**= Dividend Portion planned), is:

$$d = 1 - \{W/[1+I]-U\}/\{[1-t][M-Sf-I]\}$$
$$= 1 - \{139,796/[1+0.9900]-64,720\}$$
$$/\{[1-0.3000][41,548-51,294$$
$$*0.5800-513]\}$$
$$= \underline{30.00\%}$$

Corporate IFRS-GAAP (B/S-I/S), ISBN 13: **978-1983566332**, ISBN 10: **1983566330**

Law-5221:

If both (**M**= Margin of Contribution= 41,548 millions USD), (**f**= L/E= Leverage or Gearing Ratio= 99.00%), (**t**= T/B= Tax Rate= 30.00%), (**S**= Sales or Revenues= 51,294 millions USD), (**d**= D/A= Dividend Portion or Payout= 30.00%), (**W**= Wealth or Total Assets= 139,796 millions USD), (**f**= F/S= Fixed Portion= 58.00%), and (**I**= Interest Expense= 513 millions USD), then it's (**U**= Utilized or Starting Capital planned), is:

$$U= W/[1+f]-[1-d][1-t][M-Sf-I]$$
$$= 139,796/[1+0.9900]-[1-0.3000]$$
$$[1-0.3000][41,548-51,294$$
$$*0.5800-513]$$
$$= 64,720 \text{ millions USD}$$

Corporate IFRS-GAAP (B/S-I/S), ISBN 13: **978-1983566332**, ISBN 10: **1983566330**

<u>Law-5222</u>:

If both (**M**= Margin of Contribution= <u>41,548</u> millions USD), (**U**= Utilized or Starting Capital= <u>64,720</u> millions USD), (**t**= T/B= Tax Rate= <u>30.00%</u>), (**S**= Sales or Revenues= <u>51,294</u> millions USD), (**d**= D/A= Dividend Portion or Payout= <u>30.00%</u>), (**W**= Wealth or Total Assets= <u>139,796</u> millions USD), (**f**= F/S=Fixed Portion= <u>58.00%</u>), and (**I**= Interest Expense= <u>513</u> millions USD), then it's (**/** = Leverage or Gearing Ratio planned), is:

$$\textbf{/} = \textbf{W}/\{\textbf{U}+[1\text{-}\textbf{d}][1\text{-}\textbf{t}][\textbf{M-Sf-I}]\}$$
$$= 139{,}796/\{64{,}720+[1\text{-}0.3000]$$
$$[1\text{-}0.3000][41{,}548\text{-}51{,}294$$
$$*0.5800\text{-}513]\}$$
$$= \underline{99.00\%}$$

Corporate IFRS-GAAP (B/S-I/S), ISBN 13: **978-1983566332**, ISBN 10: **1983566330**

Law-5223:

If both (**M**= Margin of Contribution= 41,548 millions USD), (**U**= Utilized or Starting Capital= 64,720 millions USD), (**t**= T/B= Tax Rate= 30.00%), (**S**= Sales or Revenues= 51,294 millions USD), (**d**= D/A= Dividend Portion or Payout= 30.00%), (**l**= L/E= Leverage or Gearing Ratio= 99.00%), (**f**= F/S= Fixed Portion= 58.00%), and (**i**= I/S= Interest Portion= 1.00%), then it's (**W**= Wealth or Total Assets planned), is:

$$W= [1+l]\{U+[1-d][1-t][M-Sf-Si]\}$$
$$= [1+l](U+[1-d][1-t]\{M-S[f+i]\})$$
$$= [1+0.9900](64,720+[1-0.3000]$$
$$[1-0.3000]\{41,548-51,294$$
$$[0.5800+0.0100]\})$$
$$= \underline{139,786} \text{ millions USD}$$

Corporate IFRS-GAAP (B/S-I/S), ISBN 13: **978-1983566332**, ISBN 10: **1983566330**

<u>Law-5224</u>:

If both (**W**= Wealth or Total Assets= <u>139,796</u> millions USD), (**U**= Utilized or Starting Capital= <u>64,720</u> millions USD), (**t**= T/B= Tax Rate= <u>30.00%</u>), (**$**= Sales or Revenues= <u>51,294</u> millions USD), (**d**= D/A= Dividend Portion or Payout= <u>30.00%</u>), (**l**= L/E= Leverage or Gearing Ratio= <u>99.00%</u>), (**f**= F/S= Fixed Portion= <u>58.00%</u>), and (**i**= I/S= Interest Portion= <u>1.00%</u>), then it's (**M**= Margin of Contribution planned), is:

$$M= \$[f+i]+\{W/[1+l]-U\}/\{[1-d][1-t]\}$$
$$= 51,294[0.5800+0.0100]$$
$$+\{139,796/[1+0.9900]$$
$$-64,720\}/\{[1-0.3000]$$
$$[1-0.3000]\}$$
$$= \underline{41,548} \text{ millions USD}$$

Corporate IFRS-GAAP (B/S-I/S), ISBN 13: **978-1983566332**, ISBN 10: **1983566330**

Law-5225:

If both (**W**= Wealth or Total Assets= 139,796 millions USD), (**U**= Utilized or Starting Capital= 64,720 millions USD), (**t**= T/B= Tax Rate= 30.00%), (**M**= Margin of Contribution= 41,548 millions USD), (**d**= D/A= Dividend Portion or Payout= 30.00%), (**l**= L/E= Leverage or Gearing Ratio= 99.00%), (**f**= F/S= Fixed Portion= 58.00%), and (**i**= I/S= Interest Portion= 1.00%), then it's (**S**= Sales or Revenues planned), is:

$$S= M-\{W/[1+l]-U\}/\{[1-d][1-t]\})/[f+i]$$
$$= 41,548-\{139,796/[1+0.9900]$$
$$-64,720\}/\{[1-0.3000]$$
$$[1-0.3000]\})$$
$$/[0.5800+0.0100]$$
$$= 51,294 \text{ millions USD}$$

Corporate IFRS-GAAP (B/S-I/S), ISBN 13: **978-1983566332**, ISBN 10: **1983566330**

<u>Law-5226</u>:

If both (**W**= Wealth or Total Assets= <u>139,796</u> millions USD), (**U**= Utilized or Starting Capital= <u>64,720</u> millions USD), (**t**= T/B= Tax Rate= <u>30.00%</u>), (**M**= Margin of Contribution= <u>41,548</u> millions USD), (**d**= D/A= Dividend Portion or Payout= <u>30.00%</u>), (**I**= L/E= Leverage or Gearing Ratio= <u>99.00%</u>), (**S**= Sales or Revenues= <u>51,294</u> millions USD), and (**i**= I/S= Interest Portion= <u>1.00%</u>), then it's (**f**= Fixed Portion planned), is:

$$f = M - \{W/[1+I] - U\}/\{[1-d][1-t]\})/S - i$$
$$= 41,548 - \{139,796/[1+0.9900]$$
$$-64,720\}/\{[1-0.3000]$$
$$[1-0.3000]\})/51,294 - 0.0100$$
$$= \underline{58.00\%}$$

Corporate IFRS-GAAP (B/S-I/S), ISBN 13: **978-1983566332**, ISBN 10: **1983566330**

Law-5227:

If both (**W**= Wealth or Total Assets= 139,796 millions USD), (**U**= Utilized or Starting Capital= 64,720 millions USD), (**t**= T/B= Tax Rate= 30.00%), (**M**= Margin of Contribution= 41,548 millions USD), (**d**= D/A= Dividend Portion or Payout= 30.00%), (**l**= L/E= Leverage or Gearing Ratio= 99.00%), (**$**= Sales or Revenues= 51,294 millions USD), and (**f**= F/S= Fixed Portion= 58.00%),then it's (**i**= Interest Portion planned), is:

$$i= M-\{W/[1+l]-U\}/\{[1-d][1-t]\})/\$-f$$
$$= 41,548-\{139,796/[1+0.9900]$$
$$-64,720\}/\{[1-0.3000]$$
$$[1-0.3000]\})/51,294-0.5800$$
$$= \underline{1.00\%}$$

Corporate IFRS-GAAP (B/S-I/S), ISBN 13: **978-1983566332**, ISBN 10: **1983566330**

Law-5228:

If both (**W**= Wealth or Total Assets= 139,796 millions USD), (**U**= Utilized or Starting Capital= 64,720 millions USD), (**i**= I/S= Interest Portion= 1.00%), (**M**= Margin of Contribution= 41,548 millions USD), (**d**= D/A= Dividend Portion or Payout= 30.00%), (**l**= L/E= Leverage or Gearing Ratio= 99.00%), (**S**= Sales or Revenues= 51,294 millions USD), and (**f**= F/S= Fixed Portion= 58.00%),then it's (**t**= Tax Rate planned), is:

$$t= 1-\{W/[1+l]-U\}/([1-d]\{M-S[f+i]\})$$
$$= 1-\{139,796/[1+0.9900]$$
$$-64,720\}/([1-0.3000]$$
$$\{41,548-51,294$$
$$[0.5800+0.0100]\})$$
$$= 30.00\%$$

Corporate IFRS-GAAP (B/S-I/S), ISBN 13: **978-1983566332**, ISBN 10: **1983566330**

Law-5229:

If both (**W**= Wealth or Total Assets= 139,796 millions USD), (**U**= Utilized or Starting Capital= 64,720 millions USD), (**i**= I/S= Interest Portion= 1.00%), (**M**= Margin of Contribution= 41,548 millions USD), (**t**= T/B= Tax Rate= 30.00%), (**f**= L/E= Leverage or Gearing Ratio= 99.00%), (**S**= Sales or Revenues= 51,294 millions USD), and (**f**= F/S= Fixed Portion= 58.00%),then it's (**d**= Dividend Portion or Payout planned), is:

$$d= 1-\{W/[1+f]-U\}/([1-t]\{M-S[f+i]\})$$
$$= 1-\{139,796/[1+0.9900]$$
$$-64,720\}/([1-0.3000]$$
$$\{41,548-51,294$$
$$[0.5800+0.0100]\})$$
$$= 30.00\%$$

Corporate IFRS-GAAP (B/S-I/S), ISBN 13: **978-1983566332**, ISBN 10: **1983566330**

Law-5230:

If both (**W**= Wealth or Total Assets= 139,796 millions USD), (**d**= D/A= Dividend Portion or Payout= 30.00%), (**i**= I/S= Interest Portion= 1.00%), (**M**= Margin of Contribution= 41,548 millions USD), (**t**= T/B= Tax Rate= 30.00%), (**I**= L/E= Leverage or Gearing Ratio= 99.00%), (**S**= Sales or Revenues= 51,294 millions USD), and (**f**= F/S= Fixed Portion= 58.00%),then it's (**U**= Utilized or Starting Capital planned), is:

$$U= W/[1+I]-[1-d][1-t]\{M-S[f+i]\}$$
$$= 139,796/[1+0.9900]-[1-0.3000]$$
$$[1-0.3000]\{41,548$$
$$-51,294[0.5800+0.0100]\}$$
$$= 64,720 \text{ millions USD}$$

Corporate IFRS-GAAP (B/S-I/S), ISBN 13: **978-1983566332**, ISBN 10: **1983566330**

Law-5231:

If both (**W**= Wealth or Total Assets= 139,796 millions USD), (**d**= D/A= Dividend Portion or Payout= 30.00%), (**i**= I/S= Interest Portion= 1.00%), (**M**= Margin of Contribution= 41,548 millions USD), (**t**= T/B= Tax Rate= 30.00%), (**U**= Utilized or Starting Capital= 64,720 millions USD), (**S**= Sales or Revenues= 51,294 millions USD), and (**f**= F/S= Fixed Portion= 58.00%),then it's (**I** = Leverage or Gearing Ratio planned), is:

$$I = W/(U+[1-d][1-t]\{M-S[f+i]\})$$
$$= 139,796/(64,720+[1-0.3000]$$
$$[1-0.3000]\{41,548$$
$$-51,294[0.5800+0.0100]\})$$
$$= 99.00\%$$

Corporate IFRS-GAAP (B/S-I/S), ISBN 13: **978-1983566332**, ISBN 10: **1983566330**

<u>Law-5232</u>:

If both (I= L/E= Leverage or Gearing Ratio= <u>99.00%</u>), (S'= Sales of Past Year= <u>48,851</u> millions USD), (s= [S/S']-1= Sales Growth= <u>5.00%</u>), (d= D/A= Dividend Portion or Payout= <u>30.00%</u>), (i= I/S= Interest Portion= <u>1.00%</u>), (M= Margin of Contribution= <u>41,548</u> millions USD), (t= T/B= Tax Rate= <u>30.00%</u>), (U= Utilized or Starting Capital= <u>64,720</u> millions USD), (S= Sales or Revenues= <u>51,294</u> millions USD), and (f= F/S= Fixed Portion= <u>58.00%</u>),then it's (W= Wealth or Total Assets planned), is:

$$W = [1+I](U+[1-d][1-t]\{M\text{-}Sf\text{-}S'i[1+s]\})$$
$$= [1+0.9900](64,720+[1-0.3000]$$
$$[1-0.3000]\{41,548\text{-}51,294$$
$$*0.5800\text{-}48,851*0.0100$$
$$[1+0.0500]\})$$
$$= \underline{139,796} \text{ millions USD}$$

Corporate IFRS-GAAP (B/S-I/S), ISBN 13: **978-1983566332**, ISBN 10: **1983566330**

Law-5233:

If both (I= L/E= Leverage or Gearing Ratio= 99.00%), (S'= Sales of Past Year= 48,851 millions USD), (s= [S/S']-1= Sales Growth= 5.00%), (d= D/A= Dividend Portion or Payout= 30.00%), (i= I/S= Interest Portion= 1.00%), (W= Wealth or Total Assets= 139,796 millions USD), (t= T/B= Tax Rate= 30.00%), (U= Utilized or Starting Capital= 64,720 millions USD), (S= Sales or Revenues= 51,294 millions USD), and (f= F/S= Fixed Portion= 58.00%),then it's (M = Margin of Contribution planned), is:

$$M = Sf + S'i[1+s] + \{W/[1+I]-U\}/\{[1-d][1-t]\}$$
$$= 51,294*0.5800 + 48,851*0.0100$$
$$[1+0.0500] + \{139,796$$
$$/[1+0.9900]-64,720\}$$
$$/\{[1-0.3000][1-0.3000]\}$$
$$= 41,548 \text{ millions USD}$$

Corporate IFRS-GAAP (B/S-I/S), ISBN 13: **978-1983566332**, ISBN 10: **1983566330**

Law-5234:

If both (**I**= L/E= Leverage or Gearing Ratio= 99.00%), (**S'**= Sales of Past Year= 48,851 millions USD), (**s**= [S/S']-1= Sales Growth= 5.00%), (**d**= D/A= Dividend Portion or Payout= 30.00%), (**i**= I/S= Interest Portion= 1.00%), (**W**= Wealth or Total Assets= 139,796 millions USD), (**t**= T/B= Tax Rate= 30.00%), (**U**= Utilized or Starting Capital= 64,720 millions USD), (**M**= Margin of Contribution= 41,548 millions USD), and (**f**= F/S= Fixed Portion= 58.00%),then it's (**S**= Sales or Revenues planned), is:

$$\mathbf{S}= (\mathbf{M}-\mathbf{S'i}[1+\mathbf{s}]-\{\mathbf{W}/[1+\mathbf{I}]-\mathbf{U}\}/\{[1-\mathbf{d}][1-\mathbf{t}]\})/\mathbf{f}$$
$$= (41,548-48,851*0.0100[1+0.0500]$$
$$-\{139,796/[1+0.9900]$$
$$-64,720\}/\{[1-0.3000]$$
$$[1-0.3000]\})/0.5800$$
$$= 51,294 \text{ millions USD}$$

Corporate IFRS-GAAP (B/S-I/S), ISBN 13: **978-1983566332**, ISBN 10: **1983566330**

Law-5235:

If both (I= L/E= Leverage or Gearing Ratio=
99.00%), (S'= Sales of Past Year= 48,851 millions
USD), (s= [S/S']-1= Sales Growth= 5.00%), (d=
D/A= Dividend Portion or Payout= 30.00%), (i= I/S=
Interest Portion= 1.00%), (W= Wealth or Total
Assets= 139,796 millions USD), (t= T/B= Tax Rate=
30.00%), (U= Utilized or Starting Capital= 64,720
millions USD), (M= Margin of Contribution= 41,548
millions USD), and then it's (f= Fixed Portion
planned), is:

$$f= (M-S'i[1+s]-\{W/[1+I]-U\}/\{[1-d][1-t]\})/S$$
$$= (41,548-48,851*0.0100[1+0.0500]$$
$$-\{139,796/[1+0.9900]$$
$$-64,720\}/\{[1-0.3000]$$
$$[1-0.3000]\})/51,294$$
$$= \underline{58.00\%}$$

Corporate IFRS-GAAP (B/S-I/S), ISBN 13: **978-1983566332**, ISBN 10: **1983566330**

Law-5236:

If both (I= L/E= Leverage or Gearing Ratio= 99.00%), (M= Margin of Contribution= 41,548 millions USD), (s= [S/S']-1= Sales Growth= 5.00%), (d= D/A= Dividend Portion or Payout= 30.00%), (i= I/S= Interest Portion= 1.00%), (W= Wealth or Total Assets= 139,796 millions USD), (t= T/B= Tax Rate= 30.00%), (U= Utilized or Starting Capital= 64,720 millions USD), (S= Sales or Revenues= 51,294 millions USD), and (f= F/S= Fixed Portion= 58.00%),then it's (S'= Sales past), must be:

$$S' = (M-Sf-\{W/[1+I]-U\}/\{[1-d][1-t]\})/\{i[1+s]\}$$
$$=(41,548-51,294*0.5800-\{139,796$$
$$/[1+0.9900]-64,720\})$$
$$/\{1-0.3000][1-0.3000]$$
$$/\{0.0100[1+0.0500]\}$$
$$= \underline{48,851} \text{ millions USD}$$

Corporate IFRS-GAAP (B/S-I/S), ISBN 13: **978-1983566332**, ISBN 10: **1983566330**

<u>Law-5237</u>:

If both (**⌐**= L/E= Leverage or Gearing Ratio=
<u>99.00%</u>), (**M**= Margin of Contribution= <u>41,548</u>
millions USD), (**s**= [S/S']-1= Sales Growth= <u>5.00%</u>),
(**d**= D/A= Dividend Portion or Payout= <u>30.00%</u>), (**S'**=
Sales of Past Year= <u>48,851</u> millions USD), (**W**=
Wealth or Total Assets= <u>139,796</u> millions USD), (**t**=
T/B= Tax Rate= <u>30.00%</u>), (**U**= Utilized or Starting
Capital= <u>64,720</u> millions USD), (**S**= Sales or
Revenues= <u>51,294</u> millions USD), and (**f**= F/S=
Fixed Portion= <u>58.00%</u>),then it's (**i** = Interest Portion
planned), is:

$$i= (M\text{-}Sf\text{-}\{W/[1+I]\text{-}U\}/\{[1\text{-}d][1\text{-}t]\})/\{S'[1+s]\}$$
$$= (41{,}548\text{-}51{,}294{*}0.5800\text{-}\{139{,}796$$
$$/[1+0.9900]\text{-}64{,}720\})$$
$$/\{1\text{-}0.3000][1\text{-}0.3000]$$
$$/\{48{,}851[1+0.0500]\}$$

$$= \underline{1.00\%}$$

Corporate IFRS-GAAP (B/S-I/S), ISBN 13: **978-1983566332**, ISBN 10: **1983566330**

Law-5238:

If both (**I**= L/E= Leverage or Gearing Ratio= 99.00%), (**M**= Margin of Contribution= 41,548 millions USD), (**i**= I/S= Interest Portion= 1.00%), (**d**= D/A= Dividend Portion or Payout= 30.00%), (**S'**= Sales of Past Year= 48,851 millions USD), (**W**= Wealth or Total Assets= 139,796 millions USD), (**t**= T/B= Tax Rate= 30.00%), (**U**= Utilized or Starting Capital= 64,720 millions USD), (**S**= Sales or Revenues= 51,294 millions USD), and (**f**= F/S= Fixed Portion= 58.00%),then it's (**s**= Sales Growth planned), is:

$$s= (M\text{-}Sf\text{-}\{W/[1+I]\text{-}U\}/\{[1\text{-}d][1\text{-}t]\})/[S'i]\text{-}1$$
$$= (41,548\text{-}51,294*0.5800\text{-}\{139,796$$
$$/[1+0.9900]\text{-}64,720\})$$
$$/\{[1\text{-}0.3000][1\text{-}0.3000]$$
$$/[48,851*0.0100]\text{-}1$$
$$= 5.00\%$$

Corporate IFRS-GAAP (B/S-I/S), ISBN 13: **978-1983566332**, ISBN 10: **1983566330**

Law-5239:

If both (I= L/E= Leverage or Gearing Ratio= 99.00%), (M= Margin of Contribution= 41,548 millions USD),   (i= I/S= Interest Portion= 1.00%), (d= D/A= Dividend Portion or Payout= 30.00%), (S'= Sales of Past Year= 48,851 millions USD), (W= Wealth or Total Assets= 139,796 millions USD), (s= [S/S']-1= Sales Growth= 5.00%), (U= Utilized or Starting Capital= 64,720 millions USD), (S= Sales or Revenues= 51,294 millions USD), and (f= F/S= Fixed Portion= 58.00%),then it's (t= Tax Rate planned), is:

$$t= 1-\{W/[1+I]-U\}/([1-d]\{M-Sf-S'i[1+s]\})$$
$$=1-\{139,796/[1+0.9900]-64,720\}$$
$$/\{[1-0.3000]\{41,548$$
$$-51,294*0.5800-48,851$$
$$*0.0100[1+0.0500]\})$$

$$= 30.00\%$$

Corporate IFRS-GAAP (B/S-I/S), ISBN 13: **978-1983566332**, ISBN 10: **1983566330**

Law-5240:

If both (I= L/E= Leverage or Gearing Ratio= 99.00%), (M= Margin of Contribution= 41,548 millions USD), (i= I/S= Interest Portion= 1.00%), (t= T/B= Tax Rate= 30.00%), (S'= Sales of Past Year= 48,851 millions USD), (W= Wealth or Total Assets= 139,796 millions USD), (s= [S/S']-1= Sales Growth= 5.00%), (U= Utilized or Starting Capital= 64,720 millions USD), (S= Sales or Revenues= 51,294 millions USD), and (f= F/S= Fixed Portion= 58.00%),then it's (d= Dividend Portion or Payout planned), is:

$$d= 1-\{W/[1+I]-U\}/([1-t]\{M-Sf-S'i[1+s]\})$$
$$= 1-\{139,796/[1+0.9900]-64,720\}$$
$$/\{[1-0.3000]\{41,548$$
$$-51,294*0.5800-48,851$$
$$*0.0100[1+0.0500]\})$$
$$= 30.00\%$$

Corporate IFRS-GAAP (B/S-I/S), ISBN 13: 978-1983566332, ISBN 10: 1983566330

Law-5241:

If both (I= L/E= Leverage or Gearing Ratio= 99.00%), (M= Margin of Contribution= 41,548 millions USD), (i= I/S= Interest Portion= 1.00%), (t= T/B= Tax Rate= 30.00%), (S'= Sales of Past Year= 48,851 millions USD), (W= Wealth or Total Assets= 139,796 millions USD), (s= [S/S']-1= Sales Growth= 5.00%), (d= D/A= Dividend Portion or Payout= 30.00%), (S= Sales or Revenues= 51,294 millions USD), and (f= F/S= Fixed Portion= 58.00%),then it's (U= Utilized or Starting Capital planned), is:

$$U = W/[1+I]-[1-d][1-t]\{M-Sf-S'i[1+s]\}$$
$$= 139,796/[1+0.9900]-[1-0.3000]$$
$$[1-0.3000]\{41,548-51,294$$
$$*0.5800-48,851*0.0100$$
$$[1+0.0500]\}$$
$$= 64,720 \text{ millions USD}$$

Corporate IFRS-GAAP (B/S-I/S), ISBN 13: **978-1983566332**, ISBN 10: **1983566330**

<u>Law-5242</u>:

If both (**U**= Utilized or Starting Capital= <u>64,720</u> millions USD), (**M**= Margin of Contribution= <u>41,548</u> millions USD), (**i**= I/S= Interest Portion= <u>1.00%</u>), (**t**= T/B= Tax Rate= <u>30.00%</u>), (**S'**= Sales of Past Year= <u>48,851</u> millions USD), (**W**= Wealth or Total Assets= <u>139,796</u> millions USD), (**s**= [S/S']-1= Sales Growth= <u>5.00%</u>), (**d**= D/A= Dividend Portion or Payout= <u>30.00%</u>), (**S**= Sales or Revenues= <u>51,294</u> millions USD), and (**f**= F/S= Fixed Portion= <u>58.00%</u>),then it's (**I** = Leverage or Gearing Ratio planned), is:

$$I = W/(U+[1-d][1-t]\{M-Sf-S'i[1+s]\})-1$$
$$= 139,796/(64,720+[1-0.3000]$$
$$[1-0.3000]\{41,548$$
$$-51,294*0.5800-48,851$$
$$*0.0100[1+0.0500]\})-1$$
$$= \underline{99.00\%}$$

Law-5243:

If both (**U**= Utilized or Starting Capital= 64,720 millions USD), (**M**= Margin of Contribution= 41,548 millions USD), (**I**= Interest Expense= 513 millions USD), (**t**= T/B= Tax Rate= 30.00%), (**S'**= Sales of Past Year= 48,851 millions USD), (**I**= L/E= Leverage or Gearing Ratio= 99.00%), (**s**= [S/S']-1= Sales Growth= 5.00%), (**d**= D/A= Dividend Portion or Payout= 30.00%), and (**f**= F/S= Fixed Portion= 58.00%),then it's (**W**= Wealth or Total Assesments planned), is:

$$\mathbf{W}= [1+\mathbf{I}](\mathbf{U}+[1-\mathbf{d}][1-\mathbf{t}]\{\mathbf{M}-\mathbf{I}-\mathbf{S'f}[1+\mathbf{s}]\})$$
$$= [1+0.9900](64,720+[1-0.3000]$$
$$[1-0.3000]\{41,548-513$$
$$-48,851*0.5800$$
$$[1+0.0500]\})$$
$$= 139,796 \text{ millions USD}$$

Corporate IFRS-GAAP (B/S-I/S), ISBN 13: **978-1983566332**, ISBN 10: **1983566330**

Law-5244:

If both (**U**= Utilized or Starting Capital= 64,720 millions USD), (**W**= Wealth or Total Assets= 139,796 millions USD), (**I**= Interest Expense= 513 millions USD), (**t**= T/B= Tax Rate= 30.00%), (**S'**= Sales of Past Year= 48,851 millions USD), (**I**= L/E= Leverage or Gearing Ratio= 99.00%), (**s**= [S/S']-1= Sales Growth= 5.00%), (**d**= D/A= Dividend Portion or Payout= 30.00%), and (**f**= F/S= Fixed Portion= 58.00%),then it's (**M**= Margin of Contribution planned), is:

$$M = I + S'f[1+s] + \{W/[1+I] - U\}/\{[1-d][1-t]\}$$
$$= 513 + 48,851 * 0.5800[1+0.0500]$$
$$+ \{139,796/[1+0.9900]$$
$$-64,720\}/\{[1-0.3000]$$
$$[1-0.3000]\}$$
$$= \underline{41,548} \text{ millions USD}$$

Corporate IFRS-GAAP (B/S-I/S), ISBN 13: **978-1983566332**, ISBN 10: **1983566330**

Law-5245:

If both (**U**= Utilized or Starting Capital= 64,720 millions USD), (**W**= Wealth or Total Assets= 139,796 millions USD), (**I**= Interest Expense= 513 millions USD), (**t**= T/B= Tax Rate= 30.00%), (**M**= Margin of Contribution= 41,548 millions USD), (**l**= L/E= Leverage or Gearing Ratio= 99.00%), (**s**= [S/S']-1= Sales Growth= 5.00%), (**d**= D/A= Dividend Portion or Payout= 30.00%), and (**f**= F/S= Fixed Portion= 58.00%),then it's (**S'** = Sales Past), must be:

$$S' = (M-I-\{W/[1+l]-U\}/\{[1-d][1-t]\})/\{f[1+s]\}$$
$$= (41,548-513-\{139,796/[1+0.9900]$$
$$-64,720\}/\{[1-0.3000]$$
$$[1-0.3000]\})$$
$$/\{0.5800[1+0.0500]\}$$
$$= 48,851 \text{ millions USD}$$

Corporate IFRS-GAAP (B/S-I/S), ISBN 13: **978-1983566332**, ISBN 10: **1983566330**

Law-5246:

If both (**U**= Utilized or Starting Capital= 64,720 millions USD), (**W**= Wealth or Total Assets= 139,796 millions USD), (**I**= Interest Expense= 513 millions USD), (**t**= T/B= Tax Rate= 30.00%), (**M**= Margin of Contribution= 41,548 millions USD), (**I**= L/E= Leverage or Gearing Ratio= 99.00%), (**s**= [S/S']-1= Sales Growth= 5.00%), (**d**= D/A= Dividend Portion or Payout= 30.00%), and (**S'**= Sales of Past Year= 48,851 millions USD), then it's (**f**= Fixed Portion planned), is:

$$\mathbf{f} = (\mathbf{M}\text{-}\mathbf{I}\text{-}\{\mathbf{W}/[1+\mathbf{I}]\text{-}\mathbf{U}\}/\{[1\text{-}\mathbf{d}][1\text{-}\mathbf{t}]\})/\{\mathbf{S'}[1+\mathbf{s}]\}$$

$$= (41{,}548\text{-}513\text{-}\{139{,}796/[1+0.9900]$$
$$-64{,}720\}/\{[1\text{-}0.3000]$$
$$[1\text{-}0.3000]\})$$
$$/\{48{,}851[1+0.0500]\}$$

$$= 58.00\%$$

Corporate IFRS-GAAP (B/S-I/S), ISBN 13: **978-1983566332**, ISBN 10: **1983566330**

<u>Law-5247</u>:

If both (**U**= Utilized or Starting Capital= <u>64,720</u> millions USD), (**W**= Wealth or Total Assets= <u>139,796</u> millions USD), (**I**= Interest Expense= <u>513</u> millions USD), (**t**= T/B= Tax Rate= <u>30.00%</u>), (**M**= Margin of Contribution= <u>41,548</u> millions USD), (**I**= L/E= Leverage or Gearing Ratio= <u>99.00%</u>), (**f**= F/S= Fixed Portion= <u>58.00%</u>), (**d**= D/A= Dividend Portion or Payout= <u>30.00%</u>), and (**$'**= Sales of Past Year= <u>48,851</u> millions USD), then it's (**s**= Sales Growth planned), is:

$$s= (M-I-\{W/[1+I]-U\}/\{[1-d][1-t]\})/[\$'f]-1$$
$$= (41,548-513-\{139,796/[1+0.9900]$$
$$-64,720\}/\{[1-0.3000]$$
$$[1-0.3000]\})/\{48,851$$
$$*0.5800]-1$$
$$= \underline{5.00\%}$$

Corporate IFRS-GAAP (B/S-I/S), ISBN 13: **978-1983566332**, ISBN 10: **1983566330**

Law-5248:

If both (**U**= Utilized or Starting Capital= 64,720 millions USD), (**W**= Wealth or Total Assets= 139,796 millions USD), (**s**= [S/S']-1= Sales Growth= 5.00%), (**t**= T/B= Tax Rate= 30.00%), (**M**= Margin of Contribution= 41,548 millions USD), (**l**= L/E= Leverage or Gearing Ratio= 99.00%), (**f**= F/S= Fixed Portion= 58.00%), (**d**= D/A= Dividend Portion or Payout= 30.00%), and (**S'**= Sales of Past Year= 48,851 millions USD), then it's (**I**= Interest Expense planned), is:

$$\mathbf{I}= (\mathbf{M}\text{-}S'\mathbf{f}[1+\mathbf{s}]\text{-}\{\mathbf{W}/[1+\mathbf{l}]\text{-}\mathbf{U}\}/\{[1\text{-}\mathbf{d}][1\text{-}\mathbf{t}]\}$$

$$= (41,548\text{-}48,851*0.5800[1+0.0500]$$
$$-\{139,796/[1+0.9900]$$
$$-64,720\}/\{[1\text{-}0.3000]$$
$$[1\text{-}0.3000]\}$$
$$= \underline{513} \text{ millions USD}$$

Corporate IFRS-GAAP (B/S-I/S), ISBN 13: **978-1983566332**, ISBN 10: **1983566330**

Law-5249:

If both (**U**= Utilized or Starting Capital= 64,720 millions USD), (**W**= Wealth or Total Assets= 139,796 millions USD), (**s**= [S/S']-1= Sales Growth= 5.00%), (**I**= Interest Expense= 513 millions USD), (**M**= Margin of Contribution= 41,548 millions USD), (**I**= L/E= Leverage or Gearing Ratio= 99.00%), (**f**= F/S= Fixed Portion= 58.00%), (**d**= D/A= Dividend Portion or Payout= 30.00%), and (**S'**= Sales of Past Year= 48,851 millions USD), then it's (**t**= Tax Rate planned), is:

$$t= 1-\{W/[1+I]-U\}/([1-d]\{M-S'f[1+s]-I\})$$
$$= 1-\{139,796/[1+0.9900]-64,720\}$$
$$/([1-0.3000]\{41,548$$
$$-48,851*0.5800$$
$$[1+0.0500]-513\})$$
$$= \underline{30.00\%}$$

Corporate IFRS-GAAP (B/S-I/S), ISBN 13: **978-1983566332**, ISBN 10: **1983566330**

<u>Law-5250</u>:

If both (**U**= Utilized or Starting Capital= <u>64,720</u> millions USD), (**W**= Wealth or Total Assets= <u>139,796</u> millions USD), (**s**= [S/S']-1= Sales Growth= <u>5.00%</u>), (**I**= Interest Expense= <u>513</u> millions USD), (**M**= Margin of Contribution= <u>41,548</u> millions USD), (**I**= L/E= Leverage or Gearing Ratio= <u>99.00%</u>), (**f**= F/S= Fixed Portion= <u>58.00%</u>), (**t**= T/B= Tax Rate= <u>30.00%</u>), and (**S'**= Sales of Past Year= <u>48,851</u> millions USD), then it's (**d**= Dividend Portion or Payout planned), is:

$$d= 1-\{W/[1+I]-U\}/([1-t]\{M-S'f[1+s]-I\})$$
$$= 1-\{139,796/[1+0.9900]-64,720\}$$
$$/([1-0.3000]\{41,548$$
$$-48,851*0.5800$$
$$[1+0.0500]-513\})$$
$$= \underline{30.00\%}$$

Corporate IFRS-GAAP (B/S-I/S), ISBN 13: **978-1983566332**, ISBN 10: **1983566330**

Law-5251:

If both (**d**= D/A= Dividend Portion or Payout= 30.00%), (**W**= Wealth or Total Assets= 139,796 millions USD), (**s**= [S/S']-1= Sales Growth= 5.00%), (**I**= Interest Expense= 513 millions USD), (**M**= Margin of Contribution= 41,548 millions USD), (**l**= L/E= Leverage or Gearing Ratio= 99.00%), (**f**= F/S= Fixed Portion= 58.00%), (**t**= T/B= Tax Rate= 30.00%), and (**S'**= Sales of Past Year= 48,851 millions USD), then it's (**U**= Utilized or Starting Capital planned), is:

$$\mathbf{U}= \mathbf{W}/[1+\mathbf{l}]-[1-\mathbf{d}][1-\mathbf{t}]\{\mathbf{M}-\mathbf{S'f}[1+\mathbf{s}]-\mathbf{I}\})$$
$$= 139,796/[1+0.9900]-[1-0.3000]$$
$$[1-0.3000]\{41,548-48,851$$
$$*0.5800[1+0.0500]-513\})$$
$$= 64,720 \text{ millions USD}$$

Corporate IFRS-GAAP (B/S-I/S), ISBN 13: **978-1983566332**, ISBN 10: **1983566330**

Law-5252:

If both (**d**= D/A= Dividend Portion or Payout= 30.00%), (**W**= Wealth or Total Assets= 139,796 millions USD), (**s**= [S/S']-1= Sales Growth= 5.00%), (**I**= Interest Expense= 513 millions USD), (**M**= Margin of Contribution= 41,548 millions USD), (**U**= Utilized or Starting Capital= 64,720 millions USD), (**f**= F/S= Fixed Portion= 58.00%), (**t**= T/B= Tax Rate= 30.00%), and (**S'**= Sales of Past Year= 48,851 millions USD), then it's (**I** = Leverage or Gearing Ratio planned), is:

$$I = W/(U+[1-d][1-t]\{M-S'f[1+s]-I\})-1$$
$$= 139,796/(64,720+[1-0.3000]$$
$$[1-0.3000]\{41,548-48,851$$
$$*0.5800[1+0.0500]-513\})-1$$
$$= 99.00\%$$

Corporate IFRS-GAAP (B/S-I/S), ISBN 13: **978-1983566332**, ISBN 10: **1983566330**

<u>Law-5253</u>:

If both (**d**= D/A= Dividend Portion or Payout= 30.00%), (**I**= L/E= Leverage or Gearing Ratio= 99.00%), (**s**= [S/S']-1= Sales Growth= 5.00%), (**i**= I/S= Interest Portion= 1.00%), (**S**= Sales or Revenues= 51,294 millions USD), (**M**= Margin of Contribution= 41,548 millions USD), (**U**= Utilized or Starting Capital= 64,720 millions USD), (**f**= F/S= Fixed Portion= 58.00%), (**t**= T/B= Tax Rate= 30.00%), and (**S'**= Sales of Past Year= 48,851 millions USD), then it's (**W**= Wealth or Total Assets planned), is:

$$W= [1+I](U+[1-d][1-t]\{M-S'f[1+s]-Si\})$$
$$= [1+0.9900](64,720+[1-0.3000]$$
$$[1-0.3000]\{41,548-48,851$$
$$*0.5800[1+0.0500]-51,294$$
$$*0.0100\})$$
$$= \underline{139,796} \text{ millions USD}$$

Corporate IFRS-GAAP (B/S-I/S), ISBN 13: **978-1983566332**, ISBN 10: **1983566330**

<u>Law-5254</u>:

If both (**d**= D/A= Dividend Portion or Payout= 30.00%), (**l**= L/E= Leverage or Gearing Ratio= 99.00%), (**s**= [S/S']-1= Sales Growth= 5.00%), (**i**= I/S= Interest Portion= 1.00%), (**S**= Sales or Revenues= 51,294 millions USD), (**W**= Wealth or Total Assets= 139,796 millions USD), (**U**= Utilized or Starting Capital= 64,720 millions USD), (**f**= F/S= Fixed Portion= 58.00%), (**t**= T/B= Tax Rate= 30.00%), and (**S'**= Sales of Past Year= 48,851 millions USD), then it's (**M**= Margin of Contribution planned), is:

$$M = S'f[1+s] + Si + \{W/[1+l] - U\} / \{[1-d][1-t]\}$$
$$= 48,851*0.5800[1+0.0500] + 51,294$$
$$*0.0100 + \{139,796$$
$$/[1+0.9900] - 64,720\}$$
$$/\{[1-0.3000][1-0.3000]\}$$
$$= \underline{41,548} \text{ millions USD}$$

Corporate IFRS-GAAP (B/S-I/S), ISBN 13: **978-1983566332**, ISBN 10: **1983566330**

Law-5255:

If both (**d**= D/A= Dividend Portion or Payout= 30.00%), (**l**= L/E= Leverage or Gearing Ratio= 99.00%), (**s**= [S/S']-1= Sales Growth= 5.00%), (**i**= I/S= Interest Portion= 1.00%), (**S**= Sales or Revenues= 51,294 millions USD), (**W**= Wealth or Total Assets= 139,796 millions USD), (**U**= Utilized or Starting Capital= 64,720 millions USD), (**f**= F/S= Fixed Portion= 58.00%), (**t**= T/B= Tax Rate= 30.00%), and (**M**= Margin of Contribution= 41,548 millions USD),then it's (**S'**= Sales Past), must be:

$$\mathbf{S'}= (\mathbf{M}\text{-}\mathbf{Si}\text{-}\{\mathbf{W}/[1+\mathbf{l}]\text{-}\mathbf{U}\}/\{[1\text{-}\mathbf{d}][1\text{-}\mathbf{t}]\})/\{\mathbf{f}[1+\mathbf{s}]\}$$
$$= (41{,}548\text{-}51{,}294*0.0100$$
$$+\{139{,}796/[1+0.9900]$$
$$-64{,}720\}/\{[1\text{-}0.3000]$$
$$[1\text{-}0.3000]\})$$
$$/\{0.5800[1+0.0500]\}$$
$$= \underline{48{,}851} \text{ millions USD}$$

Corporate IFRS-GAAP (B/S-I/S), ISBN 13: **978-1983566332**, ISBN 10: **1983566330**

Law-5256:

If both (**d**= D/A= Dividend Portion or Payout= 30.00%), (**f**= L/E= Leverage or Gearing Ratio= 99.00%), (**s**= [S/S']-1= Sales Growth= 5.00%), (**i**= I/S= Interest Portion= 1.00%), (**S**= Sales or Revenues= 51,294 millions USD), (**W**= Wealth or Total Assets= 139,796 millions USD), (**U**= Utilized or Starting Capital= 64,720 millions USD), (**S'**= Sales of Past Year= 48,851 millions USD), (**t**= T/B= Tax Rate= 30.00%), and (**M**= Margin of Contribution= 41,548 millions USD),then it's (**f**= Fixed Portion planned), is:

$$\mathbf{f}= (\mathbf{M}\text{-}\mathbf{S}\mathbf{i}\text{-}\{\mathbf{W}/[1+\mathbf{f}]\text{-}\mathbf{U}\}/\{[1\text{-}\mathbf{d}][1\text{-}\mathbf{t}]\})/\{\mathbf{S'}[1+\mathbf{s}]\}$$

$$= (41,548\text{-}51,294*0.0100$$
$$+\{139,796/[1+0.9900]$$
$$-64,720\}/\{[1\text{-}0.3000]$$
$$[1\text{-}0.3000]\})$$
$$/\{48,851[1+0.0500]\}$$

$$= 58.00\%$$

Corporate IFRS-GAAP (B/S-I/S), ISBN 13: **978-1983566332**, ISBN 10: **1983566330**

Law-5257:

If both (**d**= D/A= Dividend Portion or Payout= 30.00%), (**I**= L/E= Leverage or Gearing Ratio= 99.00%), (**f**= F/S= Fixed Portion= 58.00%), (**i**= I/S= Interest Portion= 1.00%), (**S**= Sales or Revenues= 51,294 millions USD), (**W**= Wealth or Total Assets= 139,796 millions USD), (**U**= Utilized or Starting Capital= 64,720 millions USD), (**S'**= Sales of Past Year= 48,851 millions USD), (**t**= T/B= Tax Rate= 30.00%), and (**M**= Margin of Contribution= 41,548 millions USD),then it's (**s**= Sales Growth planned), is:

$$s= (M-Si-\{W/[1+I]-U\}/\{[1-d][1-t]\})/[S'f]-1$$
$$= (41,548-51,294*0.0100$$
$$+\{139,796/[1+0.9900]$$
$$-64,720\}/\{[1-0.3000]$$
$$[1-0.3000]\})/[48,851$$
$$*0.5800]-1$$
$$= 5.00\%$$

Corporate IFRS-GAAP (B/S-I/S), ISBN 13: **978-1983566332**, ISBN 10: **1983566330**

Law-5258:

If both (**d**= D/A= Dividend Portion or Payout= 30.00%), (**l**= L/E= Leverage or Gearing Ratio= 99.00%), (**f**= F/S= Fixed Portion= 58.00%), (**i**= I/S= Interest Portion= 1.00%), (**s**= [S/S']-1= Sales Growth= 5.00%), (**W**= Wealth or Total Assets= 139,796 millions USD), (**U**= Utilized or Starting Capital= 64,720 millions USD), (**S'**= Sales of Past Year= 48,851 millions USD), (**t**= T/B= Tax Rate= 30.00%), and (**M**= Margin of Contribution= 41,548 millions USD),then it's (**S**= Sales or Revenues planned), is:

$$S= (M-S'f[1+s]-\{W/[1+l]-U\}/\{[1-d][1-t]\})/i$$
$$= (41,548-48,851*0.5800$$
$$[1+0.0500]-\{139,796$$
$$/[1+0.9900]-64,720\}$$
$$/\{[1-0.3000][1-0.3000]\})$$
$$/0.0100$$
$$= 51,294 \text{ millions USD}$$

Corporate IFRS-GAAP (B/S-I/S), ISBN 13: **978-1983566332**, ISBN 10: **1983566330**

<u>Law-5259</u>:

If both (**d**= D/A= Dividend Portion or Payout= 30.00%), (**l**= L/E= Leverage or Gearing Ratio= 99.00%), (**f**= F/S= Fixed Portion= 58.00%), (**$**= Sales or Revenues= 51,294 millions USD), (**s**= [S/S']-1= Sales Growth= 5.00%), (**W**= Wealth or Total Assets= 139,796 millions USD), (**U**= Utilized or Starting Capital= 64,720 millions USD), (**$'**= Sales of Past Year= 48,851 millions USD), (**t**= T/B= Tax Rate= 30.00%), and (**M**= Margin of Contribution= 41,548 millions USD),then it's (**i**= Interest Portion planned), is:

$$i= (M-\$'f[1+s]-\{W/[1+l]-U\}/\{[1-d][1-t]\})/\$$$
$$= (41,548-48,851*0.5800$$
$$[1+0.0500]-\{139,796$$
$$/[1+0.9900]-64,720\}$$
$$/\{[1-0.3000][1-0.3000]\})$$
$$/51,294$$

$$= \underline{1.00\%}$$

Corporate IFRS-GAAP (B/S-I/S), ISBN 13: **978-1983566332**, ISBN 10: **1983566330**

Law-5260:

If both (**d**= D/A= Dividend Portion or Payout= 30.00%), (**I**= L/E= Leverage or Gearing Ratio= 99.00%), (**f**= F/S= Fixed Portion= 58.00%), (**S**= Sales or Revenues= 51,294 millions USD), (**s**= [S/S']-1= Sales Growth= 5.00%), (**W**= Wealth or Total Assets= 139,796 millions USD), (**U**= Utilized or Starting Capital= 64,720 millions USD), (**S'**= Sales of Past Year= 48,851 millions USD), (**i**= I/S= Interest Portion= 1.00%), and (**M**= Margin of Contribution= 41,548 millions USD),then it's (**t**= Tax Rate planned), is:

$$t= 1-\{W/[1+I]-U\}/([1-d]\{M-S'f[1+s]-Si\})$$
$$= 1-\{139,796/[1+0.9900]-64,720\}$$
$$/([1-0.3000]\{41,548-48,851$$
$$*0.5800[1+0.0500]$$
$$-51,294*0.0100\})$$
$$= 30.00\%$$

Law-5261:

If both (**t**= T/B= Tax Rate= 30.00%), (**I**= L/E= Leverage or Gearing Ratio= 99.00%), (**f**= F/S= Fixed Portion= 58.00%), (**S**= Sales or Revenues= 51,294 millions USD), (**s**= [S/S']-1= Sales Growth= 5.00%), (**W**= Wealth or Total Assets= 139,796 millions USD), (**U**= Utilized or Starting Capital= 64,720 millions USD), (**S'**= Sales of Past Year= 48,851 millions USD), (**i**= I/S= Interest Portion= 1.00%), and (**M**= Margin of Contribution= 41,548 millions USD),then it's (**d**= Dividend Portion or Payout planned), is:

$$\mathbf{d}= 1-\{\mathbf{W}/[1+\mathbf{I}]-\mathbf{U}\}/([1-\mathbf{t}]\{\mathbf{M}-\mathbf{S'f}[1+\mathbf{s}]-\mathbf{Si}\})$$

$$= 1-\{139,796/[1+0.9900]-64,720\}$$
$$/([1-0.3000]\{41,548-48,851$$
$$*0.5800[1+0.0500]$$
$$-51,294*0.0100\})$$

$$= 30.00\%$$

Corporate IFRS-GAAP (B/S-I/S), ISBN 13: **978-1983566332**, ISBN 10: **1983566330**

Law-5262:

If both (**t**= T/B= Tax Rate= 30.00%), (**l**= L/E= Leverage or Gearing Ratio= 99.00%), (**f**= F/S= Fixed Portion= 58.00%), (**S**= Sales or Revenues= 51,294 millions USD), (**s**= [S/S']-1= Sales Growth= 5.00%), (**W**= Wealth or Total Assets= 139,796 millions USD), (**d**= D/A= Dividend Portion or Payout= 30.00%), (**S'**= Sales of Past Year= 48,851 millions USD), (**i**= I/S= Interest Portion= 1.00%), and (**M**= Margin of Contribution= 41,548 millions USD),then it's (**U**= Utilized or Starting Capital planned), is:

$$U= W/[1+l]-[1-d][1-t]\{M-S'f[1+s]-Si\}$$
$$= 139,796/[1+0.9900]-[1-0.3000]$$
$$[1-0.3000]\{41,548-48,851$$
$$*0.5800[1+0.0500]$$
$$-51,294*0.0100\}$$
$$= 64,720 \text{ millions USD}$$

Corporate IFRS-GAAP (B/S-I/S), ISBN 13: **978-1983566332**, ISBN 10: **1983566330**

<u>Law-5263</u>:

If both (**t**= T/B= Tax Rate= <u>30.00%</u>), (**U**= Utilized or Starting Capital= <u>64,720</u> millions USD), (**f**= F/S= Fixed Portion= <u>58.00%</u>), (**$**= Sales or Revenues= <u>51,294</u> millions USD), (**s**= [S/S']-1= Sales Growth= <u>5.00%</u>), (**W**= Wealth or Total Assets= <u>139,796</u> millions USD), (**d**= D/A= Dividend Portion or Payout= <u>30.00%</u>), (**$'**= Sales of Past Year= <u>48,851</u> millions USD), (**i**= I/S= Interest Portion= <u>1.00%</u>), and (**M**= Margin of Contribution= <u>41,548</u> millions USD),then it's (**I** = Leverage or Gearing Ratio planned), is:

$$I = W/(U+[1-d][1-t]\{M-S'f[1+s]-Si\})-1$$
$$= 139,796/(64,720+[1-0.3000]$$
$$[1-0.3000]\{41,548-48,851$$
$$*0.5800[1+0.0500]$$
$$-51,294*0.0100\})-1$$
$$= \underline{99.00\%}$$

Corporate IFRS-GAAP (B/S-I/S), ISBN 13: **978-1983566332**, ISBN 10: **1983566330**

Law-5264:

If both (**t**= T/B= Tax Rate= 30.00%), (**U**= Utilized or Starting Capital= 64,720 millions USD), (**f**= F/S= Fixed Portion= 58.00%), (**s**= [S/S']-1= Sales Growth= 5.00%), (**l**= L/E= Leverage or Gearing Ratio= 99.00%), (**d**= D/A= Dividend Portion or Payout= 30.00%), (**S'**= Sales of Past Year= 48,851 millions USD), (**i**= I/S= Interest Portion= 1.00%), and (**M**= Margin of Contribution= 41,548 millions USD),then it's (**W**= Wealth or Total Assets planned), is:

$$\mathbf{W}= [1+\mathbf{l}](\mathbf{U}+[1-\mathbf{d}][1-\mathbf{t}]\{\mathbf{M}-\mathbf{S'f}[1+\mathbf{s}]-\mathbf{S'i}[1+\mathbf{s}]\})$$
$$= [1+\mathbf{l}](\mathbf{U}+[1-\mathbf{d}][1-\mathbf{t}]\{\mathbf{M}-\mathbf{S'}[1+\mathbf{s}][\mathbf{f}+\mathbf{i}]\})$$
$$= [1+0.9900](64,720+[1-0.3000]$$
$$[1-0.3000]\{41,548-48,851$$
$$[1+0.0500]$$
$$[0.5800+0.0100]\})$$
$$= 139,796 \text{ millions USD}$$

Corporate IFRS-GAAP (B/S-I/S), ISBN 13: **978-1983566332**, ISBN 10: **1983566330**

Law-5265:

If both (**t**= T/B= Tax Rate= 30.00%), (**U**= Utilized or Starting Capital= 64,720 millions USD), (**f**= F/S= Fixed Portion= 58.00%), (**s**= [S/S']-1= Sales Growth= 5.00%), (**l**= L/E= Leverage or Gearing Ratio= 99.00%), (**d**= D/A= Dividend Portion or Payout= 30.00%), (**S'**= Sales of Past Year= 48,851 millions USD), (**i**= I/S= Interest Portion= 1.00%), and (**W**= Wealth or Total Assets= 139,796 millions USD), then it's (**M**= Margin of Contribution planned), is:

$$M= S'[1+s][f+i]+\{W/[1+l]-U]\}/\{[1-d][1-t]\}$$
$$= 48,851[1+0.0500][0.5800+0.0100]$$
$$+\{139,796/[1+0.9900]$$
$$-64,720]\}/\{[1-0.3000]$$
$$[1-0.3000]\}$$
$$= 41,548 \text{ millions USD}$$

Corporate IFRS-GAAP (B/S-I/S), ISBN 13: **978-1983566332**, ISBN 10: **1983566330**

Law-5266:

If both (**t**= T/B= Tax Rate= 30.00%), (**U**= Utilized or Starting Capital= 64,720 millions USD), (**f**= F/S= Fixed Portion= 58.00%), (**s**= [S/S']-1= Sales Growth= 5.00%), (**l**= L/E= Leverage or Gearing Ratio= 99.00%), (**d**= D/A= Dividend Portion or Payout= 30.00%), (**M**= Margin of Contribution= 41,548 millions USD), (**i**= I/S= Interest Portion= 1.00%), and (**W**= Wealth or Total Assets= 139,796 millions USD), then it's (**S'**= Sales Past), must be:

$$S' = (M - \{W/[1+l]-U]\}/\{[1-d][1-t]\})$$
$$/\{[1+s][f+i]\}$$
$$= (41,548 - \{139,796/[1+0.9900]$$
$$-64,720]\}/\{[1-0.3000]$$
$$[1-0.3000]\})/\{[1+0.0500]$$
$$[0.5800+0.0100]\}$$
$$= 48,851 \text{ millions USD}$$

Corporate IFRS-GAAP (B/S-I/S), ISBN 13: **978-1983566332**, ISBN 10: **1983566330**

Law-5267:

If both (**t**= T/B= Tax Rate= 30.00%), (**U**= Utilized or Starting Capital= 64,720 millions USD), (**f**= F/S= Fixed Portion= 58.00%), (**$'**= Sales of Past Year= 48,851 millions USD), (**l**= L/E= Leverage or Gearing Ratio= 99.00%), (**d**= D/A= Dividend Portion or Payout= 30.00%), (**M**= Margin of Contribution= 41,548 millions USD), (**i**= I/S= Interest Portion= 1.00%), and (**W**= Wealth or Total Assets= 139,796 millions USD), then it's (**s**= Sales Growth planned), is:

$$s = (M-\{W/[1+l]-U]\}/\{[1-d][1-t]\})$$
$$/\{\$'[f+i]\}-1$$
$$= (41,548-\{139,796/[1+0.9900]$$
$$-64,720]\}/\{[1-0.3000]$$
$$[1-0.3000]\})/\{48,851$$
$$[0.5800+0.0100]\}-1$$
$$= 5.00\%$$

Corporate IFRS-GAAP (B/S-I/S), ISBN 13: **978-1983566332**, ISBN 10: **1983566330**

<u>Law-5268</u>:

If both (**t**= T/B= Tax Rate= <u>30.00%</u>), (**U**= Utilized or Starting Capital= <u>64,720</u> millions USD), (**s**= [S/S']-1= Sales Growth= <u>5.00%</u>), (**S'**= Sales of Past Year= <u>48,851</u> millions USD), (**l**= L/E= Leverage or Gearing Ratio= <u>99.00%</u>), (**d**= D/A= Dividend Portion or Payout= <u>30.00%</u>), (**M**= Margin of Contribution= <u>41,548</u> millions USD), (**i**= I/S= Interest Portion= <u>1.00%</u>), and (**W**= Wealth or Total Assets= <u>139,796</u> millions USD), then it's (**f**= Fixed Portion planned), is:

$$\mathbf{f}= (\mathbf{M}-\{\mathbf{W}/[1+\mathbf{l}]-\mathbf{U}]\}/\{[1-\mathbf{d}][1-\mathbf{t}]\})$$
$$/\{\mathbf{S'}[1+\mathbf{s}]\}-\mathbf{i}$$
$$= (41,548-\{139,796/[1+0.9900]$$
$$-64,720]\}/\{[1-0.3000]$$
$$[1-0.3000]\})/\{48,851$$
$$[1+0.0500]\}-0.0100$$
$$= \underline{58.00\%}$$

Corporate IFRS-GAAP (B/S-I/S), ISBN 13: **978-1983566332**, ISBN 10: **1983566330**

<u>Law-5269</u>:

If both (**t**= T/B= Tax Rate= <u>30.00%</u>), (**U**= Utilized or Starting Capital= <u>64,720</u> millions USD), (**s**= [S/S']-1= Sales Growth= <u>5.00%</u>), (**S'**= Sales of Past Year= <u>48,851</u> millions USD), (**l**= L/E= Leverage or Gearing Ratio= <u>99.00%</u>), (**d**= D/A= Dividend Portion or Payout= <u>30.00%</u>), (**M**= Margin of Contribution= <u>41,548</u> millions USD), (**f**= F/S=Fixed Portion= <u>58.00%</u>), and (**W**= Wealth or Total Assets= <u>139,796</u> millions USD), then it's (**f**= Fixed Portion planned), is:

$$f= (M-\{W/[1+l]-U]\}/\{[1-d][1-t]\})$$
$$/\{S'[1+s]\}-f$$
$$= (41,548-\{139,796/[1+0.9900]$$
$$-64,720]\}/\{[1-0.3000]$$
$$[1-0.3000]\})/\{48,851$$
$$[1+0.0500]\}-0.5800$$
$$= \underline{1.00\%}$$

Corporate IFRS-GAAP (B/S-I/S), ISBN 13: **978-1983566332**, ISBN 10: **1983566330**

<u>Law-5270</u>:

If both (**i**= I/S= Interest Portion= <u>1.00%</u>), (**U**= Utilized or Starting Capital= <u>64,720</u> millions USD), (**s**= [S/S']-1= Sales Growth= <u>5.00%</u>), (**S'**= Sales of Past Year= <u>48,851</u> millions USD), (**l**= L/E= Leverage or Gearing Ratio= <u>99.00%</u>), (**d**= D/A= Dividend Portion or Payout= <u>30.00%</u>), (**M**= Margin of Contribution= <u>41,548</u> millions USD), (**f**= F/S= Fixed Portion= <u>58.00%</u>), and (**W**= Wealth or Total Assets= <u>139,796</u> millions USD), then it's (**t**= Tax Rate planned), is:

$$
\begin{aligned}
\mathbf{t} &= 1 - \{\mathbf{W}/[1+\mathbf{l}]-\mathbf{U}]\}/([1-\mathbf{d}]\{\mathbf{M}-\mathbf{S'}[1+\mathbf{s}][\mathbf{f}+\mathbf{i}]\}) \\
&= 1 - \{139{,}796/[1+0.9900]-64{,}720]\} \\
&\qquad /([1-0.3000]\{41{,}548-48{,}851 \\
&\qquad [1+0.0500] \\
&\qquad [0.5800+0.0100]\}) \\
&= \underline{30.00\%}
\end{aligned}
$$

Corporate IFRS-GAAP (B/S-I/S), ISBN 13: **978-1983566332**, ISBN 10: **1983566330**

<u>Law-5271</u>:

If both (**i**= I/S= Interest Portion= <u>1.00%</u>), (**U**= Utilized or Starting Capital= <u>64,720</u> millions USD), (**s**= [S/S']-1= Sales Growth= <u>5.00%</u>), (**S'**= Sales of Past Year= <u>48,851</u> millions USD), (**l**= L/E= Leverage or Gearing Ratio= <u>99.00%</u>), (**t**= T/B= Tax Rate= <u>30.00%</u>), (**M**= Margin of Contribution= <u>41,548</u> millions USD), (**f**= F/S= Fixed Portion= <u>58.00%</u>), and (**W**= Wealth or Total Assets= <u>139,796</u> millions USD), then it's (**d**= Dividend Portion or Payout planned), is:

$$\mathbf{d}= 1-\{\mathbf{W}/[1+\mathbf{\textit{l}}]-\mathbf{U}]\}/([1-\mathbf{t}]\{\mathbf{M}-\mathbf{S'}[1+\mathbf{s}][\mathbf{f}+\mathbf{i}]\})$$

$$= 1-\{139,796/[1+0.9900]-64,720]\}$$
$$/([1-0.3000]\{41,548-48,851$$
$$[1+0.0500][0.5800$$
$$+0.0100]\})$$

$$= \underline{30.00\%}$$

Corporate IFRS-GAAP (B/S-I/S), ISBN 13: **978-1983566332**, ISBN 10: **1983566330**

<u>Law-5272</u>:

If both (i= I/S= Interest Portion= <u>1.00%</u>), (d= D/A= Dividend Portion or Payout= <u>30.00%</u>), (s= [S/S']-1= Sales Growth= <u>5.00%</u>), (S'= Sales of Past Year= <u>48,851</u> millions USD), (I= L/E= Leverage or Gearing Ratio= <u>99.00%</u>), (t= T/B= Tax Rate= <u>30.00%</u>), (M= Margin of Contribution= <u>41,548</u> millions USD), (f= F/S= Fixed Portion= <u>58.00%</u>), and (W= Wealth or Total Assets= <u>139,796</u> millions USD), then it's (U= Utilized or Starting Capital planned), is:

$$U = W/[1+I]-[1-d][1-t]\{M-S'[1+s][f+i]\}$$
$$= 139{,}796/[1+0.9900]-[1-0.3000]$$
$$[1-0.3000]\{41{,}548-48{,}851$$
$$[1+0.0500][0.5800$$
$$+0.0100]\}$$
$$= \underline{64{,}720} \text{ millions USD}$$

Corporate IFRS-GAAP (B/S-I/S), ISBN 13: **978-1983566332**, ISBN 10: **1983566330**

<u>Law-5273</u>:

If both (**i**= I/S= Interest Portion= <u>1.00%</u>), (**d**= D/A= Dividend Portion or Payout= <u>30.00%</u>), (**s**= [S/S']-1= Sales Growth= <u>5.00%</u>), (**S'**= Sales of Past Year= <u>48,851</u> millions USD), (**U**= Utilized or Starting Capital= <u>64,720</u> millions USD), (**t**= T/B= Tax Rate= <u>30.00%</u>), (**M**= Margin of Contribution= <u>41,548</u> millions USD), (**f**= F/S= Fixed Portion= <u>58.00%</u>), and (**W**= Wealth or Total Assets= <u>139,796</u> millions USD), then it's (**I** = Leverage or Gearing Ratio planned), is:

$$I = W/(U+[1-d][1-t]\{M-S'[1+s][f+i]\})-1$$
$$= 139{,}796/(64{,}720+[1-0.3000]$$
$$[1-0.3000]\{41{,}548-48{,}851$$
$$[1+0.0500]$$
$$[0.5800+0.0100]\})-1$$
$$= \underline{99.00\%}$$

Corporate IFRS-GAAP (B/S-I/S), ISBN 13: **978-1983566332**, ISBN 10: **1983566330**

Law-5274:

If both (**I**= Interest Expense= 513 millions USD), (**F**= Fixed Cost= 29,740 millions USD), (**V**= Variable Cost= 9,746 millions USD), (**$**= Sales or Revenues= 51,294 millions USD), (**d**= D/A= Dividend Portion or Payout= 30.00%), (**U**= Utilized or Starting Capital= 64,720 millions USD), (**t**= T/B= Tax Rate= 30.00%), and (**I**= L/E= Leverage or Gearing Ratio= 99.00%), then it's (**W**= Wealth or Total Assets planned), is:

$$\mathbf{W}= [1+\mathbf{I}]\{\mathbf{U}+[1-\mathbf{d}][1-\mathbf{t}][\mathbf{\$}-\mathbf{V}-\mathbf{F}-\mathbf{I}]\}$$
$$= [1+0.9900]\{64,720+[1-0.3000]$$
$$[1-0.3000][51,294-9,746$$
$$-29,750-513]\}$$
$$= \underline{139,796} \text{ millions USD}$$

Corporate IFRS-GAAP (B/S-I/S), ISBN 13: **978-1983566332**, ISBN 10: **1983566330**

<u>Law-5275</u>:

If both (**I**= Interest Expense= <u>513</u> millions USD), (**F**= Fixed Cost= <u>29,740</u> millions USD), (**V**= Variable Cost= <u>9,746</u> millions USD), (**W**= Wealth or Total Assets= <u>139,796</u> millions USD), (**d**= D/A= Dividend Portion or Payout= <u>30.00%</u>), (**U**= Utilized or Starting Capital= <u>64,720</u> millions USD), (**t**= T/B= Tax Rate= <u>30.00%</u>), and (**l**= L/E= Leverage or Gearing Ratio= <u>99.00%</u>), then it's (**S**= Sales or Revenues planned), is:

$$S = V+F+I+\{W/[1+l]-U\}/\{[1-d][1-t]\}$$
$$= 9,746+29,750+513\{139,796$$
$$/[1+0.9900]-64,720\}$$
$$/\{[1-0.3000][1-0.3000]\}$$
$$= \underline{51,294} \text{ millions USD}$$

Corporate IFRS-GAAP (B/S-I/S), ISBN 13: **978-1983566332**, ISBN 10: **1983566330**

<u>Law-5276</u>:

If both (**I**= Interest Expense= <u>513</u> millions USD), (**F**= Fixed Cost= <u>29,740</u> millions USD), (**S**= Sales or Revenues= <u>51,294</u> millions USD), (**W**= Wealth or Total Assets= <u>139,796</u> millions USD), (**d**= D/A= Dividend Portion or Payout= <u>30.00%</u>), (**U**= Utilized or Starting Capital= <u>64,720</u> millions USD), (**t**= T/B= Tax Rate= <u>30.00%</u>), and (**I**= L/E= Leverage or Gearing Ratio= <u>99.00%</u>), then it's (**V**= Variable Cost planned), is:

$$V= S-F-I-\{W/[1+I]-U\}/\{[1-d][1-t]\}$$
$$= 51,294-29,750-513- \{139,796$$
$$/[1+0.9900]-64,720\}$$
$$/\{[1-0.3000][1-0.3000]\}$$
$$= \underline{9,746} \text{ millions USD}$$

Corporate IFRS-GAAP (B/S-I/S), ISBN 13: **978-1983566332**, ISBN 10: **1983566330**

Law-5277:

If both (**I**= Interest Expense= 513 millions USD), (**V**= Variable Cost= 9,746 millions USD), (**S**= Sales or Revenues= 51,294 millions USD), (**W**= Wealth or Total Assets= 139,796 millions USD), (**d**= D/A= Dividend Portion or Payout= 30.00%), (**U**= Utilized or Starting Capital= 64,720 millions USD), (**t**= T/B= Tax Rate= 30.00%), and (**I**= L/E= Leverage or Gearing Ratio= 99.00%), then it's (**F**= Fixed Cost planned), is:

$$F= S\text{-}V\text{-}I\text{-}\{W/[1+I]\text{-}U\}/\{[1\text{-}d][1\text{-}t]\}$$
$$= 51,294\text{-}9,746\text{-}513\text{-} \{139,796$$
$$/[1+0.9900]\text{-}64,720\}$$
$$/\{[1\text{-}0.3000][1\text{-}0.3000]\}$$
$$= 29,750 \text{ millions USD}$$

Corporate IFRS-GAAP (B/S-I/S), ISBN 13: **978-1983566332**, ISBN 10: **1983566330**

<u>Law-5278</u>:

If both (**F**= Fixed Cost= <u>29,740</u> millions USD), (**V**= Variable Cost= <u>9,746</u> millions USD), (**S**= Sales or Revenues= <u>51,294</u> millions USD), (**W**= Wealth or Total Assets= <u>139,796</u> millions USD), (**d**= D/A= Dividend Portion or Payout= <u>30.00%</u>), (**U**= Utilized or Starting Capital= <u>64,720</u> millions USD), (**t**= T/B= Tax Rate= <u>30.00%</u>), and (**I**= L/E= Leverage or Gearing Ratio= <u>99.00%</u>), then it's (**I**= Interest Expense planned), is:

$$\mathbf{I= S\text{-}V\text{-}F\text{-}\{W/[1+I]\text{-}U\}/\{[1\text{-}d][1\text{-}t]\}}$$

$$= 51,294\text{-}9,746\text{-}29,750\text{-} \{139,796$$
$$/[1+0.9900]\text{-}64,720\}$$
$$/\{[1\text{-}0.3000][1\text{-}0.3000]\}$$

$$= \underline{513} \text{ millions USD}$$

Corporate IFRS-GAAP (B/S-I/S), ISBN 13: **978-1983566332**, ISBN 10: **1983566330**

Law-5279:

If both (**F**= Fixed Cost= 29,740 millions USD), (**V**= Variable Cost= 9,746 millions USD), (**S**= Sales or Revenues= 51,294 millions USD), (**W**= Wealth or Total Assets= 139,796 millions USD), (**d**= D/A= Dividend Portion or Payout= 30.00%), (**U**= Utilized or Starting Capital= 64,720 millions USD), (**I**= Interest Expense= 513 millions USD), and (**I**= L/E= Leverage or Gearing Ratio= 99.00%), then it's (**t**= Tax Rate planned), is:

$$\textbf{t}= 1-\{\textbf{W}/[1+\textbf{I}]-\textbf{U}\}/\{[1-\textbf{d}][\textbf{S-V-F-I}]\}$$
$$= 1-\{139,796/[1+0.9900]-64,720\}$$
$$/\{[1-0.3000]\{51,294-9,746$$
$$-29,750-513]\}$$
$$= 30.00\%$$

Corporate IFRS-GAAP (B/S-I/S), ISBN 13: **978-1983566332**, ISBN 10: **1983566330**

Law-5280:

If both (**F**= Fixed Cost= 29,740 millions USD), (**V**= Variable Cost= 9,746 millions USD), (**S**= Sales or Revenues= 51,294 millions USD), (**W**= Wealth or Total Assets= 139,796 millions USD), (**t**= T/B= Tax Rate= 30.00%), (**U**= Utilized or Starting Capital= 64,720 millions USD), (**I**= Interest Expense= 513 millions USD), and (**I**= L/E= Leverage or Gearing Ratio= 99.00%), then it's (**d**= Dividend Portion or Payout planned), is:

$$\mathbf{d}= 1-\{\mathbf{W}/[1+\textit{I}-\mathbf{U}\}/\{[1-\mathbf{t}][\mathbf{S}-\mathbf{V}-\mathbf{F}-\mathbf{I}]\}$$
$$= 1-\{139,796/[1+0.9900]-64,720\}$$
$$/\{[1-0.3000]\{51,294-9,746$$
$$-29,750-513]\}$$
$$= 30.00\%$$

Corporate IFRS-GAAP (B/S-I/S), ISBN 13: **978-1983566332**, ISBN 10: **1983566330**

Law-5281:

If both (**F**= Fixed Cost= 29,740 millions USD), (**V**= Variable Cost= 9,746 millions USD), (**S**= Sales or Revenues= 51,294 millions USD), (**W**= Wealth or Total Assets= 139,796 millions USD), (**t**= T/B= Tax Rate= 30.00%), (**d**= D/A= Dividend Portion or Payout= 30.00%), (**I**= Interest Expense= 513 millions USD), and (**I**= L/E= Leverage or Gearing Ratio= 99.00%), then it's (**U**= Utilized or Starting Capital planned), is:

$$U= W/[1+I]-[1-d][1-t][S-V-F-I]\}$$
$$= 139{,}796/[1+0.9900]-[1-0.3000]$$
$$\{[1-0.3000]\{51{,}294-9{,}746$$
$$-29{,}750-513]\}$$
$$= 64{,}720 \text{ millions USD}$$

Corporate IFRS-GAAP (B/S-I/S), ISBN 13: **978-1983566332**, ISBN 10: **1983566330**

Law-5282:

If both (**F**= Fixed Cost= <u>29,740</u> millions USD), (**V**= Variable Cost= <u>9,746</u> millions USD), (**S**= Sales or Revenues= <u>51,294</u> millions USD), (**W**= Wealth or Total Assets= <u>139,796</u> millions USD), (**t**= T/B= Tax Rate= <u>30.00%</u>), (**d**= D/A= Dividend Portion or Payout= <u>30.00%</u>), (**I**= Interest Expense= <u>513</u> millions USD), and (**U**= Utilized or Starting Capital= <u>64,720</u> millions USD), then it's (**l** = Leverage or Gearing Ratio planned), is:

$$l = W/\{U+[1-d][1-t][S-V-F-I]\}-1$$
$$= 139,796/[\{64,720+[1-0.3000]$$
$$[1-0.3000][51,294-9,746$$
$$-29,750-513]\}-1$$
$$= \underline{99.00\%}$$

Corporate IFRS-GAAP (B/S-I/S), ISBN 13: **978-1983566332**, ISBN 10: **1983566330**

<u>Law-5283</u>:

If both (**F**= Fixed Cost= <u>29,740</u> millions USD), (**V**= Variable Cost= <u>9,746</u> millions USD), (**S**= Sales or Revenues= <u>51,294</u> millions USD), (**W**= Wealth or Total Assets= <u>139,796</u> millions USD), (**t**= T/B= Tax Rate= <u>30.00%</u>), (**d**= D/A= Dividend Portion or Payout= <u>30.00%</u>), (**l**= L/E= Leverage or Gearing Ratio= <u>99.00%</u>), (**i**= I/S= Interest Portion= <u>1.00%</u>), and (**U**= Utilized or Starting Capital= <u>64,720</u> millions USD), then it's (**W**= Wealth or Total Assets planned), is:

$$\mathbf{W}= [1+\mathbf{\textit{l}}]\{\mathbf{U}+[1\text{-}\mathbf{d}][1\text{-}\mathbf{t}][\mathbf{S}\text{-}\mathbf{V}\text{-}\mathbf{F}\text{-}\mathbf{Si}]\}$$

$$= [1+\mathbf{\textit{l}}](\mathbf{U}+[1\text{-}\mathbf{d}][1\text{-}\mathbf{t}]\{\mathbf{S}[1\text{-}\mathbf{i}]\text{-}\mathbf{V}\text{-}\mathbf{F}\})$$

$$= [1+0.9900](64,720+[1\text{-}0.3000]$$
$$[1\text{-}0.3000]\{51,294[1\text{-}0.0100]$$
$$-9,746\text{-}29,750\})$$

$$= \underline{139,796} \text{ millions USD}$$

Corporate IFRS-GAAP (B/S-I/S), ISBN 13: **978-1983566332**, ISBN 10: **1983566330**

Law-5284:

If both (**F**= Fixed Cost= 29,740 millions USD), (**V**= Variable Cost= 9,746 millions USD), (**W**= Wealth or Total Assets= 139,796 millions USD), (**t**= T/B= Tax Rate= 30.00%), (**d**= D/A= Dividend Portion or Payout= 30.00%), (**l**= L/E= Leverage or Gearing Ratio= 99.00%), (**i**= I/S= Interest Portion= 1.00%), and (**U**= Utilized or Starting Capital= 64,720 millions USD), then it's (**S**= Sales or Revenues planned), is:

$$\mathbf{S} = (\mathbf{V}+\mathbf{F}+\{\mathbf{W}/[1+\mathbf{l}]-\mathbf{U}\})/\{[1-\mathbf{d}][1-\mathbf{t}]\}/[1-\mathbf{i}]$$
$$= (9,746+29,750+\{139,796$$
$$/[1+0.9900]-64,720$$
$$+[1-0.3000][1-0.3000]\})$$
$$/[1-0.0100]$$
$$= 51,294 \text{ millions USD}$$

Corporate IFRS-GAAP (B/S-I/S), ISBN 13: **978-1983566332**, ISBN 10: **1983566330**

<u>Law-5285</u>:

If both (**F**= Fixed Cost= <u>29,740</u> millions USD), (**V**= Variable Cost= <u>9,746</u> millions USD), (**W**= Wealth or Total Assets= <u>139,796</u> millions USD), (**t**= T/B= Tax Rate= <u>30.00%</u>), (**d**= D/A= Dividend Portion or Payout= <u>30.00%</u>), (**⫲**= L/E= Leverage or Gearing Ratio= <u>99.00%</u>), (**$**= Sales or Revenues= <u>51,294</u> millions USD), and (**U**= Utilized or Starting Capital= <u>64,720</u> millions USD), then it's (**i**= Interest Portion planned), is:

$$\mathbf{i}= 1-(\mathbf{V}+\mathbf{F}+\{\mathbf{W}/[1+\mathbf{⫲}]-\mathbf{U}\}/\{[1-\mathbf{d}][1-\mathbf{t}]\})/\mathbf{\$}$$
$$= 1-(9,746+29,750+\{139,796$$
$$/[1+0.9900]-64,720$$
$$+[1-0.3000][1-0.3000]\})$$
$$/51,294$$

$$= \underline{1.00\%}$$

Corporate IFRS-GAAP (B/S-I/S), ISBN 13: **978-1983566332**, ISBN 10: **1983566330**

Law-5286:

If both (**F**= Fixed Cost= 29,740 millions USD), (**i**= I/S= Interest Portion= 1.00%), (**W**= Wealth or Total Assets= 139,796 millions USD), (**t**= T/B= Tax Rate= 30.00%), (**d**= D/A= Dividend Portion or Payout= 30.00%), (**l**= L/E= Leverage or Gearing Ratio= 99.00%), (**S**= Sales or Revenues= 51,294 millions USD), and (**U**= Utilized or Starting Capital= 64,720 millions USD), then it's (**V**= Variable Cost planned), is:

$$V= S[1-i]-F-\{W/[1+l]-U\}/\{[1-d][1-t]\}$$
$$= (51,294[1-0.0100]-29,750$$
$$-\{139,796/[1+0.9900]$$
$$-64,720\}/\{[1-0.3000]$$
$$[1-0.3000]\}$$
$$= \underline{9,746} \text{ millions USD}$$

Corporate IFRS-GAAP (B/S-I/S), ISBN 13: **978-1983566332**, ISBN 10: **1983566330**

Law-5287:

If both (**V**= Variable Cost= 9,746 millions USD), (**i**= I/S= Interest Portion= 1.00%), (**W**= Wealth or Total Assets= 139,796 millions USD), (**t**= T/B= Tax Rate= 30.00%), (**d**= D/A= Dividend Portion or Payout= 30.00%), (**I**= L/E= Leverage or Gearing Ratio= 99.00%), (**$**= Sales or Revenues= 51,294 millions USD), and (**U**= Utilized or Starting Capital= 64,720 millions USD), then it's (**F**= Fix Cost planned), is:

$$F= \$[1-i]-V-\{W/[1+I]-U\}/\{[1-d][1-t]\}$$
$$= (51,294[1-0.0100]-9,746$$
$$-\{139,796/[1+0.9900]$$
$$-64,720\}/\{[1-0.3000]$$
$$[1-0.3000]\}$$
$$= 29,750 \text{ millions USD}$$

Corporate IFRS-GAAP (B/S-I/S), ISBN 13: **978-1983566332**, ISBN 10: **1983566330**

Law-5288:

If both (V= Variable Cost= 9,746 millions USD), (i= I/S= Interest Portion= 1.00%), (W= Wealth or Total Assets= 139,796 millions USD), (F= Fixed Cost= 29,740 millions USD), (d= D/A= Dividend Portion or Payout= 30.00%), (l= L/E= Leverage or Gearing Ratio= 99.00%), (S= Sales or Revenues= 51,294 millions USD), and (U= Utilized or Starting Capital= 64,720 millions USD), then it's (t= Tax Rate planned), is:

$$t = 1-\{W/[1+l]-U\}/([1-d]\{S[1-i]-V-F\})$$
$$= 1-\{139,796/[1+0.9900]-64,720\}$$
$$/([1-0.3000]\{51,294$$
$$[1-0.0100]-9,746-29,750\})$$
$$= 30.00\%$$

Corporate IFRS-GAAP (B/S-I/S), ISBN 13: **978-1983566332**, ISBN 10: **1983566330**

Law-5289:
  If both (**V**= Variable Cost= 9,746 millions USD), (**i**=
  I/S= Interest Portion= 1.00%), (**W**= Wealth or Total
  Assets= 139,796 millions USD), (**F**= Fixed Cost=
  29,740 millions USD), (**I**= L/E= Leverage or Gearing
  Ratio= 99.00%), (**t**= T/B= Tax Rate= 30.00%), (**S**=
  Sales or Revenues= 51,294 millions USD), and (**U**=
  Utilized or Starting Capital= 64,720 millions USD),
  then it's (**d**= Dividend Portion or Payout planned), is:

$$d= 1-\{W/[1+I]-U\}/([1-t]\{S[1-i]-V-F\})$$
$$= 1-\{139,796/[1+0.9900]-64,720\}$$
$$/([1-0.3000]\{51,294$$
$$[1-0.0100]-9,746-29,750\})$$
$$= 30.00\%$$

Law-5290:

If both (**V**= Variable Cost= 9,746 millions USD), (**i**= I/S= Interest Portion= 1.00%), (**W**= Wealth or Total Assets= 139,796 millions USD), (**F**= Fixed Cost= 29,740 millions USD), (**I**= L/E= Leverage or Gearing Ratio= 99.00%), (**t**= T/B= Tax Rate= 30.00%), (**$**= Sales or Revenues= 51,294 millions USD), and (**d**= D/A= Dividend Portion or Payout= 30.00%), then it's (**U**= Utilized or Starting Capital planned), is:

$$\mathbf{U} = \mathbf{W}/[1+\mathbf{I}]-[1-\mathbf{d}][1-\mathbf{t}]\{\mathbf{\$}[1-\mathbf{i}]-\mathbf{V}-\mathbf{F}\}$$

$$= 139,796/[1+0.9900]-[1-0.3000]$$
$$[1-0.3000]\{51,294[1-0.0100]$$
$$-9,746-29,750\}$$

$$= 64,720 \text{ millions USD}$$

Corporate IFRS-GAAP (B/S-I/S), ISBN 13: **978-1983566332**, ISBN 10: **1983566330**

Law-5291:

If both (**V**= Variable Cost= 9,746 millions USD), (**i**= I/S= Interest Portion= 1.00%), (**W**= Wealth or Total Assets= 139,796 millions USD), (**F**= Fixed Cost= 29,740 millions USD), (**U**= Utilized or Starting Capital= 64,720 millions USD), (**t**= T/B= Tax Rate= 30.00%), (**$**= Sales or Revenues= 51,294 millions USD), and (**d**= D/A= Dividend Portion or Payout= 30.00%), then it's (**I** = Leverage or Gearing Ratio planned), is:

$$I = W/(U+[1-d][1-t]\{\$[1-i]-V-F\})-1$$
$$= 139,796/(64,720+[1-0.3000]$$
$$[1-0.3000]\{51,294$$
$$[1-0.0100]-9,746$$
$$-29,750\})-1$$
$$= 99.00\%$$

Corporate IFRS-GAAP (B/S-I/S), ISBN 13: **978-1983566332**, ISBN 10: **1983566330**

Law-5292:

If both (**$'**= Sales of Past Year= 48,851 millions USD), (**s**= [S/S']-1= Sales Growth= 5.00%), (**V**= Variable Cost= 9,746 millions USD), (**i**= I/S= Interest Portion= 1.00%), (**l**= L/E= Leverage or Gearing Ratio= 99.00%), (**F**= Fixed Cost= 29,740 millions USD), (**U**= Utilized or Starting Capital= 64,720 millions USD), (**t**= T/B= Tax Rate= 30.00%), (**$**= Sales or Revenues= 51,294 millions USD), and (**d**= D/A= Dividend Portion or Payout= 30.00%), then it's (**W**= Wealth or Total Assets planned), is:

$$W= [1+l](U+\{1-d\}[1-t]\{\$-V-F-\$'i[1+s]\})$$
$$= [1+0.9900](64,720+[1-0.3000]$$
$$[1-0.3000]\{51,294-9,746$$
$$-29,750-48,851*0.0100$$
$$[1+0.0500]\})$$
$$= \underline{139,796} \text{ millions USD}$$

Corporate IFRS-GAAP (B/S-I/S), ISBN 13: **978-1983566332**, ISBN 10: **1983566330**

Law-5293:

If both (S'= Sales of Past Year= 48,851 millions USD), (s= [S/S']-1= Sales Growth= 5.00%), (V= Variable Cost= 9,746 millions USD), (i= I/S= Interest Portion= 1.00%), (l= L/E= Leverage or Gearing Ratio= 99.00%), (F= Fixed Cost= 29,740 millions USD), (U= Utilized or Starting Capital= 64,720 millions USD), (t= T/B= Tax Rate= 30.00%), (W= Wealth or Total Assets= 139,796 millions USD), and (d= D/A= Dividend Portion or Payout= 30.00%), then it's (S= Sales or Revenues planned), is:

$$S= V+F+S'i[1+s]+\{W/[1+l]-U]\}/\{1-d][1-t]\}$$
$$= 9{,}746+29{,}750+48{,}851*0.0100$$
$$[1+0.0500]+\{139{,}796$$
$$/[1+0.9900]-64{,}720]\}$$
$$/\{[1-0.3000][1-0.3000]\}$$
$$= 51{,}294 \text{ millions USD}$$

Corporate IFRS-GAAP (B/S-I/S), ISBN 13: **978-1983566332**, ISBN 10: **1983566330**

Law-5294:

If both (**S'**= Sales of Past Year= 48,851 millions USD), (**s**= [S/S']-1= Sales Growth= 5.00%), (**S**= Sales or Revenues= 51,294 millions USD), (**i**= I/S= Interest Portion= 1.00%), (**l**= L/E= Leverage or Gearing Ratio= 99.00%), (**F**= Fixed Cost= 29,740 millions USD), (**U**= Utilized or Starting Capital= 64,720 millions USD), (**t**= T/B= Tax Rate= 30.00%), (**W**= Wealth or Total Assets= 139,796 millions USD), and (**d**= D/A= Dividend Portion or Payout= 30.00%), then it's (**V**= Variable Cost planned), is:

$$V = S - F - S'i[1+s] - \{W/[1+l] - U\} / \{1-d\}[1-t]\}$$

$$= 51,294 - 29,750 - 48,851 * 0.0100$$
$$[1+0.0500] - \{139,796$$
$$/[1+0.9900] - 64,720]\}$$
$$/\{[1-0.3000][1-0.3000]\}$$

$$= 9,746 \text{ millions USD}$$

Corporate IFRS-GAAP (B/S-I/S), ISBN 13: **978-1983566332**, ISBN 10: **1983566330**

Law-5295:

If both (**$'**= Sales of Past Year= <u>48,851</u> millions USD), (**$**= [S/S']-1= Sales Growth= <u>5.00%</u>), (**$**= Sales or Revenues= <u>51,294</u> millions USD), (**i**= I/S= Interest Portion= <u>1.00%</u>), (**l**= L/E= Leverage or Gearing Ratio= <u>99.00%</u>), (**V**= Variable Cost= <u>9,746</u> millions USD), (**U**= Utilized or Starting Capital= <u>64,720</u> millions USD), (**t**= T/B= Tax Rate= <u>30.00%</u>), (**W**= Wealth or Total Assets= <u>139,796</u> millions USD), and (**d**= D/A= Dividend Portion or Payout= <u>30.00%</u>), then it's (**F**= Fixed Cost planned), is:

$$\mathbf{F= \$-V-\$'i[1+s]-\{W/[1+\mathit{l}]-U]\}/\{1-d][1-t]\}}$$

$$= 51,294-9,746-48,851*0.0100$$
$$[1+0.0500] -\{139,796$$
$$/[1+0.9900]-64,720]\}$$
$$/\{[1-0.3000][1-0.3000]\}$$

$$= \underline{29,750} \text{ millions USD}$$

Corporate IFRS-GAAP (B/S-I/S), ISBN 13: **978-1983566332**, ISBN 10: **1983566330**

Law-5296:

If both (**F**= Fixed Cost= 29,740 millions USD), (**s**= [S/S']-1= Sales Growth= 5.00%), (**S**= Sales or Revenues= 51,294 millions USD), (**i**= I/S= Interest Portion= 1.00%), (**l**= L/E= Leverage or Gearing Ratio= 99.00%), (**V**= Variable Cost= 9,746 millions USD), (**U**= Utilized or Starting Capital= 64,720 millions USD), (**t**= T/B= Tax Rate= 30.00%), (**W**= Wealth or Total Assets= 139,796 millions USD), and (**d**= D/A= Dividend Portion or Payout= 30.00%), then it's (**S'**= Sales Past), must be:

$$\mathbf{S'= (S\text{-}V\text{-}F\text{-}\{W/[1+\mathbf{\mathit{l}}]\text{-}U\}/\{1\text{-}d][1\text{-}t]\})/\{i[1+s]\}}$$
$$= 51,294\text{-}9,746\text{-}29,750\text{-}\{139,796$$
$$/[1+0.9900]\text{-}64,720]\}$$
$$/\{[1\text{-}0.3000][1\text{-}0.3000]\})$$
$$/\{0.0100[1+0.0500]\}$$
$$= \underline{48,851} \text{ millions USD}$$

Corporate IFRS-GAAP (B/S-I/S), ISBN 13: **978-1983566332**, ISBN 10: **1983566330**

Law-5297:

If both (**F**= Fixed Cost= 29,740 millions USD),
(**s**= [S/S']-1= Sales Growth= 5.00%), (**S**= Sales or
Revenues= 51,294 millions USD), (**S'**= Sales of Past
Year= 48,851 millions USD), (**l**= L/E= Leverage or
Gearing Ratio= 99.00%), (**V**= Variable Cost= 9,746
millions USD), (**U**= Utilized or Starting Capital= 64,720
millions USD), (**t**= T/B= Tax Rate= 30.00%), (**W**=
Wealth or Total Assets= 139,796 millions USD), and
(**d**= D/A= Dividend Portion or Payout= 30.00%), then
it's (**i**= Interest Portion planned), is:

$$\mathbf{i}= (\mathbf{S}\text{-}\mathbf{V}\text{-}\mathbf{F}\text{-}\{\mathbf{W}/[1+\mathbf{l}]\text{-}\mathbf{U}]\}/\{1\text{-}\mathbf{d}][1\text{-}\mathbf{t}]\})/\{\mathbf{S'}[1+\mathbf{s}]\}$$

$$= 51,294\text{-}9,746\text{-}29,750\text{-}\{139,796$$
$$/[1+0.9900]\text{-}64,720]\}$$
$$/\{[1\text{-}0.3000][1\text{-}0.3000]\})$$
$$/\{48,851[1+0.0500]\}$$

$$= 1.00\%$$

Corporate IFRS-GAAP (B/S-I/S), ISBN 13: **978-1983566332**, ISBN 10: **1983566330**

Law-5298:

If both (**F**= Fixed Cost= 29,740 millions USD), (**i**= I/S= Interest Portion= 1.00%), (**S**= Sales or Revenues= 51,294 millions USD), (**S'**= Sales of Past Year= 48,851 millions USD), (**l**= L/E= Leverage or Gearing Ratio= 99.00%), (**V**= Variable Cost= 9,746 millions USD), (**U**= Utilized or Starting Capital= 64,720 millions USD), (**t**= T/B= Tax Rate= 30.00%), (**W**= Wealth or Total Assets= 139,796 millions USD), and (**d**= D/A= Dividend Portion or Payout= 30.00%), then it's (**s**= Sales Growth planned), is:

$$s = (S\text{-}V\text{-}F\text{-}\{W/[1+l]\text{-}U]\}/\{1\text{-}d][1\text{-}t]\})/[S'i]\text{-}1$$
$$= 51{,}294\text{-}9{,}746\text{-}29{,}750\text{-}\{139{,}796$$
$$/[1+0.9900]\text{-}64{,}720]\}$$
$$/\{[1\text{-}0.3000][1\text{-}0.3000]\})$$
$$/[48{,}851*0.0100]\text{-}1$$
$$= \underline{5.00\%}$$

Corporate IFRS-GAAP (B/S-I/S), ISBN 13: **978-1983566332**, ISBN 10: **1983566330**

Law-5299:

If both (**F**= Fixed Cost= 29,740 millions USD), (**i**= I/S= Interest Portion= 1.00%), (**S**= Sales or Revenues= 51,294 millions USD), (**S'**= Sales of Past Year= 48,851 millions USD), (**l**= L/E= Leverage or Gearing Ratio= 99.00%), (**V**= Variable Cost= 9,746 millions USD), (**U**= Utilized or Starting Capital= 64,720 millions USD), (**s**= [S/S']-1= Sales Growth= 5.00%), (**W**= Wealth or Total Assets= 139,796 millions USD), and (**d**= D/A= Dividend Portion or Payout= 30.00%), then it's (**t**= Tax Rate planned), is:

$$t= 1-\{W/[1+l]-U]\}/([1-d]\{S-V-F-S'i[1+s]\})$$
$$= 1-\{139,796/[1+0.9900]-64,720]\}$$
$$/([1-0.3000]\{51,294-9,746$$
$$-29,750-48,851*0.0100$$
$$[1+0.0500]\})$$
$$= 30.00\%$$

Corporate IFRS-GAAP (B/S-I/S), ISBN 13: **978-1983566332**, ISBN 10: **1983566330**

Law-5300:

If both (**F**= Fixed Cost= <u>29,740</u> millions USD), (**i**= I/S= Interest Portion= <u>1.00%</u>), (**S**= Sales or Revenues= <u>51,294</u> millions USD), (**S'**= Sales of Past Year= <u>48,851</u> millions USD), (**l**= L/E= Leverage or Gearing Ratio= <u>99.00%</u>), (**V**= Variable Cost= <u>9,746</u> millions USD), (**U**= Utilized or Starting Capital= <u>64,720</u> millions USD), (**s**= [S/S']-1= Sales Growth= <u>5.00%</u>), (**W**= Wealth or Total Assets= <u>139,796</u> millions USD), and (**t**= T/B= Tax Rate= <u>30.00%</u>), then it's (**d**= Dividend Portion or Payout planned), is:

$$\mathbf{d}= 1-\{\mathbf{W}/[1+\mathbf{l}]-\mathbf{U}]\}/([1-\mathbf{t}]\{\mathbf{S}-\mathbf{V}-\mathbf{F}-\mathbf{S'i}[1+\mathbf{s}]\})$$
$$= 1-\{139,796/[1+0.9900]-64,720]\}$$
$$/([1-0.3000]\{51,294-9,746$$
$$-29,750-48,851*0.0100$$
$$[1+0.0500]\})$$
$$= \underline{30.00\%}$$

<u>Law-5301</u>:

If both (**F**= Fixed Cost= <u>29,740</u> millions USD), (**i**= I/S= Interest Portion= <u>1.00%</u>), (**S**= Sales or Revenues= <u>51,294</u> millions USD), (**S'**= Sales of Past Year= <u>48,851</u> millions USD), (**l**= L/E= Leverage or Gearing Ratio= <u>99.00%</u>), (**V**= Variable Cost= <u>9,746</u> millions USD), (**d**= D/A= Dividend Portion or Payout= <u>30.00%</u>), (**s**= [S/S']-1= Sales Growth= <u>5.00%</u>), (**W**= Wealth or Total Assets= <u>139,796</u> millions USD), and (**t**= T/B= Tax Rate= <u>30.00%</u>), then it's (**U**= Utilized or Starting Capital planned), is:

$$U= W/[1+l]-[1-d][1-t]\{S-V-F-S'i[1+s]\}$$
$$= 139,796/[1+0.9900]-[1-0.3000]$$
$$[1-0.3000]\{51,294-9,746$$
$$-29,750-48,851*0.0100$$
$$[1+0.0500]\}$$
$$= \underline{64,720} \text{ millions USD}$$

Corporate IFRS-GAAP (B/S-I/S), ISBN 13: **978-1983566332**, ISBN 10: **1983566330**

Law-5302:

If both (**F**= Fixed Cost= 29,740 millions USD), (**i**= I/S= Interest Portion= 1.00%), (**$**= Sales or Revenues= 51,294 millions USD), (**$'**= Sales of Past Year= 48,851 millions USD), (**U**= Utilized or Starting Capital= 64,720 millions USD), (**V**= Variable Cost= 9,746 millions USD), (**d**= D/A= Dividend Portion or Payout= 30.00%), (**s**= [S/S']-1= Sales Growth= 5.00%), (**W**= Wealth or Total Assets= 139,796 millions USD), and (**t**= T/B= Tax Rate= 30.00%), then it's (**l** = Leverage or Gearing Ratio planned), is:

$$l = W/(U+[1-d][1-t]\{\$-V-F-\$'i[1+s]\})-1$$
$$= 139,796/(64,720+[1-0.3000]$$
$$[1-0.3000]\{51,294-9,746$$
$$-29,750-48,851*0.0100$$
$$[1+0.0500]\})-1$$
$$= 99.00\%$$

Corporate IFRS-GAAP (B/S-I/S), ISBN 13: **978-1983566332**, ISBN 10: **1983566330**

<u>Law-5303</u>:

If both (**f**= F/S= Fixed Portion= <u>58.00%</u>), (**I**= Interest Expense= <u>513</u> millions USD), (**$**= Sales or Revenues= <u>51,294</u> millions USD), (**U**= Utilized or Starting Capital= <u>64,720</u> millions USD), (**V**= Variable Cost= <u>9,746</u> millions USD), (**d**= D/A= Dividend Portion or Payout= <u>30.00%</u>), (**I**= L/E= Leverage or Gearing Ratio= <u>99.00%</u>), and (**t**= T/B= Tax Rate= <u>30.00%</u>), then it's (**W**= Wealth or Total Assets planned), is:

$$
\begin{aligned}
\mathbf{W} &= [1+\mathbf{I}]\{\mathbf{U}+[1-\mathbf{d}][1-\mathbf{t}][\mathbf{\$}-\mathbf{V}-\mathbf{\$f}-\mathbf{I}]\} \\
&= [1+\mathbf{I}](\mathbf{U}+[1-\mathbf{d}][1-\mathbf{t}]\{\mathbf{\$}[1-\mathbf{f}]-\mathbf{V}-\mathbf{I}]\}) \\
&= [1+0.9900](64{,}720+[1-0.3000] \\
&\qquad [1-0.3000]\{51{,}294 \\
&\qquad [1-0.5800]-9{,}746-513\}) \\
&= \underline{139{,}796} \text{ millions USD}
\end{aligned}
$$

Corporate IFRS-GAAP (B/S-I/S), ISBN 13: **978-1983566332**, ISBN 10: **1983566330**

<u>Law-5304</u>:

If both (f= F/S= Fixed Portion= <u>58.00%</u>), (I= Interest Expense= <u>513</u> millions USD), (W= Wealth or Total Assets= <u>139,796</u> millions USD), (U= Utilized or Starting Capital= <u>64,720</u> millions USD), (V= Variable Cost= <u>9,746</u> millions USD), (d= D/A= Dividend Portion or Payout= <u>30.00%</u>), (I= L/E= Leverage or Gearing Ratio= <u>99.00%</u>), and (t= T/B= Tax Rate= <u>30.00%</u>), then it's (S= Sales or Revenues planned), is:

$$S= (V+Sf+I+\{W/[1+I]-U\}/\{[1-d][1-t]\})/[1-f]$$
$$= (9,746+513+\{139,796/[1+0.9900]$$
$$-64,720\}/\{[1-0.3000]$$
$$[1-0.3000]\})/[1-0.5800]$$
$$= \underline{51,294} \text{ millions USD}$$

Corporate IFRS-GAAP (B/S-I/S), ISBN 13: **978-1983566332**, ISBN 10: **1983566330**

Law-5305:

If both (**S**= Sales or Revenues= 51,294 millions
USD), (**I**= Interest Expense= 513 millions USD),
(**W**= Wealth or Total Assets= 139,796 millions
USD), (**U**= Utilized or Starting Capital= 64,720
millions USD), (**V**= Variable Cost= 9,746 millions
USD), (**d**= D/A= Dividend Portion or Payout=
30.00%), (**I**= L/E= Leverage or Gearing Ratio=
99.00%), and (**t**= T/B= Tax Rate= 30.00%), then it's
(**f**= Fixed Portion planned), is:

$$f = 1-(V+Sf+I+\{W/[1+I]-U\}/\{[1-d][1-t]\})/S$$
$$= 1-(9,746+513+\{139,796$$
$$/[1+0.9900]-64,720\}$$
$$/\{[1-0.3000][1-0.3000]\})$$
$$/51,294$$
$$= 58.00\%$$

Corporate IFRS-GAAP (B/S-I/S), ISBN 13: **978-1983566332**, ISBN 10: **1983566330**

<u>Law-5306</u>:

If both (**$**= Sales or Revenues= <u>51,294</u> millions USD), (**I**= Interest Expense= <u>513</u> millions USD), (**W**= Wealth or Total Assets= <u>139,796</u> millions USD), (**U**= Utilized or Starting Capital= <u>64,720</u> millions USD), (**f**= F/S= Fixed Portion= <u>58.00%</u>), (**d**= D/A= Dividend Portion or Payout= <u>30.00%</u>), (**I**= L/E= Leverage or Gearing Ratio= <u>99.00%</u>), and (**t**= T/B= Tax Rate= <u>30.00%</u>), then it's (**V**= Variable Cost planned), is:

$$V = \$[1-f]-I-\{W/[1+I]-U\}/\{[1-d][1-t]\}$$
$$= 51{,}294[1-0.5800]-513-\{139{,}796$$
$$/[1+0.9900]-64{,}720\}$$
$$/\{[1-0.3000][1-0.3000]\}$$
$$= \underline{9{,}746} \text{ millions USD}$$

Corporate IFRS-GAAP (B/S-I/S), ISBN 13: **978-1983566332**, ISBN 10: **1983566330**

Law-5307:

If both (**$**= Sales or Revenues= 51,294 millions
USD), (**V**= Variable Cost= 9,746 millions USD),
(**W**= Wealth or Total Assets= 139,796 millions
USD), (**U**= Utilized or Starting Capital= 64,720
millions USD), (**f**= F/S= Fixed Portion= 58.00%),
(**d**= D/A= Dividend Portion or Payout= 30.00%), (**l**=
L/E= Leverage or Gearing Ratio= 99.00%), and (**t**=
T/B= Tax Rate= 30.00%), then it's (**I**= Interest
Portion planned), is:

$$\mathbf{I}= \mathbf{S}[1-\mathbf{f}]-\mathbf{V}-\{\mathbf{W}/[1+\mathbf{l}]-\mathbf{U}\}/\{[1-\mathbf{d}][1-\mathbf{t}]\}$$
$$= 51,294[1-0.5800]-9,746-\{139,796$$
$$/[1+0.9900]-64,720\}$$
$$/\{[1-0.3000][1-0.3000]\}$$
$$= \underline{513} \text{ millions USD}$$

Corporate IFRS-GAAP (B/S-I/S), ISBN 13: **978-1983566332**, ISBN 10: **1983566330**

Law-5308:

If both (**$**= Sales or Revenues= 51,294 millions USD), (**V**= Variable Cost= 9,746 millions USD), (**W**= Wealth or Total Assets= 139,796 millions USD), (**U**= Utilized or Starting Capital= 64,720 millions USD), (**f**= F/S= Fixed Portion= 58.00%), (**d**= D/A= Dividend Portion or Payout= 30.00%), (**l**= L/E= Leverage or Gearing Ratio= 99.00%), and (**I**= Interest Expense= 513 millions USD), then it's (**t**= Tax Rate planned), is:

$$t= 1-\{W/[1+l]-U\}/([1-d]\{S[1-f]-V-I\})$$

$$= 1-\{139,796/[1+0.9900]-64,720\}$$
$$/([1-0.3000]\{51,294[1-0.5800]$$
$$-9,746-513\})$$

$$= 30.00\%$$

Corporate IFRS-GAAP (B/S-I/S), ISBN 13: **978-1983566332**, ISBN 10: **1983566330**

Law-5309:

If both (**$**= Sales or Revenues= 51,294 millions USD), (**V**= Variable Cost= 9,746 millions USD), (**W**= Wealth or Total Assets= 139,796 millions USD), (**U**= Utilized or Starting Capital= 64,720 millions USD), (**f**= F/S= Fixed Portion= 58.00%), (**t**= T/B= Tax Rate= 30.00%), (**l**= L/E= Leverage or Gearing Ratio= 99.00%), and (**I**= Interest Expense= 513 millions USD), then it's (**d**= Dividend Portion or Payout planned), is:

$$d= 1-\{\mathbf{W}/[1+\mathbf{l}]-\mathbf{U}\}/([1-\mathbf{t}]\{\mathbf{\$}[1-\mathbf{f}]-\mathbf{V}-\mathbf{I}\})$$
$$= 1-\{139,796/[1+0.9900]-64,720\}$$
$$/([1-0.3000]\{51,294$$
$$[1-0.5800]-9,746-513\})$$
$$= 30.00\%$$

Corporate IFRS-GAAP (B/S-I/S), ISBN 13: **978-1983566332**, ISBN 10: **1983566330**

Law-5310:

If both (**$**= Sales or Revenues= 51,294 millions
USD), (**V**= Variable Cost= 9,746 millions USD),
(**W**= Wealth or Total Assets= 139,796 millions
USD), (**d**= D/A= Dividend Portion or Payout=
30.00%), (**f**= F/S= Fixed Portion= 58.00%), (**t**= T/B=
Tax Rate= 30.00%), (**l**= L/E= Leverage or Gearing
Ratio= 99.00%), and (**I**= Interest Expense= 513
millions USD), then it's (**U**= Utilized or Starting
Capital planned), is:

$$U = W/[1+l]-[1-d][1-t]\{\$[1-f]-V-I\}$$
$$= 139{,}796/[1+0.9900]-[1-0.3000]$$
$$[1-0.3000]\{51{,}294$$
$$[1-0.5800]-9{,}746-513\}$$
$$= 64{,}720 \text{ millions USD}$$

Corporate IFRS-GAAP (B/S-I/S), ISBN 13: **978-1983566332**, ISBN 10: **1983566330**

<u>Law-5311</u>:

If both (**$**= Sales or Revenues= <u>51,294</u> millions USD), (**V**= Variable Cost= <u>9,746</u> millions USD), (**W**= Wealth or Total Assets= <u>139,796</u> millions USD), (**d**= D/A= Dividend Portion or Payout= <u>30.00%</u>), (**f**= F/S= Fixed Portion= <u>58.00%</u>), (**t**= T/B= Tax Rate= <u>30.00%</u>), (**U**= Utilized or Starting Capital= <u>64,720</u> millions USD), and (**I**= Interest Expense= <u>513</u> millions USD), then it's (**I** = Leverage or Gearing Ratio planned), is:

$$I = W/(U+[1-d][1-t]\{\$[1-f]-V-I\})-1$$
$$= 139{,}796/(64{,}720+[1-0.3000]$$
$$[1-0.3000]\{51{,}294[1-0.5800]$$
$$-9{,}746-513\})-1$$
$$= \underline{99.00\%}$$

Corporate IFRS-GAAP (B/S-I/S), ISBN 13: **978-1983566332**, ISBN 10: **1983566330**

<u>Law-5312</u>:

If both (**S**= Sales or Revenues= <u>51,294</u> millions USD), (**V**= Variable Cost= <u>9,746</u> millions USD), (**f**= L/E= Leverage or Gearing Ratio= <u>99.00%</u>), (**d**= D/A= Dividend Portion or Payout= <u>30.00%</u>), (**f**= F/S= Fixed Portion= <u>58.00%</u>), (**t**= T/B= Tax Rate= <u>30.00%</u>), (**U**= Utilized or Starting Capital= <u>64,720</u> millions USD), and (**i**= I/S= Interest Portion= <u>1.00%</u>), then it's (**W**= Wealth or Total Assets planned), is:

$$W = [1+I](U+[1-d][1-t]\{S-V-Sf-Si\})$$
$$= [1+I](U+[1-d][1-t]\{S[1-f-i]-V\})$$
$$= [1+0.9900](64,720+[1-0.3000]$$
$$[1-0.3000]\{51,294[1-0.5800$$
$$-0.0100]-9,746\})$$
$$= \underline{139,796} \text{ millions USD}$$

Corporate IFRS-GAAP (B/S-I/S), ISBN 13: 978-1983566332, ISBN 10: 1983566330

Law-5313:

If both (**W**= Wealth or Total Assets= 139,796 millions USD), (**V**= Variable Cost= 9,746 millions USD), (**I**= L/E= Leverage or Gearing Ratio= 99.00%), (**d**= D/A= Dividend Portion or Payout= 30.00%), (**f**= F/S= Fixed Portion= 58.00%), (**t**= T/B= Tax Rate= 30.00%), (**U**= Utilized or Starting Capital= 64,720 millions USD), and (**i**= I/S= Interest Portion= 1.00%), then it's (**$**= Sales or Revenues planned), is:

$$\$= (V + \{W/[1+I]-U\}/\{[1-d][1-t]\})/[1-f-i]$$

$$= (9,746 + \{139,796/[1+0.9900]$$
$$-64,720\}/\{[1-0.3000]$$
$$[1-0.3000]\})/[1-0.5800$$
$$-0.0100]$$

$$= 51,294 \text{ millions USD}$$

Corporate IFRS-GAAP (B/S-I/S), ISBN 13: **978-1983566332**, ISBN 10: **1983566330**

Law-5314:

If both (**W**= Wealth or Total Assets= 139,796 millions USD), (**V**= Variable Cost= 9,746 millions USD), (**I**= L/E= Leverage or Gearing Ratio= 99.00%), (**d**= D/A= Dividend Portion or Payout= 30.00%), (**S**= Sales or Revenues= 51,294 millions USD), (**t**= T/B= Tax Rate= 30.00%), (**U**= Utilized or Starting Capital= 64,720 millions USD), and (**i**= I/S= Interest Portion= 1.00%),   then it's (**f**= Fixed Portion planned), is:

$$\mathbf{f}= 1-\mathbf{i}-(\mathbf{V}+\{\mathbf{W}/[1+\mathbf{I}]-\mathbf{U}\}/\{[1-\mathbf{d}][1-\mathbf{t}]\})/\mathbf{S}$$

$$= 1-0.0100-(9{,}746+\{139{,}796$$
$$/[1+0.9900]-64{,}720\}$$
$$/\{[1-0.3000][1-0.3000]\})$$
$$/51{,}294$$

$$= 58.00\%$$

Corporate IFRS-GAAP (B/S-I/S), ISBN 13: **978-1983566332**, ISBN 10: **1983566330**

Law-5315:

If both (**W**= Wealth or Total Assets= 139,796 millions USD), (**V**= Variable Cost= 9,746 millions USD), (**f**= L/E= Leverage or Gearing Ratio= 99.00%), (**d**= D/A= Dividend Portion or Payout= 30.00%), (**$**= Sales or Revenues= 51,294 millions USD), (**t**= T/B= Tax Rate= 30.00%), (**U**= Utilized or Starting Capital= 64,720 millions USD), and (**f**= F/S= Fixed Portion= 58.00%), then it's (**i**= Interest Portion planned), is:

$$i = 1\text{-}\mathbf{f}\text{-}(\mathbf{V}+\{\mathbf{W}/[1+\mathbf{f}]\text{-}\mathbf{U}\}/\{[1\text{-}\mathbf{d}][1\text{-}\mathbf{t}]\})/\mathbf{\$}$$

$$= 1\text{-}0.5800\text{-}(9,746+\{139,796$$
$$/[1+0.9900]\text{-}64,720\}$$
$$/\{[1\text{-}0.3000][1\text{-}0.3000]\})$$
$$/51,294$$

$$= 1.00\%$$

Corporate IFRS-GAAP (B/S-I/S), ISBN 13: **978-1983566332**, ISBN 10: **1983566330**

<u>Law-5316</u>:

If both (**W**= Wealth or Total Assets= <u>139,796</u> millions USD), (**i**= I/S= Interest Portion= <u>1.00%</u>), (**l**= L/E= Leverage or Gearing Ratio= <u>99.00%</u>), (**d**= D/A= Dividend Portion or Payout= <u>30.00%</u>), (**S**= Sales or Revenues= <u>51,294</u> millions USD), (**t**= T/B= Tax Rate= <u>30.00%</u>), (**U**= Utilized or Starting Capital= <u>64,720</u> millions USD), and (**f**= F/S= Fixed Portion= <u>58.00%</u>), then it's (**V**= Variable Cost planned), is:

$$V = S[1\text{-}f\text{-}i]\text{-}\{W/[1\text{+}l]\text{-}U\}/\{[1\text{-}d][1\text{-}t]\}$$
$$= 51,294[1\text{-}0.5800\text{-}0.0100]$$
$$-\{139,796/[1\text{+}0.9900]$$
$$-64,720\}/\{[1\text{-}0.3000]$$
$$[1\text{-}0.3000]\}$$
$$= \underline{9,746} \text{ millions USD}$$

Corporate IFRS-GAAP (B/S-I/S), ISBN 13: **978-1983566332**, ISBN 10: **1983566330**

Law-5317:

If both (**W**= Wealth or Total Assets= 139,796 millions USD), (**i**= I/S= Interest Portion= 1.00%), (**F**= L/E= Leverage or Gearing Ratio= 99.00%), (**d**= D/A= Dividend Portion or Payout= 30.00%), (**S**= Sales or Revenues= 51,294 millions USD), (**V**= Variable Cost= 9,746 millions USD), (**U**= Utilized or Starting Capital= 64,720 millions USD), and (**f**= F/S= Fixed Portion= 58.00%), then it's (**t**= Tax Rate planned), is:

$$\mathbf{t}= 1-\{\mathbf{W}/[1+\mathbf{I}]-\mathbf{U}\}/([1-\mathbf{d}]\{\mathbf{S}[1-\mathbf{f}-\mathbf{i}]-\mathbf{V}\})$$
$$= 1-\{139{,}796/[1+0.9900]-64{,}720\}$$
$$/\{[1-0.3000]\{51{,}294$$
$$[1-0.5800-0.0100]-9{,}746\})$$
$$= 30.00\%$$

Corporate IFRS-GAAP (B/S-I/S), ISBN 13: **978-1983566332**, ISBN 10: **1983566330**

Law-5318:

If both (**W**= Wealth or Total Assets= 139,796 millions USD), (**i**= I/S= Interest Portion= 1.00%), (**f**= L/E= Leverage or Gearing Ratio= 99.00%), (**t**= T/B= Tax Rate= 30.00%), (**S**= Sales or Revenues= 51,294 millions USD), (**V**= Variable Cost= 9,746 millions USD), (**U**= Utilized or Starting Capital= 64,720 millions USD), and (**f**= F/S= Fixed Portion= 58.00%), then it's (**d**= Dividend Portion or Payout planned), is:

$$\mathbf{d} = 1 - \{\mathbf{W}/[1+\mathbf{f}] - \mathbf{U}\}/([1-\mathbf{d}]\{\mathbf{S}[1-\mathbf{f}-\mathbf{i}] - \mathbf{V}\})$$
$$= 1 - \{139{,}796/[1+0.9900] - 64{,}720\}$$
$$/\{[1-0.3000]\{51{,}294$$
$$[1-0.5800-0.0100] - 9{,}746\})$$
$$= 30.00\%$$

Corporate IFRS-GAAP (B/S-I/S), ISBN 13: **978-1983566332**, ISBN 10: **1983566330**

Law-5319:

If both (**W**= Wealth or Total Assets= 139,796 millions USD), (**i**= I/S= Interest Portion= 1.00%), (**f**= L/E= Leverage or Gearing Ratio= 99.00%), (**t**= T/B= Tax Rate= 30.00%), (**$**= Sales or Revenues= 51,294 millions USD), (**V**= Variable Cost= 9,746 millions USD), (**d**= D/A=vDividend Portion or Payout= 30.00%), and (**f**= F/S=vFixed Portion= 58.00%), then it's (**U**= Utilized or Starting Capital planned), is:

$$U= \{W/[1+i]\}-[1-d][1-t]\{S[1-f-i]-V\})$$
$$= \{139,796/[1+0.9900]-[1-0.3000]$$
$$[1-0.3000]\{51,294[1-0.5800$$
$$-0.0100]-9,746\})$$
$$= \underline{64,720} \text{ millions USD}$$

Corporate IFRS-GAAP (B/S-I/S), ISBN 13: **978-1983566332**, ISBN 10: **1983566330**

Law-5320:

If both (**W**= Wealth or Total Assets= 139,796 millions USD), (**i**= I/S= Interest Portion= 1.00%), (**U**= Utilized or Starting Capital= 64,720 millions USD), (**t**= T/B= Tax Rate= 30.00%), (**S**= Sales or Revenues= 51,294 millions USD), (**V**= Variable Cost= 9,746 millions USD), (**d**= D/A= Dividend Portion or Payout= 30.00%), and (**f**= F/S= Fixed Portion= 58.00%), then it's (**I** = Leverage or Gearing Ratio planned), is:

$$I = \{W/(U+[1-d][1-t]\{S[1-f-i]-V\})-1$$
$$= \{139,796/(64,720+[1-0.3000]$$
$$[1-0.3000]\{51,294[1-0.5800$$
$$-0.0100]-9,746\})-1$$
$$= \underline{99.00\%}$$

Corporate IFRS-GAAP (B/S-I/S), ISBN 13: **978-1983566332**, ISBN 10: **1983566330**

Law-5321:

If both ($\textbf{S'}$= Sales of Past Year= <u>48,851</u> millions USD), ($\textbf{s}$= [S/S']-1= Sales Growth= <u>5.00%</u>), ($\textbf{l}$= L/E= Leverage or Gearing Ratio= <u>99.00%</u>), ($\textbf{i}$= I/S= Interest Portion= <u>1.00%</u>), ($\textbf{U}$= Utilized or Starting Capital= <u>64,720</u> millions USD), ($\textbf{t}$= T/B= Tax Rate= <u>30.00%</u>), ($\textbf{S}$= Sales or Revenues= <u>51,294</u> millions USD), ($\textbf{V}$= Variable Cost= <u>9,746</u> millions USD), ($\textbf{d}$= D/A= Dividend Portion or Payout= <u>30.00%</u>), and ($\textbf{f}$= F/S= Fixed Portion= <u>58.00%</u>), then it's ($\textbf{W}$= Wealth or Total Assets planned), is:

$$\textbf{W}= [1+\textbf{l}](\textbf{U}+[1-\textbf{d}][1-\textbf{t}]\{\textbf{S}-\textbf{V}-\textbf{Sf}-\textbf{S'i}[1+\textbf{s}]\})$$

$$= [1+\textbf{l}](\textbf{U}+[1-\textbf{d}][1-\textbf{t}]\{\textbf{S}[1-\textbf{f}]-\textbf{V}-\textbf{S'i}[1+\textbf{s}]\})$$

$$= [1+0.9900](64,720+[1-0.3000][1-0.3000]\{51,294[1-0.5800]-9,746-48,851*0.0100[1+0.0500]\})$$

$$= \underline{139,796} \text{ millions USD}$$

Corporate IFRS-GAAP (B/S-I/S), ISBN 13: **978-1983566332**, ISBN 10: **1983566330**

Law-5322:

If both (**$'$**= Sales of Past Year= 48,851 millions USD), (**s**= [S/S']-1= Sales Growth= 5.00%), (**f**= L/E= Leverage or Gearing Ratio= 99.00%), (**i**= I/S= Interest Portion= 1.00%), (**U**= Utilized or Starting Capital= 64,720 millions USD), (**t**= T/B= Tax Rate= 30.00%), (**W**= Wealth or Total Assets= 139,796 millions USD), (**V**= Variable Cost= 9,746 millions USD), (**d**= D/A= Dividend Portion or Payout= 30.00%), and (**f**= F/S= Fixed Portion= 58.00%), then it's (**$**= Sales or Revenues planned), is:

$$S= (V+S'i[1+s]+\{W/[1+f]-U\}/\{[1-d][1-t]\})/[1-f]$$

$$= (9,746+48,851*0.0100[1+0.0500]$$
$$+\{139,796/[1+0.9900]$$
$$-64,720\}/\{[1-0.3000]$$
$$[1-0.3000]\})/[1-0.5800]$$
$$= \underline{51,294} \text{ millions USD}$$

Corporate IFRS-GAAP (B/S-I/S), ISBN 13: **978-1983566332**, ISBN 10: **1983566330**

Law-5323:

If both (**$'**= Sales of Past Year= <u>48,851</u> millions USD), (**s**= [S/S']-1= Sales Growth= <u>5.00%</u>), (**l**= L/E= Leverage or Gearing Ratio= <u>99.00%</u>), (**i**= I/S= Interest Portion= <u>1.00%</u>), (**U**= Utilized or Starting Capital= <u>64,720</u> millions USD), (**t**= T/B= Tax Rate= <u>30.00%</u>), (**W**= Wealth or Total Assets= <u>139,796</u> millions USD), (**$**= Sales or Revenues= <u>51,294</u> millions USD), (**d**= D/A= Dividend Portion or Payout= <u>30.00%</u>), and (**f**= F/S=Fixed Portion= <u>58.00%</u>), then it's (**V**= Variable Cost planned), is:

$$V = \$[1-f] - \$'i[1+s] - \{W/[1+l] - U\}$$
$$/\{[1-d][1-t]\}$$
$$= (51{,}294[1-0.5800] - 48{,}851$$
$$*0.0100[1+0.0500]$$
$$-\{139{,}796/[1+0.9900]$$
$$-64{,}720\}/\{[1-0.3000]$$
$$[1-0.3000]\}$$
$$= \underline{9{,}746} \text{ millions USD}$$

Corporate IFRS-GAAP (B/S-I/S), ISBN 13: **978-1983566332**, ISBN 10: **1983566330**

<u>Law-5324</u>:

If both (**V**= Variable Cost= <u>9,746</u> millions USD), (**s**= [S/S']-1= Sales Growth= <u>5.00%</u>), (**f**= L/E= Leverage or Gearing Ratio= <u>99.00%</u>), (**i**= I/S= Interest Portion= <u>1.00%</u>), (**U**= Utilized or Starting Capital= <u>64,720</u> millions USD), (**t**= T/B= Tax Rate= <u>30.00%</u>), (**W**= Wealth or Total Assets= <u>139,796</u> millions USD), (**\$**= Sales or Revenues= <u>51,294</u> millions USD), (**d**= D/A= Dividend Portion or Payout= <u>30.00%</u>), and (**f**= F/S= Fixed Portion= <u>58.00%</u>), then it's (**\$'**= Sales Past), must be:

$$\$' = (\$[1-f]-V-\{W/[1+f]-U\}/\{[1-d][1-t]\})/\{i[1+s]\}$$

$$= (51{,}294[1-0.5800]-9{,}746-\{139{,}796/[1+0.9900]-64{,}720\}/\{[1-0.3000][1-0.3000]\})/\{0.0100[1+0.0500]\}$$

$$= \underline{48{,}851} \text{ millions USD}$$

Corporate IFRS-GAAP (B/S-I/S), ISBN 13: **978-1983566332**, ISBN 10: **1983566330**

<u>Law-5325</u>:

If both (**V**= Variable Cost= <u>9,746</u> millions USD), (**s**= [S/S']-1= Sales Growth= <u>5.00%</u>), (**I**= L/E= Leverage or Gearing Ratio= <u>99.00%</u>), (**i**= I/S= Interest Portion= <u>1.00%</u>), (**U**= Utilized or Starting Capital= <u>64,720</u> millions USD), (**t**= T/B= Tax Rate= <u>30.00%</u>), (**W**= Wealth or Total Assets= <u>139,796</u> millions USD), (**S**= Sales or Revenues= <u>51,294</u> millions USD), (**d**= D/A= Dividend Portion or Payout= <u>30.00%</u>), and (**S'**= Sales of Past Year= <u>48,851</u> millions USD), then it's (**f**= Fixed Portion planned), must be:

$$f = 1 - (S'i[1+s] + V + \{W/[1+I] - U\} / \{[1-d][1-t]\})/S$$

$$= 1 - (48,851 * 0.0100[1+0.0500]$$
$$+ 9,746 + \{139,796$$
$$/[1+0.9900] - 64,720\}$$
$$/ \{[1-0.3000][1-0.3000]\})$$
$$/ 51,294$$

$$= \underline{58.00\%}$$

Corporate IFRS-GAAP (B/S-I/S), ISBN 13: **978-1983566332**, ISBN 10: **1983566330**

<u>Law-5326</u>:

If both (V= Variable Cost= <u>9,746</u> millions USD), (S'= Sales of Past Year= <u>48,851</u> millions USD), (I= L/E= Leverage or Gearing Ratio= <u>99.00%</u>), (i= I/S= Interest Portion= <u>1.00%</u>), (U= Utilized or Starting Capital= <u>64,720</u> millions USD), (t= T/B= Tax Rate= <u>30.00%</u>), (W= Wealth or Total Assets= <u>139,796</u> millions USD), (S= Sales or Revenues= <u>51,294</u> millions USD), (d= D/A= Dividend Portion or Payout= <u>30.00%</u>), and (f= F/S= Fixed Portion= <u>58.00%</u>), then it's (s= Sales Growth planned), is:

$$s = (S[1-f]-V-\{W/[1+I]-U\}$$
$$/\{[1-d][1-t]\})/[S'i]-1$$
$$= (51,294[1-0.5800]-9,746$$
$$-\{139,796/[1+0.9900]$$
$$-64,720\}/\{[1-0.3000]$$
$$[1-0.3000]\})$$
$$/\{[48,851*0.0100]-1$$
$$= \underline{5.00\%}$$

Corporate IFRS-GAAP (B/S-I/S), ISBN 13: **978-1983566332**, ISBN 10: **1983566330**

Law-5327:

If both (**V**= Variable Cost= 9,746 millions USD), (**S'**= Sales of Past Year= 48,851 millions USD), (**I**= L/E= Leverage or Gearing Ratio= 99.00%), (**s**= [S/S']-1= Sales Growth= 5.00%), (**U**= Utilized or Starting Capital= 64,720 millions USD), (**t**= T/B= Tax Rate= 30.00%), (**W**= Wealth or Total Assets= 139,796 millions USD), (**S**= Sales or Revenues= 51,294 millions USD), (**d**= D/A= Dividend Portion or Payout= 30.00%), and (**f**= F/S= Fixed Portion= 58.00%), then it's (**i**= Interest Portion planned), is:

$$i= (S[1-f]-V-\{W/[1+I]-U\}$$
$$/\{[1-d][1-t]\})/\{S'[1+s]\}$$
$$= (51,294[1-0.5800]-9,746-\{139,796$$
$$/[1+0.9900]-64,720\}$$
$$/\{[1-0.3000][1-0.3000]\})$$
$$/\{48,851[1+0.0500]\}$$
$$= 1.00\%$$

<u>Law-5328</u>:

If both (**V**= Variable Cost= <u>9,746</u> millions USD), (**$'**= Sales of Past Year= <u>48,851</u> millions USD), (**f**= L/E= Leverage or Gearing Ratio= <u>99.00%</u>), (**s**= [S/S']-1= Sales Growth= <u>5.00%</u>), (**U**= Utilized or Starting Capital= <u>64,720</u> millions USD), (**i**= I/S= Interest Portion= <u>1.00%</u>), (**W**= Wealth or Total Assets= <u>139,796</u> millions USD), (**$**= Sales or Revenues= <u>51,294</u> millions USD), (**d**= D/A= Dividend Portion or Payout= <u>30.00%</u>), and (**f**= F/S= Fixed Portion= <u>58.00%</u>), then it's (**t**= Tax Rate planned), is:

$$t= 1-\{W/[1+f]-U\}/([1-d]\{\$[1-f]-V-\$'i[1+s]\})$$
$$= 1-\{139,796/[1+0.9900]-64,720\}$$
$$/([1-0.3000]\{51,294$$
$$[1-0.5800]-9,746$$
$$-48,851*0.0100$$
$$[1+0.0500]\})$$
$$= \underline{30.00\%}$$

Corporate IFRS-GAAP (B/S-I/S), ISBN 13: **978-1983566332**, ISBN 10: **1983566330**

<u>Law-5329</u>:

If both (**V**= Variable Cost= <u>9,746</u> millions USD), (**S'**= Sales of Past Year= <u>48,851</u> millions USD), (**I**= L/E= Leverage or Gearing Ratio= <u>99.00%</u>), (**s**= [S/S']-1= Sales Growth= <u>5.00%</u>), (**U**= Utilized or Starting Capital= <u>64,720</u> millions USD), (**i**= I/S= Interest Portion= <u>1.00%</u>), (**W**= Wealth or Total Assets= <u>139,796</u> millions USD), (**S**= Sales or Revenues= <u>51,294</u> millions USD), (**t**= T/B= Tax Rate= <u>30.00%</u>), and (**f**= F/S= Fixed Portion= <u>58.00%</u>), then it's (**d**= Dividend Portion or Payout planned), is:

$$d= 1-\{\mathbf{W}/[1+\mathbf{I}]-\mathbf{U}\}/([1-\mathbf{t}]\{\mathbf{S}[1-\mathbf{f}]-\mathbf{V}-\mathbf{S'i}[1+\mathbf{s}]\})$$

$$= 1-\{139,796/[1+0.9900]-64,720\}$$
$$/([1-0.3000]\{51,294$$
$$[1-0.5800]-9,746-48,851$$
$$*0.0100[1+0.0500]\})$$

$$= \underline{30.00\%}$$

Corporate IFRS-GAAP (B/S-I/S), ISBN 13: **978-1983566332**, ISBN 10: **1983566330**

<u>Law-5330</u>:

If both (**V**= Variable Cost= <u>9,746</u> millions USD), (**S'**= Sales of Past Year= <u>48,851</u> millions USD), (**I**= L/E= Leverage or Gearing Ratio= <u>99.00%</u>), (**s**= [S/S']-1= Sales Growth= <u>5.00%</u>), (**d**= D/A= Dividend Portion or Payout= <u>30.00%</u>), (**i**= I/S= Interest Portion= <u>1.00%</u>), (**W**= Wealth or Total Assets= <u>139,796</u> millions USD), (**S**= Sales or Revenues= <u>51,294</u> millions USD), (**t**= T/B= Tax Rate= <u>30.00%</u>), and (**f**= F/S= Fixed Portion= <u>58.00%</u>), then it's (**U**= Utilized or Starting Capital planned), is:

$$U= W/[1+I]-[1-d][1-t]\{S[1-f]-V-S'i[1+s]\}$$
$$= 139{,}796/\,[1+0.9900]-[1-0.3000]$$
$$[1-0.3000]\{51{,}294[1-0.5800]$$
$$-9{,}746-48{,}851*0.0100$$
$$[1+0.0500]\}$$
$$= \underline{64{,}720} \text{ millions USD}$$

Corporate IFRS-GAAP (B/S-I/S), ISBN 13: **978-1983566332**, ISBN 10: **1983566330**

<u>Law-5331</u>:

If both (**V**= Variable Cost= <u>9,746</u> millions USD), (**S'**= Sales of Past Year= <u>48,851</u> millions USD), (**U**= Utilized or Starting Capital= <u>64,720</u> millions USD), (**s**= [S/S']-1= Sales Growth= <u>5.00%</u>), (**d**= D/A= Dividend Portion or Payout= <u>30.00%</u>), (**i**= I/S= Interest Portion= <u>1.00%</u>), (**W**= Wealth or Total Assets= <u>139,796</u> millions USD), (**S**= Sales or Revenues= <u>51,294</u> millions USD), (**t**= T/B= Tax Rate= <u>30.00%</u>), and (**f**= F/S= Fixed Portion= <u>58.00%</u>), then it's (**I** = Leverage or Gearing Ratio planned), is:

$$I = W/(U+[1-d][1-t]\{S[1-f]-V-S'i[1+s]\})-1$$
$$= 139,796/(64,720+[[1-0.3000]$$
$$[1-0.3000]\{51,294$$
$$[1-0.5800]-9,746-48,851$$
$$*0.0100[1+0.0500]\})-1$$

$$= \underline{99.00\%}$$

Corporate IFRS-GAAP (B/S-I/S), ISBN 13: **978-1983566332**, ISBN 10: **1983566330**

<u>Law-5332</u>:

If both (**V**= Variable Cost= <u>9,746</u> millions USD), (**S'**= Sales of Past Year= <u>48,851</u> millions USD), (**U**= Utilized or Starting Capital= <u>64,720</u> millions USD), (**s**= [S/S']-1= Sales Growth= <u>5.00%</u>), (**d**= D/A= Dividend Portion or Payout= <u>30.00%</u>), (**I**= Interest Expense= <u>513</u> millions USD), (**f**= L/E= Leverage or Gearing Ratio= <u>99.00%</u>), (**S**= Sales or Revenues= <u>51,294</u> millions USD), (**t**= T/B= Tax Rate= <u>30.00%</u>), and (**f**= F/S=Fixed Portion= <u>58.00%</u>), then it's (**W**= Wealth or Total Assets planned), is:

$$W= [1+I](U+[1-d][1-t]\{S-V-I-S'f[1+s]\})$$
$$= [1+0.9900](64,720+[[1-0.3000]$$
$$[1-0.3000]\{51,294-9,746$$
$$-513-48,851*0.5800$$
$$[1+0.0500]\})$$
$$= \underline{139,796} \text{ millions USD}$$

Corporate IFRS-GAAP (B/S-I/S), ISBN 13: **978-1983566332**, ISBN 10: **1983566330**

Law-5333:

If both (**V**= Variable Cost= 9,746 millions USD), (**S'**= Sales of Past Year= 48,851 millions USD), (**U**= Utilized or Starting Capital= 64,720 millions USD), (**s**= [S/S']-1= Sales Growth= 5.00%), (**d**= D/A= Dividend Portion or Payout= 30.00%), (**I**= Interest Expense= 513 millions USD), (**l**= L/E= Leverage or Gearing Ratio= 99.00%), (**W**= Wealth or Total Assets= 139,796 millions USD), (**t**= T/B= Tax Rate= 30.00%), and (**f**= F/S= Fixed Portion= 58.00%), then it's (**S**= Sales or Revenues planned), is:

$$S= V+I+S'f[1+s]+\{W/[1+l]-U\}/\{[1-d][1-t]\}$$
$$= 9,746+513+48,851*0.5800$$
$$[1+0.0500] +\{139,796$$
$$/[1+0.9900]-64,720\}$$
$$/\{[[1-0.3000][1-0.3000]]\}$$
$$= 51,294 \text{ millions USD}$$

Corporate IFRS-GAAP (B/S-I/S), ISBN 13: **978-1983566332**, ISBN 10: **1983566330**

<u>Law-5334</u>:

If both (**$**= Sales or Revenues= <u>51,294</u> millions USD), (**$'**= Sales of Past Year= <u>48,851</u> millions USD), (**U**= Utilized or Starting Capital= <u>64,720</u> millions USD), (**s**= [S/S']-1= Sales Growth= <u>5.00%</u>), (**d**= D/A= Dividend Portion or Payout= <u>30.00%</u>),(**I**= Interest Expense= <u>513</u> millions USD), (**l**= L/E= Leverage or Gearing Ratio= <u>99.00%</u>), (**W**= Wealth or Total Assets= <u>139,796</u> millions USD), (**t**= T/B= Tax Rate= <u>30.00%</u>), and (**f**= F/S= Fixed Portion= <u>58.00%</u>), then it's (**V**= Variable Cost planned), is:

$$V= S-I-S'f[1+s]-\{W/[1+l]-U\}/\{[1-d][1-t]\}$$
$$= 51,294-513-48,851*0.5800$$
$$[1+0.0500]-\{139,796$$
$$/[1+0.9900]-64,720\}$$
$$/\{[[1-0.3000][1-0.3000]\}$$
$$= \underline{9,746} \text{ millions USD}$$

<u>Law-5336</u>:

If both (**$**= Sales or Revenues= <u>51,294</u> millions USD), (**U**= Utilized or Starting Capital= <u>64,720</u> millions USD), (**V**= Variable Cost= <u>9,746</u> millions USD), (**s**= [S/S']-1= Sales Growth= <u>5.00%</u>), (**d**= D/A= Dividend Portion or Payout= <u>30.00%</u>), (**I**= Interest Expense= <u>513</u> millions USD), (**f**= L/E= Leverage or Gearing Ratio= <u>99.00%</u>), (**W**= Wealth or Total Assets= <u>139,796</u> millions USD), (**t**= T/B= Tax Rate= <u>30.00%</u>), and (**$'**= Sales of Past Year= <u>48,851</u> millions USD), then it's (**f**= Fixed Portion planned), is:

$$f= (\$-I-V-\{W/[1+I]-U\}/\{[1-d][1-t]\})/\{\$'[1+s]\}$$
$$= (51,294-513-9,746-\{139,796$$
$$/[1+0.9900]-64,720\}$$
$$/\{[[1-0.3000][1-0.3000]\})$$
$$/\{48,851[1+0.0500]\}$$
$$= \underline{58.00\%}$$

Corporate IFRS-GAAP (B/S-I/S), ISBN 13: **978-1983566332**, ISBN 10: **1983566330**

<u>Law-5337</u>:

If both (**$**= Sales or Revenues= <u>51,294</u> millions USD), (**U**= Utilized or Starting Capital= <u>64,720</u> millions USD), (**V**= Variable Cost= <u>9,746</u> millions USD), (**f**= F/S= Fixed Portion= <u>58.00%</u>), (**d**= D/A= Dividend Portion or Payout= <u>30.00%</u>), (**I**= Interest Expense= <u>513</u> millions USD), (**⌐**= L/E= Leverage or Gearing Ratio= <u>99.00%</u>), (**W**= Wealth or Total Assets= <u>139,796</u> millions USD), (**t**= T/B= Tax Rate= <u>30.00%</u>), and (**$'**= Sales of Past Year= <u>48,851</u> millions USD), then it's (**$**= Sales Growth planned), is:

$$s= (\mathbf{\$}\text{-}\mathbf{I}\text{-}\mathbf{V}\text{-}\{\mathbf{W}/[1+\mathbf{I}]\text{-}\mathbf{U}\}/\{[1\text{-}\mathbf{d}][1\text{-}\mathbf{t}]\})/[\mathbf{\$'f}]\text{-}1$$

$$= (51,294\text{-}513\text{-}9,746\text{-}\{139,796$$
$$/[1+0.9900]\text{-}64,720\}$$
$$/\{[[1\text{-}0.3000][1\text{-}0.3000]\})$$
$$/[48,851*0.5800]\text{-}1$$

$$= \underline{5.00\%}$$

Corporate IFRS-GAAP (B/S-I/S), ISBN 13: **978-1983566332**, ISBN 10: **1983566330**

<u>Law-5338</u>:

If both (**$**= Sales or Revenues= <u>51,294</u> millions USD), (**U**= Utilized or Starting Capital= <u>64,720</u> millions USD), (**V**= Variable Cost= <u>9,746</u> millions USD), (**s**= [S/S']-1= Sales Growth= <u>5.00%</u>), (**d**= D/A= Dividend Portion or Payout= <u>30.00%</u>), (**f**= F/S= Fixed Portion= <u>58.00%</u>), (**I**= L/E= Leverage or Gearing Ratio= <u>99.00%</u>), (**W**= Wealth or Total Assets= <u>139,796</u> millions USD), (**t**= T/B= Tax Rate= <u>30.00%</u>), and (**$'**= Sales of Past Year= <u>48,851</u> millions USD), then it's (**I**= Interest Expense planned), is:

$$\textbf{I}= \textbf{\$}-\textbf{\$'f}[1+\textbf{s}]-\textbf{V}-\{\textbf{W}/[1+\textbf{I}]-\textbf{U}\}/\{[1-\textbf{d}][1-\textbf{t}]\}$$
$$= 51,294-48,851*0.5800[1+0.0500]$$
$$-9,746-\{139,796/[1+0.9900]$$
$$-64,720\}/\{[[1-0.3000]$$
$$[1-0.3000]\}$$
$$= \underline{513} \text{ millions USD}$$

Corporate IFRS-GAAP (B/S-I/S), ISBN 13: **978-1983566332**, ISBN 10: **1983566330**

Law-5339:

If both (**$**= Sales or Revenues= 51,294 millions USD), (**U**= Utilized or Starting Capital= 64,720 millions USD), (**V**= Variable Cost= 9,746 millions USD), (**s**= [S/S']-1= Sales Growth= 5.00%), (**d**= D/A= Dividend Portion or Payout= 30.00%), (**f**= F/S= Fixed Portion= 58.00%), (**l**= L/E= Leverage or Gearing Ratio= 99.00%), (**W**= Wealth or Total Assets= 139,796 millions USD), (**I**= Interest Expense= 513 millions USD), and (**$'**= Sales of Past Year= 48,851 millions USD), then it's (**t**= Tax Rate planned), is:

$$t= 1-\{W/[1+l]-U\}/([1-d]\{S-S'f[1+s]-V-I\})$$
$$= 1-\{139,796/[1+0.9900]-64,720\}$$
$$/([1-0.3000]\{51,294$$
$$-48,851*0.5800[1+0.0500]$$
$$-9,746-513\})$$

$$= 30.00\%$$

Corporate IFRS-GAAP (B/S-I/S), ISBN 13: **978-1983566332**, ISBN 10: **1983566330**

Law-5340:

If both (**S**= Sales or Revenues= 51,294 millions USD), (**U**= Utilized or Starting Capital= 64,720 millions USD), (**V**= Variable Cost= 9,746 millions USD), (**s**= [S/S']-1= Sales Growth= 5.00%), (**t**= T/B= Tax Rate= 30.00%), (**f**= F/S= Fixed Portion= 58.00%), (**l**= L/E= Leverage or Gearing Ratio= 99.00%), (**W**= Wealth or Total Assets= 139,796 millions USD), (**l**= Interest Expense= 513 millions USD), and (**S'**= Sales of Past Year= 48,851 millions USD), then it's (**d**= Dividend Portion or Payout planned), is:

$$d = 1-\{W/[1+l]-U\}/([1-t]\{S-S'f[1+s]-V-l\})$$
$$= 1-\{139,796/[1+0.9900]-64,720\}$$
$$/([1-0.3000]\{51,294-48,851$$
$$*0.5800[1+0.0500]$$
$$-9,746-513\})$$

$$= 30.00\%$$

Corporate IFRS-GAAP (B/S-I/S), ISBN 13: **978-1983566332**, ISBN 10: **1983566330**

Law-5341:

If both ($= Sales or Revenues= 51,294 millions USD), (d= D/A= Dividend Portion or Payout= 30.00%), (V= Variable Cost= 9,746 millions USD), (s= [S/S']-1= Sales Growth= 5.00%), (t= T/B= Tax Rate= 30.00%), (f= F/S= Fixed Portion= 58.00%), (I= L/E= Leverage or Gearing Ratio= 99.00%), (W= Wealth or Total Assets= 139,796 millions USD), (I= Interest Expense= 513 millions USD), and ($'= Sales of Past Year= 48,851 millions USD), then it's (U= Utilized or Starting Capital planned), is:

$$U= W/[1+I]-[1-d][1-t]\{S-S'f[1+s]-V-I\}$$
$$= 139,796/[1+0.9900]-[1-0.3000]$$
$$[1-0.3000]\{51,294-48,851$$
$$*0.5800[1+0.0500]$$
$$-9,746-513\}$$
$$= 64,720 \text{ millions USD}$$

Corporate IFRS-GAAP (B/S-I/S), ISBN 13: **978-1983566332**, ISBN 10: **1983566330**

Law-5342:

If both (**$**= Sales or Revenues= 51,294 millions USD), (**d**= D/A= Dividend Portion or Payout= 30.00%), (**V**= Variable Cost= 9,746 millions USD), (**s**= [S/S']-1= Sales Growth= 5.00%), (**t**= T/B= Tax Rate= 30.00%), (**f**= F/S= Fixed Portion= 58.00%), (**U**= Utilized or Starting Capital= 64,720 millions USD), (**W**= Wealth or Total Assets= 139,796 millions USD), (**l**= Interest Expense= 513 millions USD), and (**$'**= Sales of Past Year= 48,851 millions USD), then it's (**l** = Leverage or Gearing Ratio planned), is:

$$l = W/(U+[1-d][1-t]\{\$-\$'f[1+s]-V-l\})-1$$
$$= 139,796/(64,720+[1-0.3000]$$
$$[1-0.3000]\{51,294-48,851$$
$$*0.5800[1+0.0500]-9,746$$
$$-513\})-1$$
$$= 99.00\%$$

Corporate IFRS-GAAP (B/S-I/S), ISBN 13: **978-1983566332**, ISBN 10: **1983566330**

<u>Law-5343</u>:

If both (**$**= Sales or Revenues= <u>51,294</u> millions USD), (**d**= D/A= Dividend Portion or Payout= <u>30.00%</u>), (**V**= Variable Cost= <u>9,746</u> millions USD), (**s**= [S/S']-1= Sales Growth= <u>5.00%</u>), (**t**= T/B= Tax Rate= <u>30.00%</u>), (**f**= F/S= Fixed Portion= <u>58.00%</u>), (**U**= Utilized or Starting Capital= <u>64,720</u> millions USD), (**l**= L/E= Leverage or Gearing Ratio= <u>99.00%</u>), (**i**= I/S= Interest Portion= <u>1.00%</u>), and (**$'**= Sales of Past Year= <u>48,851</u> millions USD), then it's (**W**= Wealth or Total Assesment planned), is:

$$W= [1+l](U+[1-d][1-t]\{\$-V-\$'f[1+s]-\$i\})$$
$$= [1+l](U+[1-d][1-t]\{\$[1-i]-V$$
$$-\$'f[1+s]\})$$
$$= [1+0.9900](64,720+[1-0.3000]$$
$$[1-0.3000]\{51,294[1-0.0100]$$
$$-9,746-48,851*0.5800$$
$$[1+0.0500]\})$$
$$= \underline{139,796} \text{ millions USD}$$

Corporate IFRS-GAAP (B/S-I/S), ISBN 13: **978-1983566332**, ISBN 10: **1983566330**

Law-5344:

If both (**W**= Wealth or Total Assets= 139,796 millions USD), (**d**= D/A= Dividend Portion or Payout= 30.00%), (**V**= Variable Cost= 9,746 millions USD), (**s**= [S/S']-1= Sales Growth= 5.00%), (**t**= T/B= Tax Rate= 30.00%), (**f**= F/S= Fixed Portion= 58.00%), (**U**= Utilized or Starting Capital= 64,720 millions USD), (**l**= L/E= Leverage or Gearing Ratio= 99.00%), (**i**= I/S= Interest Portion= 1.00%), and (**S'**= Sales of Past Year= 48,851 millions USD), then it's (**S**= Sales or Revenues planned), is:

$$S= (V+S'f[1+s]+\{W/[1+l]-U\}$$
$$/\{[1-d][1-t]\})/[1-i]$$
$$= (9,746+48,851*0.5800$$
$$[1+0.0500] +\{139,796$$
$$/[1+0.9900]-64,720\}$$
$$/\{[1-0.3000][1-0.3000]\})$$
$$/[1-0.0100]$$
$$= \underline{51,294} \text{ millions USD}$$

Corporate IFRS-GAAP (B/S-I/S), ISBN 13: **978-1983566332**, ISBN 10: **1983566330**

Law-5345:

If both (**W**= Wealth or Total Assets= 139,796 millions USD), (**d**= D/A= Dividend Portion or Payout= 30.00%), (**V**= Variable Cost= 9,746 millions USD), (**s**= [S/S']-1= Sales Growth= 5.00%), (**t**= T/B= Tax Rate= 30.00%), (**f**= F/S= Fixed Portion= 58.00%), (**U**= Utilized or Starting Capital= 64,720 millions USD), (**I**= L/E= Leverage or Gearing Ratio= 99.00%), (**S**= Sales or Revenues= 51,294 millions USD), and (**S'**= Sales of Past Year= 48,851 millions USD), then it's (**i**= Interest Portion planned), is:

$$\mathbf{i}= 1-(\mathbf{V}+\mathbf{S'f}[1+\mathbf{s}]+\{\mathbf{W}/[1+\mathbf{I}]-\mathbf{U}\}/\{[1-\mathbf{d}][1-\mathbf{t}]\})/\mathbf{S}$$

$$= 1-(9{,}746+48{,}851*0.5800$$
$$[1+0.0500]+\{139{,}796$$
$$/[1+0.9900]-64{,}720\}$$
$$/\{[1-0.3000][1-0.3000]\})$$
$$/51{,}294$$

$$= \underline{1.00\%}$$

Corporate IFRS-GAAP (B/S-I/S), ISBN 13: **978-1983566332**, ISBN 10: **1983566330**

Law-5346:

If both (**W**= Wealth or Total Assets= 139,796 millions USD), (**d**= D/A= Dividend Portion or Payout= 30.00%), (**f**= F/S= Fixed Portion= 58.00%), (**s**= [S/S']-1= Sales Growth= 5.00%), (**t**= T/B= Tax Rate= 30.00), (**i**= I/S= Interest Portion= 1.00%), (**U**= Utilized or Starting Capital= 64,720 millions USD), (**l**= L/E= Leverage or Gearing Ratio= 99.00%), (**S**= Sales or Revenues= 51,294 millions USD), and (**S'**= Sales of Past Year= 48,851 millions USD), then it's (**V**= Variable Cost planned), is:

$$\mathbf{V}= \mathbf{S}[1-\mathbf{i}]-\mathbf{S'f}[1+\mathbf{s}]-\{\mathbf{W}/[1+\mathbf{l}]-\mathbf{U}\}/\{[1-\mathbf{d}][1-\mathbf{t}]\}$$
$$= 51,294[1-0.0100]-48,851*0.5800$$
$$[1+0.0500]-\{139,796$$
$$/[1+0.9900]-64,720\}$$
$$/\{[1-0.3000][1-0.3000]\}$$
$$= 9,746 \text{ millions USD}$$

Corporate IFRS-GAAP (B/S-I/S), ISBN 13: **978-1983566332**, ISBN 10: **1983566330**

Law-5347:

If both (**W**= Wealth or Total Assets= 139,796 millions USD), (**d**= D/A= Dividend Portion or Payout= 30.00%), (**f**= F/S= Fixed Portion= 58.00%), (**s**= [S/S']-1= Sales Growth= 5.00%), (**t**= T/B= Tax Rate= 30.00), (**i**= I/S= Interest Portion= 1.00%), (**U**= Utilized or Starting Capital= 64,720 millions USD), (**l**= L/E= Leverage or Gearing Ratio= 99.00%), (**S**= Sales or Revenues= 51,294 millions USD), and (**V**= Variable Cost= 9,746 millions USD), then it's (**S'**= Sales Past), must be:

$$\mathbf{S'} = (\mathbf{S}[1\text{-}\mathbf{i}]\text{-}\mathbf{V}\text{-}\{\mathbf{W}/[1\text{+}\mathbf{l}]\text{-}\mathbf{U}\}/\{[1\text{-}\mathbf{d}][1\text{-}\mathbf{t}]\})$$
$$/\{\mathbf{f}[1\text{+}\mathbf{s}]\}$$
$$= (51{,}294[1\text{-}0.0100]\text{-}9{,}746$$
$$-\{139{,}796/[1\text{+}0.9900]$$
$$-64{,}720\}/\{[1\text{-}0.3000]$$
$$[1\text{-}0.3000]\})$$
$$/\{0.5800[1\text{+}0.0500]\}$$
$$= \underline{48{,}851} \text{ millions USD}$$

Corporate IFRS-GAAP (B/S-I/S), ISBN 13: **978-1983566332**, ISBN 10: **1983566330**

<u>Law-5348</u>:

If both (**W**= Wealth or Total Assets= <u>139,796</u> millions USD), (**d**= D/A= Dividend Portion or Payout= <u>30.00%</u>), (**S'**= Sales of Past Year= <u>48,851</u> millions USD), (**s**= [S/S']-1= Sales Growth= <u>5.00%</u>), (**t**= T/B= Tax Rate= <u>30.00</u>), (**i**= I/S= Interest Portion= <u>1.00%</u>), (**U**= Utilized or Starting Capital= <u>64,720</u> millions USD), (**l**= L/E= Leverage or Gearing Ratio= <u>99.00%</u>), (**S**= Sales or Revenues= <u>51,294</u> millions USD), and (**V**= Variable Cost= <u>9,746</u> millions USD), then it's (**f**= Fixed Portion planned), is:

$$f = (S[1-i]-V-\{W/[1+l]-U\}/\{[1-d][1-t]\})$$
$$/\{S'[1+s]\}$$
$$= (51{,}294[1-0.0100]-9{,}746$$
$$-\{139{,}796/[1+0.9900]$$
$$-64{,}720\}/\{[1-0.3000]$$
$$[1-0.3000]\})$$
$$/\{48{,}851[1+0.0500]\}$$
$$= \underline{58.00\%}$$

Corporate IFRS-GAAP (B/S-I/S), ISBN 13: **978-1983566332**, ISBN 10: **1983566330**

<u>Law-5349</u>:

If both (**W**= Wealth or Total Assets= <u>139,796</u> millions USD), (**d**= D/A= Dividend Portion or Payout= <u>30.00%</u>), (**$'**= Sales of Past Year= <u>48,851</u> millions USD), (**f**= F/S= Fixed Portion= <u>58.00%</u>), (**t**= T/B= Tax Rate= <u>30.00</u>), (**i**= I/S= Interest Portion= <u>1.00%</u>), (**U**= Utilized or Starting Capital= <u>64,720</u> millions USD), (**l**= L/E= Leverage or Gearing Ratio= <u>99.00%</u>), (**$**= Sales or Revenues= <u>51,294</u> millions USD), and (**V**= Variable Cost= <u>9,746</u> millions USD), then it's (**s**= Sales Growth planned), is:

$$s = (\$[1-i]-V-\{W/[1+l]-U\}/\{[1-d][1-t]\})/[\$'f]-1$$
$$= (51{,}294[1-0.0100]-9{,}746-\{139{,}796$$
$$/[1+0.9900]-64{,}720\}$$
$$/\{[1-0.3000][1-0.3000]\})$$
$$/[48{,}851*0.5800]-1$$
$$= \underline{5.00\%}$$

Corporate IFRS-GAAP (B/S-I/S), ISBN 13: **978-1983566332**, ISBN 10: **1983566330**

<u>Law-5350</u>:

If both (**W**= Wealth or Total Assets= <u>139,796</u> millions USD), (**d**= D/A= Dividend Portion or Payout= <u>30.00%</u>), (**S'**= Sales of Past Year= <u>48,851</u> millions USD), (**f**= F/S= Fixed Portion= <u>58.00%</u>), (**s**= [S/S']-1= Sales Growth= <u>5.00%</u>), (**i**= I/S= Interest Portion= <u>1.00%</u>), (**U**= Utilized or Starting Capital= <u>64,720</u> millions USD), (**I**= L/E= Leverage or Gearing Ratio= <u>99.00%</u>), (**S**= Sales or Revenues= <u>51,294</u> millions USD), and (**V**= Variable Cost= <u>9,746</u> millions USD), then it's (**t**= Tax Rate planned), is:

$$t= 1-\{W/[1+I]-U\}/([1-d]\{S[1-i]-V-S'f[1+s]\})$$
$$= 1-\{139,796/[1+0.9900]-64,720\}$$
$$/([1-0.3000]\{51,294$$
$$[1-0.0100]-9,746-48,851$$
$$*0.5800[1+0.0500]\})$$

$$= \underline{30.00\%}$$

Corporate IFRS-GAAP (B/S-I/S), ISBN 13: **978-1983566332**, ISBN 10: **1983566330**

Law-5351:

If both (**W**= Wealth or Total Assets= 139,796 millions USD), (**t**= T/B= Tax Rate= 30.00%), (**S'**= Sales of Past Year= 48,851 millions USD), (**f**= F/S= Fixed Portion= 58.00%), (**s**= [S/S']-1= Sales Growth= 5.00%), (**i**= I/S= Interest Portion= 1.00%), (**U**= Utilized or Starting Capital= 64,720 millions USD), (**l**= L/E= Leverage or Gearing Ratio= 99.00%), (**S**= Sales or Revenues= 51,294 millions USD), and (**V**= Variable Cost= 9,746 millions USD), then it's (**d**= Dividend Portion or Payout planned), is:

$$d= 1-\{W/[1+l]-U\}/([1-t]\{S[1-i]-V-S'f[1+s]\})$$
$$= 1-\{139,796/[1+0.9900]-64,720\}$$
$$/([1-0.3000]\{51,294$$
$$[1-0.0100]-9,746-48,851$$
$$*0.5800[1+0.0500]\})$$
$$= 30.00\%$$

Corporate IFRS-GAAP (B/S-I/S), ISBN 13: **978-1983566332**, ISBN 10: **1983566330**

Law-5352:

If both (**W**= Wealth or Total Assets= 139,796 millions USD), (**t**= T/B= Tax Rate= 30.00%), (**S'**= Sales of Past Year= 48,851 millions USD), (**f**= F/S= Fixed Portion= 58.00%), (**s**= [S/S']-1= Sales Growth= 5.00%), (**i**= I/S= Interest Portion= 1.00%), (**d**= D/A= Dividend Portion or Payout= 30.00%), (**I**= L/E= Leverage or Gearing Ratio= 99.00%), (**S**= Sales or Revenues= 51,294 millions USD), and (**V**= Variable Cost= 9,746 millions USD), then it's (**U**= Utilized or Starting Capital planned), is:

$$\mathbf{U}= \mathbf{W}/[1+\mathbf{I}]-[1-\mathbf{d}][1-\mathbf{t}]\{\mathbf{S}[1-\mathbf{i}]-\mathbf{V}-\mathbf{S'}\mathbf{f}[1+\mathbf{s}]\}$$

$$= 139{,}796/[1+0.9900]-[1-0.3000]$$
$$[1-0.3000]\{51{,}294$$
$$[1-0.0100]-9{,}746-48{,}851$$
$$*0.5800[1+0.0500]\}$$

$$= \underline{64{,}720} \text{ millions USD}$$

Corporate IFRS-GAAP (B/S-I/S), ISBN 13: **978-1983566332**, ISBN 10: **1983566330**

Law-5353:

If both (**W**= Wealth or Total Assets= 139,796 millions USD), (**t**= T/B= Tax Rate= 30.00%), (**S'**= Sales of Past Year= 48,851 millions USD), (**f**= F/S= Fixed Portion= 58.00%), (**s**= [S/S']-1= Sales Growth= 5.00%), (**i**= I/S= Interest Portion= 1.00%), (**d**= D/A= Dividend Portion or Payout= 30.00%), (**U**= Utilized or Starting Capital= 64,720 millions USD), (**S**= Sales or Revenues= 51,294 millions USD), and (**V**= Variable Cost= 9,746 millions USD), then it's (**I** = Leverage or Gearing Ratio planned), is:

$$I = W/(U+[1-d][1-t]\}/\{S[1-i]-V-S'f[1+s]\})-1$$
$$= 139,796/(64,720+[1-0.3000]$$
$$[1-0.3000]\{51,294$$
$$[1-0.0100]-9,746-48,851$$
$$*0.5800[1+0.0500]\})-1$$
$$= 99.00\%$$

Corporate IFRS-GAAP (B/S-I/S), ISBN 13: **978-1983566332**, ISBN 10: **1983566330**

<u>Law-5354</u>:

If both (**I**= L/E= Leverage or Gearing Ratio= 99.00%), (**t**= T/B= Tax Rate= 30.00%), (**S'**= Sales of Past Year= 48,851 millions USD), (**f**= F/S= Fixed Portion= 58.00%), (**s**= [S/S']-1= Sales Growth= 5.00%), (**i**= I/S= Interest Portion= 1.00%), (**d**= D/A= Dividend Portion or Payout= 30.00%), (**U**= Utilized or Starting Capital= 64,720 millions USD), (**S**= Sales or Revenues= 51,294 millions USD), and (**V**= Variable Cost= 9,746 millions USD), then it's (**W**= Wealth or Total Assets planned), is:

$$\mathbf{W}= [1+\mathbf{I}](\mathbf{U}+[1-\mathbf{d}][1-\mathbf{t}]\}/\{\mathbf{S}-\mathbf{V}-\mathbf{S'f}[1+\mathbf{s}]$$
$$-\mathbf{S'i}[1+\mathbf{s}]\})$$
$$= [1+\mathbf{I}](\mathbf{U}+[1-\mathbf{d}][1-\mathbf{t}]\}$$
$$/\{\mathbf{S}-\mathbf{V}-\mathbf{S'}[1+\mathbf{s}][\mathbf{f}+\mathbf{i}]\})$$
$$= [1+0.9900](64,720+[1-0.3000]$$
$$[1-0.3000]\{51,294-9,746$$
$$-48,851[1+0.0500]$$
$$[0.5800+0.0100]\})$$
$$= \underline{139,796} \text{ millions USD}$$

Corporate IFRS-GAAP (B/S-I/S), ISBN 13: **978-1983566332**, ISBN 10: **1983566330**

<u>Law-5355</u>:

If both (I= L/E= Leverage or Gearing Ratio= <u>99.00%</u>), (t= T/B= Tax Rate= <u>30.00%</u>), (S'= Sales of Past Year= <u>48,851</u> millions USD), (f= F/S= Fixed Portion= <u>58.00%</u>), (s= [S/S']-1= Sales Growth= <u>5.00%</u>), (i= I/S= Interest Portion= <u>1.00%</u>), (d= D/A= Dividend Portion or Payout= <u>30.00%</u>), (U= Utilized or Starting Capital= <u>64,720</u> millions USD), (W= Wealth or Total Assets= <u>139,796</u> millions USD), and (V= Variable Cost= <u>9,746</u> millions USD), then it's (S= Sales or Revenues planned), is:

$$S= V+S'[1+s][f+i]+\{W/[1+I]-U\}/\{[1-d][1-t]\}$$
$$= 9,746+48,851[1+0.0500][0.5800$$
$$+0.0100] +\{139,796$$
$$/[1+0.9900]-64,720\}$$
$$/\{[1-0.3000][1-0.3000]\}$$
$$= \underline{51,294} \text{ millions USD}$$

Corporate IFRS-GAAP (B/S-I/S), ISBN 13: **978-1983566332**, ISBN 10: **1983566330**

<u>Law-5356</u>:

If both (**I**= L/E= Leverage or Gearing Ratio= <u>99.00%</u>), (**t**= T/B= Tax Rate= <u>30.00%</u>), (**S'**= Sales of Past Year= <u>48,851</u> millions USD), (**f**= F/S= Fixed Portion= <u>58.00%</u>), (**s**= [S/S']-1= Sales Growth= <u>5.00%</u>), (**i**= I/S= Interest Portion= <u>1.00%</u>), (**d**= D/A= Dividend Portion or Payout= <u>30.00%</u>), (**U**= Utilized or Starting Capital= <u>64,720</u> millions USD), (**W**= Wealth or Total Assets= <u>139,796</u> millions USD), and (**S**= Sales or Revenues= <u>51,294</u> millions USD), then it's (**V**= Variable Cost planned), is:

$$\mathbf{V}= (\mathbf{S}\text{-}\mathbf{S'}[1+\mathbf{s}][\mathbf{f}+\mathbf{i}]\text{-}\{\mathbf{W}/[1+\mathbf{I}]\text{-}\mathbf{U}\}/\{[1\text{-}\mathbf{d}][1\text{-}\mathbf{t}]\}$$

$$= (51,294\text{-}48,851 \ [1+0.0500]$$
$$[0.5800+0.0100]\text{-}\{139,796$$
$$/[1+0.9900]\text{-}64,720\}$$
$$/\{[1\text{-}0.3000][1\text{-}0.3000]\}$$

$$= \underline{9,746} \text{ millions USD}$$

Corporate IFRS-GAAP (B/S-I/S), ISBN 13: **978-1983566332**, ISBN 10: **1983566330**

<u>Law-5357</u>:

If both (I= L/E= Leverage or Gearing Ratio= 99.00%), (t= T/B= Tax Rate= 30.00%), (V= Variable Cost= 9,746 millions USD), (f= F/S= Fixed Portion= 58.00%), (s= [S/S']-1= Sales Growth= 5.00%), (i= I/S= Interest Portion= 1.00%), (d= D/A= Dividend Portion or Payout= 30.00%), (U= Utilized or Starting Capital= 64,720 millions USD), (W= Wealth or Total Assets= 139,796 millions USD), and (S= Sales or Revenues= 51,294 millions USD), then it's (S'= Sales Past), must be:

$$S' = (S-V-\{W/[1+I]-U\}/\{[1-d][1-t]\})$$
$$/\{[1+s][f+i]\}$$
$$= (51,294-9,746-\{139,796$$
$$/[1+0.9900]-64,720\}$$
$$/\{[1-0.3000][1-0.3000]\}$$
$$/\{[1+0.0500][0.5800+0.0100]\}$$
$$= \underline{48,851} \text{ millions USD}$$

Corporate IFRS-GAAP (B/S-I/S), ISBN 13: **978-1983566332**, ISBN 10: **1983566330**

Law-5358:

If both (I= L/E= Leverage or Gearing Ratio= 99.00%), (t= T/B= Tax Rate= 30.00%), (V= Variable Cost= 9,746 millions USD), (f= F/S= Fixed Portion= 58.00%), ($\$'$= Sales of Past Year= 48,851 millions USD), (i= I/S= Interest Portion= 1.00%), (d= D/A= Dividend Portion or Payout= 30.00%), (U= Utilized or Starting Capital= 64,720 millions USD), (W= Wealth or Total Assets= 139,796 millions USD), and ($\$$= Sales or Revenues= 51,294 millions USD), then it's (s= Sales Growth planned), is:

$$s = (\$-V-\{W/[1+I]-U\}/\{[1-d][1-t]\})$$
$$/\{\$'[f+i]\}-1$$
$$= (51{,}294-9{,}746-\{139{,}796$$
$$/[1+0.9900]-64{,}720\}$$
$$/\{[1-0.3000][1-0.3000]\}$$
$$/\{48{,}851[0.5800+0.0100]\}-1$$
$$= 5.00\%$$

Corporate IFRS-GAAP (B/S-I/S), ISBN 13: **978-1983566332**, ISBN 10: **1983566330**

<u>Law-5359</u>:

If both (**I**= L/E= Leverage or Gearing Ratio= <u>99.00%</u>), (**t**= T/B= Tax Rate= <u>30.00%</u>), (**V**= Variable Cost= <u>9,746</u> millions USD), (**s**= [S/S']-1= Sales Growth= <u>5.00%</u>), (**S'**= Sales of Past Year= <u>48,851</u> millions USD), (**i**= I/S= Interest Portion= <u>1.00%</u>), (**d**= D/A= Dividend Portion or Payout= <u>30.00%</u>), (**U**= Utilized or Starting Capital= <u>64,720</u> millions USD), (**W**= Wealth or Total Assets= <u>139,796</u> millions USD), and (**S**= Sales or Revenues= <u>51,294</u> millions USD), then it's (**f**= Fixed Portion planned), is:

$$f = (S-V-\{W/[1+I]-U\}/\{[1-d][1-t]\})$$
$$/\{S'[1+s]\}-I$$
$$= (51,294-9,746-\{139,796$$
$$/[1+0.9900]-64,720\}$$
$$/\{[1-0.3000][1-0.3000]\}$$
$$/\{48,851[1+0.0500]\}-0.0100$$
$$= \underline{58.00\%}$$

Corporate IFRS-GAAP (B/S-I/S), ISBN 13: **978-1983566332**, ISBN 10: **1983566330**

<u>Law-5360</u>:

If both (I= L/E= Leverage or Gearing Ratio= <u>99.00%</u>), (t= T/B= Tax Rate= <u>30.00%</u>), (V= Variable Cost= <u>9,746</u> millions USD), (s= [S/S']-1= Sales Growth= <u>5.00%</u>), (S'= Sales of Past Year= <u>48,851</u> millions USD), (f= F/S= Fixed Portion= <u>58.00%</u>), (d= D/A= Dividend Portion or Payout= <u>30.00%</u>), (U= Utilized or Starting Capital= <u>64,720</u> millions USD), (W= Wealth or Total Assets= <u>139,796</u> millions USD), and (S= Sales or Revenues= <u>51,294</u> millions USD), then it's (i= Interest Portion planned), is:

$$i = (S\text{-}V\text{-}\{W/[1+I]\text{-}U\}/\{[1\text{-}d][1\text{-}t]\})$$
$$/\{S'[1+s]\}\text{-}f$$
$$= (51,294\text{-}9,746\text{-}\{139,796$$
$$/[1+0.9900]\text{-}64,720\}$$
$$/\{[1\text{-}0.3000][1\text{-}0.3000]\}$$
$$/\{48,851[1+0.0500]\}\text{-}0.5800$$
$$= \underline{1.00}\%$$

Corporate IFRS-GAAP (B/S-I/S), ISBN 13: **978-1983566332**, ISBN 10: **1983566330**

Law-5361:

If both (I= L/E= Leverage or Gearing Ratio=
99.00%), (i= I/S= Interest Portion= 1.00%), (V=
Variable Cost= 9,746 millions USD), (s= [S/S']-1=
Sales Growth= 5.00%), (S'= Sales of Past Year=
48,851 millions USD), (f= F/S= Fixed Portion=
58.00%), (d= D/A= Dividend Portion or Payout=
30.00%), (U= Utilized or Starting Capital= 64,720
millions USD), (W= Wealth or Total Assets=
139,796 millions USD), and (S= Sales or Revenues=
51,294 millions USD), then it's (t= Tax Rate
planned), is:

$$t = 1-\{W/[1+I]-U\}/([1-d]\{S-V-S'[1+s][f+i]\})$$
$$= 1-\{139,796/[1+0.9900]-64,720\}$$
$$/([1-0.3000]\{51,294-9,746$$
$$-48,851[1+0.0500]$$
$$[0.5800+0.0100]\})$$
$$= 30.00\%$$

Corporate IFRS-GAAP (B/S-I/S), ISBN 13: **978-1983566332**, ISBN 10: **1983566330**

<u>Law-5362</u>:

If both ($\textbf{\textit{f}}$= L/E= Leverage or Gearing Ratio= 99.00%), ($\textbf{i}$= I/S= Interest Portion= 1.00%), ($\textbf{V}$= Variable Cost= 9,746 millions USD), ($\textbf{s}$= [S/S']-1= Sales Growth= 5.00%), ($\textbf{S'}$= Sales of Past Year= 48,851 millions USD), ($\textbf{f}$= F/S= Fixed Portion= 58.00%), ($\textbf{t}$= T/B= Tax Rate= 30.00%), ($\textbf{U}$= Utilized or Starting Capital= 64,720 millions USD), ($\textbf{W}$= Wealth or Total Assets= 139,796 millions USD), and ($\textbf{S}$= Sales or Revenues= 51,294 millions USD), then it's ($\textbf{d}$= Dividend Portion or Payout planned), is:

$$\textbf{d}= 1-\{\textbf{W}/[1+\textbf{\textit{f}}]-\textbf{U}\}/([1-\textbf{t}]\{\textbf{S}-\textbf{V}-\textbf{S'}[1+\textbf{s}][\textbf{f}+\textbf{i}]\})$$
$$= 1-\{139,796/[1+0.9900]-64,720\}$$
$$/([1-0.3000]\{51,294-9,746$$
$$-48,851[1+0.0500]$$
$$[0.5800+0.0100]\})$$
$$= 30.00\%$$

Corporate IFRS-GAAP (B/S-I/S), ISBN 13: **978-1983566332**, ISBN 10: **1983566330**

<u>Law-5363</u>:

If both (I= L/E= Leverage or Gearing Ratio= <u>99.00%</u>), (i= I/S= Interest Portion= <u>1.00%</u>), (V= Variable Cost= <u>9,746</u> millions USD), (s= [S/S']-1= Sales Growth= <u>5.00%</u>), (S'= Sales of Past Year= <u>48,851</u> millions USD), (f= F/S= Fixed Portion= <u>58.00%</u>), (t= T/B= Tax Rate= <u>30.00%</u>), (d= D/A= Dividend Portion or Payout= <u>30.00%</u>), (W= Wealth or Total Assets= <u>139,796</u> millions USD), and (S= Sales or Revenues= <u>51,294</u> millions USD), then it's (U= Utilized or Starting Capital planned), is:

$$U= W/[1+I]-[1-d][1-t]\{S-V-S'[1+s][f+i]\}$$
$$= 139,796/[1+0.9900]-[1-0.3000]$$
$$[1-0.3000]\{51,294-9,746$$
$$-48,851[1+0.0500][0.5800$$
$$+0.0100]\}$$
$$= \underline{64,720} \text{ millions USD}$$

<u>Law-5364</u>:

If both (**U**= Utilized or Starting Capital= <u>64,720</u> millions USD), (**i**= I/S= Interest Portion= <u>1.00%</u>), (**V**= Variable Cost= <u>9,746</u> millions USD), (**s**= [S/S']-1= Sales Growth= <u>5.00%</u>), (**S'**= Sales of Past Year= <u>48,851</u> millions USD), (**f**= F/S= Fixed Portion= <u>58.00%</u>), (**t**= T/B= Tax Rate= <u>30.00%</u>), (**d**= D/A= Dividend Portion or Payout= <u>30.00%</u>), (**W**= Wealth or Total Assets= <u>139,796</u> millions USD), and (**S**= Sales or Revenues= <u>51,294</u> millions USD), then it's (**I** = Leverage or Starting Capital planned), is:

$$I = W/(U+[1-d][1-t]\{S-V-S'[1+s][f+i]\})-1$$
$$= 139,796/(64,720+[1-0.3000]$$
$$[1-0.3000]\{51,294-9,746$$
$$-48,851[1+0.0500]$$
$$[0.5800+0.0100]\})-1$$
$$= \underline{99.00\%}$$

Corporate IFRS-GAAP (B/S-I/S), ISBN 13: **978-1983566332**, ISBN 10: **1983566330**

Law-5365:

  If both (**U**= Utilized or Starting Capital= 64,720
  millions USD), (**I**= Interest Expense= 513 millions
  USD), (**F**= Fixed Cost= 29,740 millions USD), (**v**=
  V/S= Variable Portion= 19.00%), (**t**= T/B= Tax
  Rate= 30.00%), (**d**= D/A= Dividend Portion or
  Payout= 30.00%), (**I**= L/E= Leverage or Gearing
  Ratio= 99.00%), and (**$**= Sales or Revenues= 51,294
  millions USD), then it's (**W**= Wealth or Total Assets
  planned), is:

$$W= [1+I]\{U+[1-d][1-t]\{\$-\$v-F-I]\}$$
$$= [1+I](U+[1-d][1-t]\{\$[1-v]-F-I]\})$$
$$= [1+0.9900](64,720+[1-0.3000]$$
$$[1-0.3000]\{51,294[1-0.5800]$$
$$-29,750-513\})$$
$$= 139,796 \text{ millions USD}$$

Corporate IFRS-GAAP (B/S-I/S), ISBN 13: **978-1983566332**, ISBN 10: **1983566330**

<u>Law-5366</u>:

If both (**U**= Utilized or Starting Capital= <u>64,720</u> millions USD), (**I**= Interest Expense= <u>513</u> millions USD), (**F**= Fixed Cost= <u>29,740</u> millions USD), (**v**= V/S= Variable Portion= <u>19.00%</u>), (**t**= T/B= Tax Rate= <u>30.00%</u>), (**d**= D/A= Dividend Portion or Payout= <u>30.00%</u>), (**I**= L/E= Leverage or Gearing Ratio= <u>99.00%</u>), and (**W**= Wealth or Total Assets= <u>139,796</u> millions USD), then it's (**S**= Sales or Revenues planned), is:

$$S= (F+I+\{W/[1+I]-U\}/\{[1-d][1-t]\})/[1-v]$$
$$= (29,750+513+\{139,796$$
$$/[1+0.9900]-64,720\}$$
$$/\{[1-0.3000][1-0.3000]\})$$
$$/[1-0.5800]$$
$$= \underline{51,294} \text{ millions USD}$$

Corporate IFRS-GAAP (B/S-I/S), ISBN 13: **978-1983566332**, ISBN 10: **1983566330**

Law-5367:

If both (**U**= Utilized or Starting Capital= 64,720 millions USD), (**I**= Interest Expense= 513 millions USD), (**F**= Fixed Cost= 29,740 millions USD), (**$**= Sales or Revenues= 51,294 millions USD), (**t**= T/B= Tax Rate= 30.00%), (**d**= D/A= Dividend Portion or Payout= 30.00%), (**l**= L/E= Leverage or Gearing Ratio= 99.00%), and (**W**= Wealth or Total Assets= 139,796 millions USD), then it's (**$**= Sales or Revenues planned), is:

$$\upsilon = 1-(\mathbf{F}+\mathbf{I}+\{\mathbf{W}/[1+\mathbf{I}]-\mathbf{U}\}/\{[1-\mathbf{d}][1-\mathbf{t}]\})/\mathbf{\$}$$
$$= 1-(29,750+513+\{139,796$$
$$/[1+0.9900]-64,720\}$$
$$/\{[1-0.3000][1-0.3000]\})$$
$$/51,294$$
$$= \underline{19.00\%}$$

Corporate IFRS-GAAP (B/S-I/S), ISBN 13: **978-1983566332**, ISBN 10: **1983566330**

Law-5368:

If both (**U**= Utilized or Starting Capital= 64,720 millions USD), (**I**= Interest Expense= 513 millions USD), (**v**= V/S= Variable Portion= 19.00%), (**$**= Sales or Revenues= 51,294 millions USD), (**t**= T/B= Tax Rate= 30.00%), (**d**= D/A= Dividend Portion or Payout= 30.00%), (**I**= L/E= Leverage or Gearing Ratio= 99.00%), and (**W**= Wealth or Total Assets= 139,796 millions USD), then it's (**F**= Fixed Portion planned), is:

$$F= \$[1\text{-}v]\text{-}I\text{-}\{W/[1\text{+}I]\text{-}U\}/\{[1\text{-}d][1\text{-}t]\}$$
$$= 51{,}294[1\text{-}0.1900]\text{-}513\text{-}\{139{,}796$$
$$/[1\text{+}0.9900]\text{-}64{,}720\}$$
$$/\{[1\text{-}0.3000][1\text{-}0.3000]\}$$
$$= 29{,}750 \text{ millions USD}$$

Corporate IFRS-GAAP (B/S-I/S), ISBN 13: **978-1983566332**, ISBN 10: **1983566330**

Law-5369:

If both (**U**= Utilized or Starting Capital= 64,720 millions USD), (**F**= Fixed Cost= 29,740 millions USD), (**v**= V/S= Variable Portion= 19.00%), (**S**= Sales or Revenues= 51,294 millions USD), (**t**= T/B= Tax Rate= 30.00%), (**d**= D/A= Dividend Portion or Payout= 30.00%), (**l**= L/E= Leverage or Gearing Ratio= 99.00%), and (**W**= Wealth or Total Assets= 139,796 millions USD), then it's (**I**= Interest Expense planned), is:

$$I = S[1-v] - F - \{W/[1+l] - U\} / \{[1-d][1-t]\}$$
$$= 51,294[1-0.1900] - 29,750 - \{139,796 /[1+0.9900] - 64,720\} /\{[1-0.3000][1-0.3000]\}$$
$$= \underline{513 \text{ millions USD}}$$

Corporate IFRS-GAAP (B/S-I/S), ISBN 13: **978-1983566332**, ISBN 10: **1983566330**

Law-5370:

If both (**U**= Utilized or Starting Capital= 64,720 millions USD), (**F**= Fixed Cost= 29,740 millions USD), (**v**= V/S= Variable Portion= 19.00%), (**$**= Sales or Revenues= 51,294 millions USD), (**I**= Interest Expense= 513 millions USD), (**d**= D/A= Dividend Portion or Payout= 30.00%), (**I**= L/E= Leverage or Gearing Ratio= 99.00%), and (**W**= Wealth or Total Assets= 139,796 millions USD), then it's (**t**= Tax Rate planned), is:

$$t = 1-\{W/[1+I]-U\}/([1-d]\{\$[1-v]-F-I\})$$
$$= 1-\{139,796/[1+0.9900]-64,720\}$$
$$/([1-0.3000]\{51,294$$
$$[1-0.1900]-29,750-513\})$$
$$= 30.00\%$$

Corporate IFRS-GAAP (B/S-I/S), ISBN 13: **978-1983566332**, ISBN 10: **1983566330**

Law-5371:

If both (**U**= Utilized or Starting Capital= 64,720 millions USD), (**F**= Fixed Cost= 29,740 millions USD), (**v**= V/S= Variable Portion= 19.00%), (**S**= Sales or Revenues= 51,294 millions USD), (**I**= Interest Expense= 513 millions USD), (**t**= T/B= Tax Rate= 30.00%), (**I**= L/E= Leverage or Gearing Ratio= 99.00%), and (**W**= Wealth or Total Assets= 139,796 millions USD), then it's (**d**= Dividend Portion or Payout planned), is:

$$d= 1-\{W/[1+I]-U\}/([1-t]\{S[1-v]-F-I\})$$
$$= 1-\{139,796/[1+0.9900]-64,720\}$$
$$/([1-0.3000]\{51,294$$
$$[1-0.1900]-29,750-513\})$$
$$= 30.00\%$$

Corporate IFRS-GAAP (B/S-I/S), ISBN 13: **978-1983566332**, ISBN 10: **1983566330**

Law-5372:

If both (**d**= D/A= Dividend Portion or Payout=
30.00%), (**F**= Fixed Cost= 29,740 millions USD), (**v**=
V/S= Variable Portion= 19.00%), (**$**= Sales or
Revenues= 51,294 millions USD), (**I**= Interest
Expense= 513 millions USD), (**t**= T/B= Tax Rate=
30.00%), (**l**= L/E= Leverage or Gearing Ratio=
99.00%), and (**W**= Wealth or Total Assets= 139,796
millions USD), then it's (**U**= Utilized or Starting
Capital planned), is:

$$U= W/[1+I]-[1-d][1-t]\{\$[1-v]-F-I\}$$
$$= 139,796/[1+0.9900]-[1-0.3000]$$
$$[1-0.3000]\{51,294$$
$$[1-0.1900]-29,750-513\}$$
$$= 64,720 \text{ millions USD}$$

Corporate IFRS-GAAP (B/S-I/S), ISBN 13: **978-1983566332**, ISBN 10: **1983566330**

Law-5373:

If both (**d**= D/A= Dividend Portion or Payout= 30.00%), (**F**= Fixed Cost= 29,740 millions USD), (**v**= V/S= Variable Portion= 19.00%), (**S**= Sales or Revenues= 51,294 millions USD), (**I**= Interest Expense= 513 millions USD), (**t**= T/B= Tax Rate= 30.00%), (**U**= Utilized or Starting Capital= 64,720 millions USD), and (**W**= Wealth or Total Assets= 139,796 millions USD), then it's (**I** = Leverage or Gearing Ratio planned), is:

$$I = W/(U+[1-d][1-t]\{S[1-v]-F-I\})-1$$
$$= 139,796/(64,720+[1-0.3000]$$
$$[1-0.3000]\{51,294$$
$$[1-0.1900]-29,750-513\})-1$$
$$= 99.00\%$$

Corporate IFRS-GAAP (B/S-I/S), ISBN 13: **978-1983566332**, ISBN 10: **1983566330**

Law-5374:

If both (**d**= D/A= Dividend Portion or Payout= 30.00%), (**F**= Fixed Cost= 29,740 millions USD), (**v**= V/S= Variable Portion= 19.00%), (**$**= Sales or Revenues= 51,294 millions USD), (**i**= I/S= Interest Portion= 1.00%), (**t**= T/B= Tax Rate= 30.00%), (**U**= Utilized or Starting Capital= 64,720 millions USD), and (**l**= L/E= Leverage or Gearing Ratio= 99.00%), then it's (**W** = Wealth or Total Assets planned), is:

$$W= [1+l](U+[1-d][1-t]\{\$-\$v-F-\$i\})$$

$$=[1+l](U+[1-d][1-t]\{\$[1-v-i]-F\})$$

$$= [1+0.9900](64,720+[1-0.3000]$$
$$[1-0.3000]\{51,294$$
$$[1-0.1900-0.0100]-29,750\})$$

$$= 139,796 \text{ millions USD}$$

Corporate IFRS-GAAP (B/S-I/S), ISBN 13: **978-1983566332**, ISBN 10: **1983566330**

Law-5375:

If both (**d**= D/A=Dividend Portion or Payout= 30.00%), (**F**= Fixed Cost= 29,740 millions USD), (**v**= V/S= Variable Portion= 19.00%), (**W**= Wealth or Total Assets= 139,796 millions USD), (**i**= I/S= Interest Portion= 1.00%), (**t**= T/B= Tax Rate= 30.00%), (**U**= Utilized or Starting Capital= 64,720 millions USD), and (**l**= L/E= Leverage or Gearing Ratio= 99.00%), then it's (**S**= Sales or Revenues planned), is:

$$S= (F+[1+l]-U\}/\{[1-d][1-t]\})/[1-v-i]$$
$$= (29,750+\{139,796/[1+0.9900]$$
$$-64,720\}/\{[1-0.3000]$$
$$[1-0.3000]\})/[1-0.1900$$
$$-0.0100]$$
$$= \underline{51,294} \text{ millions USD}$$

Corporate IFRS-GAAP (B/S-I/S), ISBN 13: **978-1983566332**, ISBN 10: **1983566330**

<u>Law-5376</u>:

If both (**d**= D/A=Dividend Portion or Payout= <u>30.00%</u>), (**F**= Fixed Cost= <u>29,740</u> millions USD), (**$**= Sales or Revenues= <u>51,294</u> millions USD), (**W**= Wealth or Total Assets= <u>139,796</u> millions USD), (**i**= I/S= Interest Portion= <u>1.00%</u>), (**t**= T/B= Tax Rate= <u>30.00%</u>), (**U**= Utilized or Starting Capital= <u>64,720</u> millions USD), and (**/**= L/E= Leverage or Gearing Ratio= <u>99.00%</u>), then it's (**v**= Variable Portion planned), is:

$$v= 1-(F+\{W/[1+/\ ]-U\}/\{[1-d][1-t]\})/\$-I$$
$$= 1-(29,750+\{139,796/[1+0.9900]$$
$$-64,720\}/\{[1-0.3000]$$
$$[1-0.3000]\})/51,294$$
$$-0.0100$$
$$= \underline{19.00\%}$$

Law-5377:

If both (**d**= D/A= Dividend Portion or Payout= 30.00%), (**F**= Fixed Cost= 29,740 millions USD), (**S**= Sales or Revenues= 51,294 millions USD), (**W**= Wealth or Total Assets= 139,796 millions USD), (**v**= V/S= Variable Portion= 19.00%), (**t**= T/B= Tax Rate= 30.00%), (**U**= Utilized or Starting Capital= 64,720 millions USD), and (**l**= L/E= Leverage or Gearing Ratio= 99.00%), then it's (**i**= Interest Portion planned), is:

$$i= 1-(F+\{W/[1+l]-U\}/\{[1-d][1-t]\})/S-v$$
$$= 1-(29,750+\{139,796/[1+0.9900]$$
$$-64,720\}/\{[1-0.3000]$$
$$[1-0.3000]\})/51,294-0.1900$$
$$= 1.00\%$$

Corporate IFRS-GAAP (B/S-I/S), ISBN 13: **978-1983566332**, ISBN 10: **1983566330**

Law-5378:

If both (**d**= D/A= Dividend Portion or Payout= 30.00%), (**i**= I/S= Interest Portion= 1.00%), (**S**= Sales or Revenues= 51,294 millions USD), (**W**= Wealth or Total Assets= 139,796 millions USD), (**v**= V/S= Variable Portion= 19.00%), (**t**= T/B= Tax Rate= 30.00%), (**U**= Utilized or Starting Capital= 64,720 millions USD), and (**I**= L/E= Leverage or Gearing Ratio= 99.00%), then it's (**F**= Fixed Cost planned), is:

$$F= S[1-v-i]-\{W/[1+I]-U\}/\{[1-d][1-t]\}$$
$$= 51,294[1-0.1900-0.0100]$$
$$-\{139,796/[1+0.9900]$$
$$-64,720\}/\{[1-0.3000]$$
$$[1-0.3000]\}$$
$$= 29,750 \text{ millions USD}$$

Corporate IFRS-GAAP (B/S-I/S), ISBN 13: **978-1983566332**, ISBN 10: **1983566330**

Law-5379:

If both (**d**= D/A=Dividend Portion or Payout= 30.00%), (**i**= I/S= Interest Portion= 1.00%), (**$**= Sales or Revenues= 51,294 millions USD), (**W**= Wealth or Total Assets= 139,796 millions USD), (**v**= V/S= Variable Portion= 19.00%), (**F**= Fixed Cost= 29,740 millions USD), (**U**= Utilized or Starting Capital= 64,720 millions USD), and (**l**= L/E= Leverage or Gearing Ratio= 99.00%), then it's (**t**= Tax Rate planned), is:

$$t = 1-\{\mathbf{W}/[1+\mathbf{l}]-\mathbf{U}\}/([1-\mathbf{d}]\{\mathbf{\$}[1-\mathbf{v}-\mathbf{i}]-\mathbf{F}\})$$
$$= 1-\{139,796/[1+0.9900]-64,720\}$$
$$/([1-0.3000]\{51,294$$
$$[1-0.1900-0.0100]$$
$$-29,750\})$$
$$= 30.00\%$$

Corporate IFRS-GAAP (B/S-I/S), ISBN 13: **978-1983566332**, ISBN 10: **1983566330**

Law-5380:

If both (**t**= T/B= Tax Rate= 30.00%), (**i**= I/S= Interest Portion= 1.00%), (**S**= Sales or Revenues= 51,294 millions USD), (**W**= Wealth or Total Assets= 139,796 millions USD), (**v**= V/S= Variable Portion= 19.00%), (**F**= Fixed Cost= 29,740 millions USD), (**U**= Utilized or Starting Capital= 64,720 millions USD), and (**l**= L/E= Leverage or Gearing Ratio= 99.00%), then it's (**d**= Dividend Portion planned), is:

$$d = 1 - \{W/[1+l]-U\}/([1-t]\{S[1-v-i]-F\})$$
$$= 1 - \{139,796/[1+0.9900]-64,720\}$$
$$/([1-0.3000]\{51,294$$
$$[1-0.1900-0.0100]$$
$$-29,750\})$$
$$= 30.00\%$$

Corporate IFRS-GAAP (B/S-I/S), ISBN 13: **978-1983566332**, ISBN 10: **1983566330**

Law-5381:

If both (**t**= T/B= Tax Rate= 30.00%), (**i**= I/S= Interest Portion= 1.00%), (**$**= Sales or Revenues= 51,294 millions USD), (**W**= Wealth or Total Assets= 139,796 millions USD), (**v**= V/S= Variable Portion= 19.00%), (**F**= Fixed Cost= 29,740 millions USD), (**d**= D/A=Dividend Portion or Payout= 30.00%), and (**l**= L/E= Leverage or Gearing Ratio= 99.00%), then it's (**U**= Utilized or Starting Capital planned), is:

$$U= W/[1+l]-[1-d][1-t]\{\$[1-v-i]-F\}$$
$$= 139,796/[1+0.9900]-[1-0.3000]$$
$$[1-0.3000]\{51,294[1-0.1900$$
$$-0.0100]-29,750\}$$
$$= 64,720 \text{ millions USD}$$

Corporate IFRS-GAAP (B/S-I/S), ISBN 13: **978-1983566332**, ISBN 10: **1983566330**

Law-5382:

If both (**t**= T/B= Tax Rate= 30.00%), (**i**= I/S= Interest Portion= 1.00%), (**$**= Sales or Revenues= 51,294 millions USD), (**W**= Wealth or Total Assets= 139,796 millions USD), (**v**= V/S= Variable Portion= 19.00%), (**F**= Fixed Cost= 29,740 millions USD), (**d**= D/A= Dividend Portion or Payout= 30.00%), and (**U**= Utilized or Starting Capital= 64,720 millions USD), then it's (**I** = Leverage or Gearing Ratio planned), is:

$$I = W/(U+[1-d][1-t]\{\$[1-v-i]-F\})-1$$
$$= 139,796/(64,720+[1-0.3000]$$
$$[1-0.3000]\{51,294$$
$$[1-0.1900-0.0100]$$
$$-29,750\})-1$$
$$= \underline{99.00\%}$$

Corporate IFRS-GAAP (B/S-I/S), ISBN 13: **978-1983566332**, ISBN 10: **1983566330**

<u>Law-5383</u>:

If both (**t**= T/B= Tax Rate= <u>30.00%</u>), (**i**= I/S= Interest Portion= <u>1.00%</u>), (**$'**= Sales of Past Year= <u>48,851</u> millions USD), (**s**= [S/S']-1= Sales Growth= <u>5.00%</u>), (**$**= Sales or Revenues= <u>51,294</u> millions USD), (**l**= L/E= Leverage or Gearing Ratio= <u>99.00%</u>), (**v**= V/S= Variable Portion= <u>19.00%</u>), (**F**= Fixed Cost= <u>29,740</u> millions USD), (**d**= D/A= Dividend Portion or Payout= <u>30.00%</u>), and (**U**= Utilized or Starting Capital= <u>64,720</u> millions USD), then it's (**W**= Wealth or Total Assets planned), is:

$$W= [1+l](U+[1-d][1-t]\{\$-\$v-i]-F-\$'i[1+s]\})$$
$$= [1+l](U+[1-d][1-t]$$
$$\{\$[1-v]-F-\$'i[1+s]\})$$
$$= [1+0.9900](64,720+[1-0.3000]$$
$$[1-0.3000]\{51,294[1-0.1900]$$
$$-29,750-48,851*0.0100$$
$$[1+0.0500]\})$$
$$= \underline{139,796} \text{ millions USD}$$

Corporate IFRS-GAAP (B/S-I/S), ISBN 13: **978-1983566332**, ISBN 10: **1983566330**

<u>Law-5384</u>:

If both (**t**= T/B= Tax Rate= <u>30.00%</u>), (**i**= I/S= Interest Portion= <u>1.00%</u>), (**$'**= Sales of Past Year= <u>48,851</u> millions USD), (**s**= [S/S']-1= Sales Growth= <u>5.00%</u>), (**W**= Wealth or Total Assets= <u>139,796</u> millions USD), (**l**= L/E= Leverage or Gearing Ratio= <u>99.00%</u>), (**v**= V/S= Variable Portion= <u>19.00%</u>), (**F**= Fixed Cost= <u>29,740</u> millions USD), (**d**= D/A=Dividend Portion or Payout= <u>30.00%</u>), and (**U**= Utilized or Starting Capital= <u>64,720</u> millions USD), then it's (**$**= Sales or Revenues planned), is:

$$\begin{aligned}
\$ &= (\mathbf{F}+\$'i[1+\mathbf{s}]+\{\mathbf{W}/[1+\mathbf{l}]-\mathbf{U}\}/\{[1-\mathbf{d}][1-\mathbf{t}]\}) \\
&\quad /[1-\mathbf{v}] \\
&= (29,750+48,851*0.0100[1+0.0500] \\
&\quad +\{139,796/[1+0.9900] \\
&\quad -64,720\}/\{[1-0.3000] \\
&\quad [1-0.3000]\})/[1-0.1900] \\
&= \underline{51,294} \text{ millions USD}
\end{aligned}$$

Corporate IFRS-GAAP (B/S-I/S), ISBN 13: **978-1983566332**, ISBN 10: **1983566330**

<u>Law-5385</u>:

If both (**t**= T/B= Tax Rate= <u>30.00%</u>), (**i**= I/S= Interest Portion= <u>1.00%</u>), (**$'**= Sales of Past Year= <u>48,851</u> millions USD), (**s**= [S/S']-1= Sales Growth= <u>5.00%</u>), (**W**= Wealth or Total Assets= <u>139,796</u> millions USD), (**l**= L/E= Leverage or Gearing Ratio= <u>99.00%</u>), (**$**= Sales or Revenues= <u>51,294</u> millions USD), (**F**= Fixed Cost= <u>29,740</u> millions USD), (**d**= D/A= Dividend Portion or Payout= <u>30.00%</u>), and (**U**= Utilized or Starting Capital= <u>64,720</u> millions USD), then it's (**v**= Variable Portion planned), is:

$$v = 1-(F+\$'i[1+s]+\{W/[1+l]-U\}$$
$$/\{[1-d][1-t]\})/\$$$
$$= 1-(29,750+48,851*0.0100$$
$$[1+0.0500]+\{139,796$$
$$/[1+0.9900]-64,720\}$$
$$/\{[1-0.3000][1-0.3000]\})$$
$$/51,294$$
$$= \underline{19.00\%}$$

Corporate IFRS-GAAP (B/S-I/S), ISBN 13: **978-1983566332**, ISBN 10: **1983566330**

<u>Law-5386</u>:

If both (**t**= T/B= Tax Rate= <u>30.00%</u>), (**i**= I/S= Interest Portion= <u>1.00%</u>), (**S'**= Sales of Past Year= <u>48,851</u> millions USD), (**s**= [S/S']-1= Sales Growth= <u>5.00%</u>), (**W**= Wealth or Total Assets= <u>139,796</u> millions USD), (**l**= L/E= Leverage or Gearing Ratio= <u>99.00%</u>), (**S**= Sales or Revenues= <u>51,294</u> millions USD), (**v**= V/S= Variable Portion= <u>19.00%</u>), (**d**= D/A= Dividend Portion or Payout= <u>30.00%</u>), and (**U**= Utilized or Starting Capital= <u>64,720</u> millions USD), then it's (**F**= Fixed Cost planned), is:

$$\mathbf{F} = \mathbf{S}[1\text{-}\mathbf{v}]\text{-}\mathbf{S'i}[1\text{+}\mathbf{s}]\text{-}\{\mathbf{W}/[1\text{+}\mathbf{l}]\text{-}\mathbf{U}\}/\{[1\text{-}\mathbf{d}][1\text{-}\mathbf{t}]\}$$

$$= 51,294[1\text{-}0.1900]\text{-} \ 48,851*0.0100$$
$$[1\text{+}0.0500]\text{-}\{139,796$$
$$/[1\text{+}0.9900]\text{-}64,720\}$$
$$/\{[1\text{-}0.3000][1\text{-}0.3000]\}$$

$$= \underline{29,750} \text{ millions USD}$$

Corporate IFRS-GAAP (B/S-I/S), ISBN 13: **978-1983566332**, ISBN 10: **1983566330**

Law-5387:

If both (**t**= T/B= Tax Rate= 30.00%), (**i**= I/S= Interest Portion= 1.00%), (**F**= Fixed Cost= 29,740 millions USD), (**s**= [S/S']-1= Sales Growth= 5.00%), (**W**= Wealth or Total Assets= 139,796 millions USD), (**l**= L/E= Leverage or Gearing Ratio= 99.00%), (**S**= Sales or Revenues= 51,294 millions USD), (**v**= V/S= Variable Portion= 19.00%), (**d**= D/A= Dividend Portion or Payout= 30.00%), and (**U**= Utilized or Starting Capital= 64,720 millions USD), then it's (**S'**= Sales Past), must be:

$$S' = (S[1-v]-F-\{W/[1+l]-U\}/\{[1-d][1-t]\})/\{i[1+s]\}$$

$$= (51,294[1-0.1900]-29,750-\{139,796/[1+0.9900]-64,720\}/\{[1-0.3000][1-0.3000]\})/\{0.0100[1+0.0500]\}$$

$$= 48,851 \text{ millions USD}$$

Corporate IFRS-GAAP (B/S-I/S), ISBN 13: **978-1983566332**, ISBN 10: **1983566330**

Law-5388:

If both (**t**= T/B= Tax Rate= 30.00%), (**$'**= Sales of Past Year= 48,851 millions USD), (**F**= Fixed Cost= 29,740 millions USD), (**s**= [S/S']-1= Sales Growth= 5.00%), (**W**= Wealth or Total Assets= 139,796 millions USD), (**l**= L/E= Leverage or Gearing Ratio= 99.00%), (**$**= Sales or Revenues= 51,294 millions USD), (**v**= V/S= Variable Portion= 19.00%), (**d**= D/A= Dividend Portion or Payout= 30.00%), and (**U**= Utilized or Starting Capital= 64,720 millions USD), then it's (**i**= Interest Portion planned), is:

$$i= (\$[1\text{-}\textbf{v}]\text{-}\textbf{F}\text{-}\{\textbf{W}/[1+\textbf{l}]\text{-}\textbf{U}\}/\{[1\text{-}\textbf{d}][1\text{-}\textbf{t}]\})$$
$$/\{\$'[1+\textbf{s}]\}$$
$$= (51,294[1\text{-}0.1900]\text{-}29,750$$
$$-\{139,796/[1+0.9900]$$
$$-64,720\}/\{[1\text{-}0.3000]$$
$$[1\text{-}0.3000]\})$$
$$/\{48,851[1+0.0500]\}$$
$$= 1.00\%$$

Corporate IFRS-GAAP (B/S-I/S), ISBN 13: **978-1983566332**, ISBN 10: **1983566330**

Law-5389:

If both (**t**= T/B= Tax Rate= 30.00%), (**$'**= Sales of Past Year= 48,851 millions USD), (**F**= Fixed Cost= 29,740 millions USD), (**i**= I/S= Interest Portion= 1.00%), (**W**= Wealth or Total Assets= 139,796 millions USD), (**l**= L/E= Leverage or Gearing Ratio= 99.00%), (**$**= Sales or Revenues= 51,294 millions USD), (**v**= V/S= Variable Portion= 19.00%), (**d**= D/A= Dividend Portion or Payout= 30.00%), and (**U**= Utilized or Starting Capital= 64,720 millions USD), then it's (**s**= Sales Growth planned), is:

$$s= (\$[1\text{-}v]\text{-}F\text{-}\{W/[1+l]\text{-}U\}/\{[1\text{-}d][1\text{-}t]\})$$
$$/[\$'i]\text{-}1$$
$$= (51{,}294[1\text{-}0.1900]\text{-}29{,}750$$
$$-\{139{,}796/[1+0.9900]$$
$$-64{,}720\}/\{[1\text{-}0.3000]$$
$$[1\text{-}0.3000]\})/[48{,}851$$
$$*0.0100]\text{-}1$$
$$= 5.00\%$$

Corporate IFRS-GAAP (B/S-I/S), ISBN 13: **978-1983566332**, ISBN 10: **1983566330**

Law-5390:

If both (**s**= [S/S']-1= Sales Growth= 5.00%), (**S'**= Sales of Past Year= 48,851 millions USD), (**F**= Fixed Cost= 29,740 millions USD), (**i**= I/S= Interest Portion= 1.00%), (**W**= Wealth or Total Assets= 139,796 millions USD), (**l**= L/E= Leverage or Gearing Ratio= 99.00%), (**S**= Sales or Revenues= 51,294 millions USD), (**v**= V/S= Variable Portion= 19.00%), (**d**= D/A=Dividend Portion or Payout= 30.00%), and (**U**= Utilized or Starting Capital= 64,720 millions USD), then it's (**t**= Tax Rate planned), is:

$$s= 1-\{W/[1+l]-U\}/([1-d]\{S[1-v]$$
$$-F-S'i[1+s]\})$$
$$= 1-\{139,796/[1+0.9900]-64,720\}$$
$$/([1-0.3000]\{51,294$$
$$[1-0.1900]-29,750-48,851$$
$$*0.0100[1+0.0500]\})$$
$$= 30.00\%$$

Corporate IFRS-GAAP (B/S-I/S), ISBN 13: **978-1983566332**, ISBN 10: **1983566330**

Law-5391:

If both (**s**= [S/S']-1= Sales Growth= 5.00%), (**S'**= Sales of Past Year= 48,851 millions USD), (**F**= Fixed Cost= 29,740 millions USD), (**i**= I/S= Interest Portion= 1.00%), (**W**= Wealth or Total Assets= 139,796 millions USD), (**l**= L/E= Leverage or Gearing Ratio= 99.00%), (**S**= Sales or Revenues= 51,294 millions USD), (**v**= V/S= Variable Portion= 19.00%), (**t**= T/B= Tax Rate= 30.00%), and (**U**= Utilized or Starting Capital= 64,720 millions USD), then it's (**d**= Dividend Portion or Payout planned), is:

$$\mathbf{d}= 1-\{\mathbf{W}/[1+\mathbf{l}]-\mathbf{U}\}/([1-\mathbf{t}]\{\mathbf{S}[1-\mathbf{v}]-\mathbf{F}-\mathbf{S'i}[1+\mathbf{s}]\})$$

$$= 1-\{139,796/[1+0.9900]-64,720\}$$
$$/([1-0.3000]\{51,294$$
$$[1-0.1900]-29,750-48,851$$
$$*0.0100[1+0.0500]\})$$

$$= 30.00\%$$

Corporate IFRS-GAAP (B/S-I/S), ISBN 13: **978-1983566332**, ISBN 10: **1983566330**

Law-5392:

If both (s= [S/S']-1= Sales Growth= 5.00%), (S'=
Sales of Past Year= 48,851 millions USD), (F= Fixed
Cost= 29,740 millions USD), (i= I/S= Interest
Portion= 1.00%), (W= Wealth or Total Assets=
139,796 millions USD), (I= L/E= Leverage or
Gearing Ratio= 99.00%), (S= Sales or Revenues=
51,294 millions USD), (v= V/S= Variable Portion=
19.00%), (t= T/B= Tax Rate= 30.00%), and (d=
D/A= Dividend Portion or Payout= 30.00%), then it's
(U = Utilized or Starting Capital planned), is:

$$U= W/[1+I]-[1-d][1-t]\{S[1-v]-F-S'i[1+s]\}$$
$$= 139,796/[1+0.9900]-[1-0.3000]$$
$$[1-0.3000]\{51,294$$
$$[1-0.1900]-29,750-48,851$$
$$*0.0100[1+0.0500]\}$$
$$= 64,720 \text{ millions USD}$$

Corporate IFRS-GAAP (B/S-I/S), ISBN 13: **978-1983566332**, ISBN 10: **1983566330**

<u>Law-5393</u>:

If both (**s**= [S/S']-1= Sales Growth= <u>5.00%</u>), (**S'**= Sales of Past Year= <u>48,851</u> millions USD), (**F**= Fixed Cost= <u>29,740</u> millions USD), (**i**= I/S= Interest Portion= <u>1.00%</u>), (**W**= Wealth or Total Assets= <u>139,796</u> millions USD), (**U**= Utilized or Starting Capital= <u>64,720</u> millions USD), (**S**= Sales or Revenues= <u>51,294</u> millions USD), (**v**= V/S= Variable Portion= <u>19.00%</u>), (**t**= T/B= Tax Rate= <u>30.00%</u>), and (**d**= D/A= Dividend Portion or Payout= <u>30.00%</u>), then it's (**l** = Leverage or Gearing Ratio planned), is:

$$\mathbf{l} = \mathbf{W}/(\mathbf{U}+[1-\mathbf{d}][1-\mathbf{t}]\{\mathbf{S}[1-\mathbf{v}]-\mathbf{F}-\mathbf{S'i}[1+\mathbf{s}]\})-1$$
$$= 139,796/(64,720+[1-0.3000]$$
$$[1-0.3000]\{51,294$$
$$[1-0.1900]-29,750-48,851$$
$$*0.0100[1+0.0500]\})-1$$
$$= \underline{99.00\%}$$

Corporate IFRS-GAAP (B/S-I/S), ISBN 13: **978-1983566332**, ISBN 10: **1983566330**

Law-5394:

If (f= F/S= Fixed Portion= 58.00%), (I= Interest
Expense= 513 millions USD), (l= L/E= Leverage or
Gearing Ratio= 99.00%), (U= Utilized or Starting
Capital= 64,720 millions USD), (S= Sales or
Revenues= 51,294 millions USD), (v= V/S= Variable
Portion= 19.00%), (t= T/B= Tax Rate= 30.00%), and
(d= D/A= Dividend Portion or Payout= 30.00%),
then it's (W= Wealth or Total Assesments planned),
is:

$$W = [1+l]\{U+[1-d][1-t][S-Sv-Sf-I]\}$$
$$= [1+l](U+[1-d][1-t]\{S[1-v-f]-I\})$$
$$= [1+0.9900](64,720+[1-0.3000]$$
$$[1-0.3000]\{51,294$$
$$[1-0.1900-0.5800]-513\})$$
$$= 139,796 \text{ millions USD}$$

Corporate IFRS-GAAP (B/S-I/S), ISBN 13: **978-1983566332**, ISBN 10: **1983566330**

Law-5395:

If (**f**= F/S= Fixed Portion= 58.00%), (**I**= Interest
Expense= 513 millions USD), (**I**= L/E= Leverage or
Gearing Ratio= 99.00%), (**U**= Utilized or Starting
Capital= 64,720 millions USD), (**W**= Wealth or Total
Assets= 139,796 millions USD), (**v**= V/S= Variable
Portion= 19.00%), (**t**= T/B= Tax Rate= 30.00%), and
(**d**= D/A= Dividend Portion or Payout= 30.00%),
then it's (**S**= Sales or Revenues planned), is:

$$S= (I+\{W/[1+I]-U\}/\{[1-d][1-t]\})/[1-v-f]$$

$$= (513+\{139,796/[1+0.9900]$$
$$-64,720\}/\{[1-0.3000]$$
$$[1-0.3000]\})$$
$$/[1-0.1900-0.5800]$$

$$= 51,294 \text{ millions USD}$$

Corporate IFRS-GAAP (B/S-I/S), ISBN 13: **978-1983566332**, ISBN 10: **1983566330**

Law-5396:

If (**f**= F/S= Fixed Portion= 58.00%), (**I**= Interest
Expense= 513 millions USD), (**I**= L/E= Leverage or
Gearing Ratio= 99.00%), (**U**= Utilized or Starting
Capital= 64,720 millions USD), (**W**= Wealth or Total
Assets= 139,796 millions USD), (**$**= Sales or
Revenues= 51,294 millions USD), (**t**= T/B= Tax
Rate= 30.00%), and (**d**= D/A= Dividend Portion or
Payout= 30.00%), then it's (**v**= Variable Portion
planned), is:

$$v= 1\text{-}f\text{-}(I+\{W/[1+I]\text{-}U\}/\{[1\text{-}d][1\text{-}t]\})/\$$$
$$= 1\text{-}0.5800\text{-}(513+\{139,796$$
$$/[1+0.9900]\text{-}64,720\}$$
$$/\{[1\text{-}0.3000][1\text{-}0.3000]\})$$
$$/51,294$$

$$= 19.00\%$$

Corporate IFRS-GAAP (B/S-I/S), ISBN 13: **978-1983566332**, ISBN 10: **1983566330**

<u>Law-5397</u>:

If (**v**= V/S= Variable Portion= <u>19.00%</u>), (**I**= Interest Expense= <u>513</u> millions USD), (**I**= L/E= Leverage or Gearing Ratio= <u>99.00%</u>), (**U**= Utilized or Starting Capital= <u>64,720</u> millions USD), (**W**= Wealth or Total Assets= <u>139,796</u> millions USD), (**S**= Sales or Revenues= <u>51,294</u> millions USD), (**t**= T/B= Tax Rate= <u>30.00%</u>), and (**d**= D/A= Dividend Portion or Payout= <u>30.00%</u>), then it's (**f**= Fixed Portion planned), is:

$$f= 1-v-(I+\{W/[1+I]-U\}/\{[1-d][1-t]\})/S$$
$$= 1-0.1900-(513+\{139,796$$
$$/[1+0.9900]-64,720\}$$
$$/\{[1-0.3000][1-0.3000]\})$$
$$/51,294$$
$$= \underline{58.00\%}$$

Corporate IFRS-GAAP (B/S-I/S), ISBN 13: **978-1983566332**, ISBN 10: **1983566330**

<u>Law-5398</u>:

If (v= V/S= Variable Portion= <u>19.00%</u>), (f= F/S= Fixed Portion= <u>58.00%</u>), (I= L/E= Leverage or Gearing Ratio= <u>99.00%</u>), (U= Utilized or Starting Capital= <u>64,720</u> millions USD), (W= Wealth or Total Assets= <u>139,796</u> millions USD), (S= Sales or Revenues= <u>51,294</u> millions USD), (t= T/B= Tax Rate= <u>30.00%</u>), and (d= D/A= Dividend Portion or Payout= <u>30.00%</u>), then it's (I= Interest Expense planned), is:

$$I = S[1-v-f]-\{W/[1+I]-U\}/\{[1-d][1-t]\}$$
$$= 51,294[1-0.1900-0.5800]$$
$$-\{139,796/[1+0.9900]$$
$$-64,720\}/\{[1-0.3000]$$
$$[1-0.3000]\}$$
$$= \underline{513} \text{ millions USD}$$

Corporate IFRS-GAAP (B/S-I/S), ISBN 13: **978-1983566332**, ISBN 10: **1983566330**

Law-5399:

If (v= V/S= Variable Portion= 19.00%), (f= F/S= Fixed Portion= 58.00%), (l= L/E= Leverage or Gearing Ratio= 99.00%), (U= Utilized or Starting Capital= 64,720 millions USD), (W= Wealth or Total Assets= 139,796 millions USD), (S= Sales or Revenues= 51,294 millions USD), (I= Interest Expense= 513 millions USD), and (d= D/A= Dividend Portion or Payout= 30.00%), then it's (t= Tax Rate planned), is:

$$t = 1 - \{W/[1+l]-U\}/([1-d]\{S[1-v-f]-I\})$$
$$= 1 - \{139{,}796/[1+0.9900]-64{,}720\}$$
$$/([1-0.3000]\{51{,}294$$
$$[1-0.1900-0.5800]-513\})$$
$$= 30.00\%$$

Corporate IFRS-GAAP (B/S-I/S), ISBN 13: **978-1983566332**, ISBN 10: **1983566330**

Law-5400:

If (v= V/S= Variable Portion= 19.00%), (f= F/S= Fixed Portion= 58.00%), (l= L/E= Leverage or Gearing Ratio= 99.00%), (U= Utilized or Starting Capital= 64,720 millions USD), (W= Wealth or Total Assets= 139,796 millions USD), (S= Sales or Revenues= 51,294 millions USD), (I= Interest Expense= 513 millions USD), and (t= T/B= Tax Rate= 30.00%), then it's (d= Dividend Portion or Payout planned), is:

$$d = 1-\{W/[1+l]-U\}/([1-t]\{S[1-v-f]-I\})$$
$$= 1-\{139,796/[1+0.9900]-64,720\}$$
$$/([1-0.3000]\{51,294$$
$$[1-0.1900-0.5800]-513\})$$
$$= 30.00\%$$

Corporate IFRS-GAAP (B/S-I/S), ISBN 13: **978-1983566332**, ISBN 10: **1983566330**

Law-5401:

If (v= V/S= Variable Portion= 19.00%), (f= F/S= Fixed Portion= 58.00%), (l= L/E= Leverage or Gearing Ratio= 99.00%), (d= D/A= Dividend Portion or Payout= 30.00%), (W= Wealth or Total Assets= 139,796 millions USD), (S= Sales or Revenues= 51,294 millions USD), (I= Interest Expense= 513 millions USD), and (t= T/B= Tax Rate= 30.00%), then it's (U= Utilized or Starting Capital planned), is:

U= W/[1+l]-[1-d][1-t]{S[1-v-f]-I}

$$= 139,796/[1+0.9900]-[1-0.3000]$$
$$[1-0.3000]\{51,294$$
$$[1-0.1900-0.5800]-513\}$$

$$= 64,720 \text{ millions USD}$$

Corporate IFRS-GAAP (B/S-I/S), ISBN 13: **978-1983566332**, ISBN 10: **1983566330**

Law-5402:

If (v= V/S= Variable Portion= 19.00%), (f= F/S= Fixed Portion= 58.00%), (U= Utilized or Starting Capital= 64,720 millions USD), (d= D/A= Dividend Portion or Payout= 30.00%), (W= Wealth or Total Assets= 139,796 millions USD), (S= Sales or Revenues= 51,294 millions USD), (I= Interest Expense= 513 millions USD), and (t= T/B= Tax Rate= 30.00%), then it's (I = Leverage or Gearing Ratio planned), is:

$$I = W/(U+[1-d][1-t]\{S[1-v-f]-I\})-1$$
$$= 139,796/(64,720+[1-0.3000]$$
$$[1-0.3000]\{51,294$$
$$[1-0.1900-0.5800]-513\})-1$$
$$= 99.00\%$$

Corporate IFRS-GAAP (B/S-I/S), ISBN 13: **978-1983566332**, ISBN 10: **1983566330**

<u>Law-5403</u>:

If (**v**= V/S= Variable Portion= <u>19.00%</u>), (**f**= F/S= Fixed Portion= <u>58.00%</u>), (**U**= Utilized or Starting Capital= <u>64,720</u> millions USD), (**d**= D/A= Dividend Portion or Payout= <u>30.00%</u>), (**l**= L/E= Leverage or Gearing Ratio= <u>99.00%</u>), (**$**= Sales or Revenues= <u>51,294</u> millions USD), (**i**= I/S= Interest Portion= <u>1.00%</u>), and (**t**= T/B= Tax Rate= <u>30.00%</u>), then it's (**W**= Wealth or Total Assets planned), is:

$$W= [1+l]\{U+[1-d][1-t][\$-\$v-\$f-\$i]\}$$
$$= [1+l](U+[1-d][1-t]\{\$[1-v-f-i]\})$$
$$= [1+0.9900](64,720+[1-0.3000]$$
$$[1-0.3000]\{51,294[1-0.1900$$
$$-0.5800-0.0100]\})$$
$$= \underline{139,796} \text{ millions USD}$$

Corporate IFRS-GAAP (B/S-I/S), ISBN 13: **978-1983566332**, ISBN 10: **1983566330**

<u>Law-5404</u>:

If (**v**= V/S= Variable Portion= <u>19.00%</u>), (**f**= F/S= Fixed Portion= <u>58.00%</u>), (**U**= Utilized or Starting Capital= <u>64,720</u> millions USD), (**d**= D/A= Dividend Portion or Payout= <u>30.00%</u>), (**l**= L/E= Leverage or Gearing Ratio= <u>99.00%</u>), (**W**= Wealth or Total Assets= <u>139,796</u> millions USD), (**i**= I/S= Interest Portion= <u>1.00%</u>), and (**t**= T/B= Tax Rate= <u>30.00%</u>), then it's (**S**= Sales or Revenues planned), is:

$$S= \{W/[1+l]-U\}/\{[1-d][1-t]\})/[1-v-f-i]$$
$$= \{139,796/[1+0.9900]-64,720\}$$
$$/\{[1-0.3000][1-0.3000]\})$$
$$/[1-0.1900-0.5800-0.0100]$$
$$= \underline{51,294} \text{ millions USD}$$

Corporate IFRS-GAAP (B/S-I/S), ISBN 13: **978-1983566332**, ISBN 10: **1983566330**

Law-5405:

If ($\$$= Sales or Revenues= 51,294 millions USD), (f= F/S= Fixed Portion= 58.00%), (U= Utilized or Starting Capital= 64,720 millions USD), (d= D/A= Dividend Portion or Payout= 30.00%), (l= L/E= Leverage or Gearing Ratio= 99.00%), (W= Wealth or Total Assets= 139,796 millions USD), (i= I/S= Interest Portion= 1.00%), and (t= T/B= Tax Rate= 30.00%), then it's (v= Variable Portion planned), is:

$$v= 1\text{-}f\text{-}i\text{-}\{W/[1+l]\text{-}U\}/\{[1\text{-}d][1\text{-}t]\})/\$$$
$$= 1\text{-}0.5800\text{-}0.0100\text{-}\{139,796$$
$$/[1+0.9900]\text{-}64,720\}$$
$$/\{[1\text{-}0.3000][1\text{-}0.3000]\})$$
$$/51,294$$
$$= 19.00\%$$

Corporate IFRS-GAAP (B/S-I/S), ISBN 13: **978-1983566332**, ISBN 10: **1983566330**

Law-5406:

If (**\$**= Sales or Revenues= 51,294 millions USD), (**v**= V/S= Variable Portion= 19.00%), (**U**= Utilized or Starting Capital= 64,720 millions USD), (**d**= D/A= Dividend Portion or Payout= 30.00%), (**l**= L/E= Leverage or Gearing Ratio= 99.00%), (**W**= Wealth or Total Assets= 139,796 millions USD), (**i**= I/S= Interest Portion= 1.00%), and (**t**= T/B= Tax Rate= 30.00%), then it's (**f**= Fixed Portion planned), is:

$$\mathbf{f} = 1 - \mathbf{v} - \mathbf{i} - \{\mathbf{W}/[1+\mathbf{l}] - \mathbf{U}\}/\{[1-\mathbf{d}][1-\mathbf{t}]\})/\$$$

$$= 1 - 0.1900 - 0.0100 - \{139,796$$
$$/[1+0.9900] - 64,720\}$$
$$/\{[1-0.3000][1-0.3000]\})$$
$$/51,294$$

$$= 58.00\%$$

Corporate IFRS-GAAP (B/S-I/S), ISBN 13: 978-1983566332, ISBN 10: 1983566330

Law-5407:

If (**$**= Sales or Revenues= 51,294 millions USD), (**v**= V/S= Variable Portion= 19.00%), (**U**= Utilized or Starting Capital= 64,720 millions USD), (**d**= D/A= Dividend Portion or Payout= 30.00%), (**f**= L/E= Leverage or Gearing Ratio= 99.00%), (**W**= Wealth or Total Assets= 139,796 millions USD), (**f**= F/S= Fixed Portion= 58.00%), and (**t**= T/B= Tax Rate= 30.00%), then it's (**i**= Interest Portion planned), is:

$$i= 1\text{-}\mathbf{v}\text{-}\mathbf{f}\text{-}\{\mathbf{W}/[1\text{+}\mathbf{f}]\text{-}\mathbf{U}\}/\{[1\text{-}\mathbf{d}][1\text{-}\mathbf{t}]\})/\mathbf{\$}$$

$$= 1\text{-}0.1900\text{-}0.5800\text{-}\{139,796$$
$$/[1\text{+}0.9900]\text{-}64,720\}$$
$$/\{[1\text{-}0.3000][1\text{-}0.3000]\})$$
$$/51,294$$

$$= 1.00\%$$

Corporate IFRS-GAAP (B/S-I/S), ISBN 13: **978-1983566332**, ISBN 10: **1983566330**

Law-5408:

If (**$**= Sales or Revenues= 51,294 millions USD), (**v**= V/S= Variable Portion= 19.00%), (**U**= Utilized or Starting Capital= 64,720 millions USD), (**d**= D/A= Dividend Portion or Payout= 30.00%), (**l**= L/E= Leverage or Gearing Ratio= 99.00%), (**W**= Wealth or Total Assets= 139,796 millions USD), (**f**= F/S= Fixed Portion= 58.00%), and (**i**= I/S= Interest Portion= 1.00%), then it's (**t**= Tax Rate planned), is:

$$t = 1-\{W/[1+l]-U\}/([1-d]\{\$[1-v-f-i]\})$$
$$= 1-\{139,796/[1+0.9900]-64,720\}$$
$$/([1-0.3000]\{51,294[1-0.1900$$
$$-0.5800-0.0100]\})$$
$$= 30.00\%$$

Corporate IFRS-GAAP (B/S-I/S), ISBN 13: **978-1983566332**, ISBN 10: **1983566330**

Law-5409:

If ($ = Sales or Revenues= 51,294 millions USD), (v= V/S= Variable Portion= 19.00%), (U= Utilized or Starting Capital= 64,720 millions USD), (t= T/B= Tax Rate= 30.00%), (l= L/E= Leverage or Gearing Ratio= 99.00%), (W= Wealth or Total Assets= 139,796 millions USD), (f= F/S= Fixed Portion= 58.00%), and (i= I/S= Interest Portion= 1.00%), then it's (d= Dividend Portion planned), is:

$$d = 1 - \{W/[1+l]-U\}/([1-t]\{$[1-v-f-i]\})$$
$$= 1 - \{139,796/[1+0.9900]-64,720\}$$
$$/([1-0.3000]\{51,294[1-0.1900$$
$$-0.5800-0.0100]\})$$
$$= \underline{30.00\%}$$

Corporate IFRS-GAAP (B/S-I/S), ISBN 13: 978-1983566332, ISBN 10: 1983566330

Law-5410:

If (**$** = Sales or Revenues= 51,294 millions USD), (**v**= V/S= Variable Portion= 19.00%), (**d**= D/A= Dividend Portion or Payout= 30.00%), (**t**= T/B= Tax Rate= 30.00%), (**l**= L/E= Leverage or Gearing Ratio= 99.00%), (**W**= Wealth or Total Assets= 139,796 millions USD), (**f**= F/S= Fixed Portion= 58.00%), and (**i**= I/S= Interest Portion= 1.00%), then it's (**U**= Utilized or Starting Capital planned), is:

$$\mathbf{U}= \mathbf{W}/[1+\mathbf{l}]-[1-\mathbf{d}][1-\mathbf{t}]\{\mathbf{\$}[1-\mathbf{v}-\mathbf{f}-\mathbf{i}]\}$$

$$= 139{,}796/[1+0.9900]-[1-0.3000]$$
$$[1-0.3000]\{51{,}294[1-0.1900$$
$$-0.5800-0.0100]\}$$

$$= 64{,}720 \text{ millions USD}$$

Corporate IFRS-GAAP (B/S-I/S), ISBN 13: **978-1983566332**, ISBN 10: **1983566330**

<u>Law-5411</u>:

If ($= Sales or Revenues= <u>51,294</u> millions USD), (**v**= V/S= Variable Portion= <u>19.00%</u>), (**d**= D/A= Dividend Portion or Payout= <u>30.00%</u>), (**t**= T/B= Tax Rate= <u>30.00%</u>), (**U**= Utilized or Starting Capital= <u>64,720</u> millions USD), (**W**= Wealth or Total Assets= <u>139,796</u> millions USD), (**f**= F/S= Fixed Portion= <u>58.00%</u>), and (**i**= I/S= Interest Portion= <u>1.00%</u>), then it's (**I** = Leverage or gearing Ratio planned), is:

$$I = W/(U+[1-d][1-t]\{S[1-v-f-i]\})-1$$
$$= 139,796/(64,720+[1-0.3000]$$
$$[1-0.3000]\{51,294[1-0.1900$$
$$-0.5800-0.0100]\})-1$$
$$= \underline{99.00\%}$$

Corporate IFRS-GAAP (B/S-I/S), ISBN 13: **978-1983566332**, ISBN 10: **1983566330**

Law-5412:

If (**$'**= Sales of Past Year= 48,851 millions USD),
(**s**= [S/S']-1= Sales Growth= 5.00%), (**$**= Sales or
Revenues= 51,294 millions USD), (**v**= V/S= Variable
Portion= 19.00%), (**d**= D/A= Dividend Portion or
Payout= 30.00%), (**t**= T/B= Tax Rate= 30.00%), (**U**=
Utilized or Starting Capital= 64,720 millions USD),
(**l**= L/E= Leverage or Gearing Ratio= 99.00%), (**f**=
F/S= Fixed Portion= 58.00%), and (**i**= I/S= Interest
Portion= 1.00%), then it's (**W**= Wealth or Total
Assets planned), is:

$$W = [1+l](U+[1-d][1-t]\{\$-\$v-\$f-\$'i[1+s]\})$$
$$= [1+l](U+[1-d][1-t]\{\$[1-v-f]$$
$$-\$'i[1+s]\})$$
$$= [1+0.9900](64{,}720+[1-0.3000]$$
$$[1-0.3000]\{51{,}294[1-0.1900$$
$$-0.5800]-48{,}851*0.0100$$
$$[1+0.0500]\})$$
$$= 139{,}796 \text{ millions USD}$$

Corporate IFRS-GAAP (B/S-I/S), ISBN 13: **978-1983566332**, ISBN 10: **1983566330**

Law-5413:

If (**$'**= Sales of Past Year= 48,851 millions USD), (**s**= [S/S']-1= Sales Growth= 5.00%), (**W**= Wealth or Total Assets= 139,796 millions USD), (**v**= V/S= Variable Portion= 19.00%), (**d**= D/A= Dividend Portion or Payout= 30.00%), (**t**= T/B= Tax Rate= 30.00%), (**U**= Utilized or Starting Capital= 64,720 millions USD), (**I**= L/E= Leverage or Gearing Ratio= 99.00%), (**f**= F/S= Fixed Portion= 58.00%), and (**i**= I/S= Interest Portion= 1.00%), then it's (**$**= Sales or Revenues planned), is:

$$\$= (\$'i[1+s]+\{W/[1+I]-U\}/\{[1-d][1-t]\})/[1-v-f]$$

$$= (48,851*0.0100[1+0.0500]$$
$$+\{139,796/[1+0.9900]$$
$$-64,720\}/\{[1-0.3000]$$
$$[1-0.3000]\})$$
$$/[1-0.1900-0.5800]$$

$$= 51,294 \text{ millions USD}$$

Corporate IFRS-GAAP (B/S-I/S), ISBN 13: **978-1983566332**, ISBN 10: **1983566330**

Law-5414:

If ($\mathbf{S'}$= Sales of Past Year= 48,851 millions USD), ($\mathbf{s}$= [S/S']-1= Sales Growth= 5.00%), ($\mathbf{W}$= Wealth or Total Assets= 139,796 millions USD), ($\mathbf{S}$= Sales or Revenues= 51,294 millions USD), ($\mathbf{d}$= D/A= Dividend Portion or Payout= 30.00%), ($\mathbf{t}$= T/B= Tax Rate= 30.00%), ($\mathbf{U}$= Utilized or Starting Capital= 64,720 millions USD), ($\mathbf{l}$= L/E= Leverage or Gearing Ratio= 99.00%), ($\mathbf{f}$= F/S= Fixed Portion= 58.00%), and ($\mathbf{i}$= I/S= Interest Portion= 1.00%), then it's ($\mathbf{v}$= Variable Portion planned), is:

$$v = 1 - f - (S'i[1+s] + \{W/[1+l] - U\} / \{[1-d][1-t]\})/S$$

$$= 1 - 0.5800 - (48,851 * 0.0100 [1+0.0500] + \{139,796 /[1+0.9900] - 64,720\} /\{[1-0.3000][1-0.3000]\}) /51,294$$

$$= 19.00\%$$

Corporate IFRS-GAAP (B/S-I/S), ISBN 13: **978-1983566332**, ISBN 10: **1983566330**

Law-5415:

If (**$'**= Sales of Past Year= 48,851 millions USD), (**s**= [S/S']-1= Sales Growth= 5.00%), (**W**= Wealth or Total Assets= 139,796 millions USD), (**$**= Sales or Revenues= 51,294 millions USD), (**d**= D/A= Dividend Portion or Payout= 30.00%), (**t**= T/B= Tax Rate= 30.00%), (**U**= Utilized or Starting Capital= 64,720 millions USD), (**I**= L/E= Leverage or Gearing Ratio= 99.00%), (**v**= V/S= Variable Portion= 19.00%), and (**i**= I/S= Interest Portion= 1.00%), then it's (**f**= Fixed Portion planned), is:

$$f= 1-v-(\$'i[1+s]+\{W/[1+I]-U\}$$
$$/\{[1-d][1-t]\})/\$$$
$$= 1-0.1900-(48,851*0.0100$$
$$[1+0.0500]+\{139,796$$
$$/[1+0.9900]-64,720\}$$
$$/\{[1-0.3000][1-0.3000]\})$$
$$/51,294$$
$$= 58.00\%$$

Corporate IFRS-GAAP (B/S-I/S), ISBN 13: **978-1983566332**, ISBN 10: **1983566330**

Law-5416:

If (**f**= F/S= Fixed Portion= 58.00%), (**s**= [S/S']-1= Sales Growth= 5.00%), (**W**= Wealth or Total Assets= 139,796 millions USD), (**$**= Sales or Revenues= 51,294 millions USD), (**d**= D/A= Dividend Portion or Payout= 30.00%), (**t**= T/B= Tax Rate= 30.00%), (**U**= Utilized or Starting Capital= 64,720 millions USD), (**⌐**= L/E= Leverage or Gearing Ratio= 99.00%), (**v**= V/S= Variable Portion= 19.00%), and (**i**= I/S= Interest Portion= 1.00%), then it's (**$'**= Sales Past), must be:

$$\$' = (\$[1\text{-}v\text{-}f]\text{-}\{W/[1\text{+}⌐]\text{-}U\}/\{[1\text{-}d][1\text{-}t]\})$$
$$/\{i[1\text{+}s]\}$$
$$= (51{,}294[1\text{-}0.1900\text{-}0.5800]$$
$$-\{139{,}796/[1\text{+}0.9900]$$
$$-64{,}720\}/\{[1\text{-}0.3000]$$
$$[1\text{-}0.3000]\})/\{0.0100$$
$$[1\text{+}0.0500]\}$$
$$= \underline{48{,}851} \text{ millions USD}$$

Corporate IFRS-GAAP (B/S-I/S), ISBN 13: **978-1983566332**, ISBN 10: **1983566330**

Law-5417:

If (f= F/S= Fixed Portion= 58.00%), (S'= Sales of
Past Year= 48,851 millions USD), (W= Wealth or
Total Assets= 139,796 millions USD), (S= Sales or
Revenues= 51,294 millions USD), (d= D/A=
Dividend Portion or Payout= 30.00%), (t= T/B= Tax
Rate= 30.00%), (U= Utilized or Starting Capital=
64,720 millions USD), (l= L/E= Leverage or Gearing
Ratio= 99.00%), (v= V/S= Variable Portion=
19.00%), and (s= [S/S']-1= Sales Growth= 5.00%),
then it's (i= Interest Portion planned), is:

$$i= (S[1-v-f]-\{W/[1+l]-U\}/\{[1-d][1-t]\})$$
$$/\{S'[1+s]\}$$
$$= (51,294[1-0.1900-0.5800]$$
$$-\{139,796/[1+0.9900]$$
$$-64,720\}/\{[1-0.3000]$$
$$[1-0.3000]\})/\{48,851$$
$$[1+0.0500]\}$$
$$= \underline{1.00\%}$$

Corporate IFRS-GAAP (B/S-I/S), ISBN 13: **978-1983566332**, ISBN 10: **1983566330**

Law-5418:

If (**f**= F/S= Fixed Portion= 58.00%), (**W**= Wealth or Total Assets= 139,796 millions USD), (**S'**= Sales of Past Year= 48,851 millions USD), (**S**= Sales or Revenues= 51,294 millions USD), (**d**= D/A= Dividend Portion or Payout= 30.00%), (**t**= T/B= Tax Rate= 30.00%), (**U**= Utilized or Starting Capital= 64,720 millions USD), (**l**= L/E= Leverage or Gearing Ratio= 99.00%), (**v**= V/S= Variable Portion= 19.00%), and (**i**= I/S= Interest Portion= 1.00%), then it's (**s**= Sales Growth planned), is:

$$s = (S[1-v-f]-\{W/[1+l]-U\}/\{[1-d][1-t]\})/[S'i]-1$$

$$= (51,294[1-0.1900-0.5800]$$
$$-\{139,796/[1+0.9900]$$
$$-64,720\}/\{[1-0.3000]$$
$$[1-0.3000]\})/[48,851$$
$$*0.0100]-1$$

$$= 5.00\%$$

Corporate IFRS-GAAP (B/S-I/S), ISBN 13: **978-1983566332**, ISBN 10: **1983566330**

Law-5419:

If (**f**= F/S= Fixed Portion= 58.00%), (**W**= Wealth or Total Assets= 139,796 millions USD), (**$'**= Sales of Past Year= 48,851 millions USD), (**$**= Sales or Revenues= 51,294 millions USD), (**d**= D/A= Dividend Portion or Payout= 30.00%), (**s**= [S/S']-1= Sales Growth= 5.00%), (**U**= Utilized or Starting Capital= 64,720 millions USD), (**I**= L/E= Leverage or Gearing Ratio= 99.00%), (**v**= V/S= Variable Portion= 19.00%), and (**i**= I/S= Interest Portion= 1.00%), then it's (**t**= Tax Rate planned), is:

$$t= 1-\{W/[1+I]-U\}/([1-d]\{\$[1-v-f] -\$'i[1+s]\})$$

$$= 1-\{139,796/[1+0.9900]-64,720\} /([1-0.3000]\{51,294 [1-0.1900-0.5800]-48,851 *0.0100[1+0.0500]\})$$

$$= 30.00\%$$

Corporate IFRS-GAAP (B/S-I/S), ISBN 13: **978-1983566332**, ISBN 10: **1983566330**

<u>Law-5420</u>:

If (**f**= F/S= Fixed Portion= <u>58.00%</u>), (**W**= Wealth or Total Assets= <u>139,796</u> millions USD), (**$'**= Sales of Past Year= <u>48,851</u> millions USD), (**$**= Sales or Revenues= <u>51,294</u> millions USD), (**t**= T/B= Tax ate= <u>30.00%</u>), (**s**= [S/S']-1= Sales Growth= <u>5.00%</u>), (**U**= Utilized or Starting Capital= <u>64,720</u> millions USD), (**l**= L/E= Leverage or Gearing Ratio= <u>99.00%</u>), (**v**= V/S= Variable Portion= <u>19.00%</u>), and (**i**= I/S= Interest Portion= <u>1.00%</u>), then it's (**d**= Dividend Portion planned), is:

$$d= 1-\{W/[1+l]-U\}/([1-t]\{\$[1-v-f]-\$'i[1+s]\})$$
$$= 1-\{139,796/[1+0.9900]-64,720\}$$
$$/([1-0.3000]\{51,294$$
$$[1-0.1900-0.5800]-48,851$$
$$*0.0100[1+0.0500]\})$$
$$= \underline{30.00\%}$$

Corporate IFRS-GAAP (B/S-I/S), ISBN 13: **978-1983566332**, ISBN 10: **1983566330**

Law-5421:

If (**f**= F/S= Fixed Portion= 58.00%), (**W**= Wealth or Total Assets= 139,796 millions USD), (**S'**= Sales of Past Year= 48,851 millions USD), (**S**= Sales or Revenues= 51,294 millions USD), (**t**= T/B= Tax Rate= 30.00%), (**s**= [S/S']-1= Sales Growth= 5.00%), (**d**= D/A= Dividend Portion or Payout= 30.00%), (**f**= L/E= Leverage or Gearing Ratio= 99.00%), (**v**= V/S= Variable Portion= 19.00%), and (**i**= I/S= Interest Portion= 1.00%), then it's (**U**= Utilized or Starting Capital planned), is:

$$\mathbf{U}= \mathbf{W}/[1+\mathbf{f}]-[1-\mathbf{d}][1-\mathbf{t}]\{\mathbf{S}[1-\mathbf{v}-\mathbf{f}]-\mathbf{S'i}[1+\mathbf{s}]\}$$

$$= \{139{,}796/[1+0.9900]-[1-0.3000]$$
$$[1-0.3000]\{51{,}294[1-0.1900$$
$$-0.5800]-48{,}851*0.0100$$
$$[1+0.0500]\}$$

$$= 64{,}720 \text{ millions USD}$$

Corporate IFRS-GAAP (B/S-I/S), ISBN 13: **978-1983566332**, ISBN 10: **1983566330**

Law-5422:

If (**f**= F/S= Fixed Portion= 58.00%), (**W**= Wealth or Total Assets= 139,796 millions USD), (**$'**= Sales of Past Year= 48,851 millions USD), (**$**= Sales or Revenues= 51,294 millions USD), (**t**= T/B= Tax Rate= 30.00%), (**s**= [S/S']-1= Sales Growth= 5.00%), (**d**= D/A= Dividend Portion or Payout= 30.00%), (**U**= Utilized or Starting Capital= 64,720 millions USD), (**v**= V/S= Variable Portion= 19.00%), and (**i**= I/S= Interest Portion= 1.00%), then it's (**I** = Leverage or Gearing Ratio planned), is:

$$I = W/(U+[1-d][1-t]\{\$[1-v-f]-\$'i[1+s]\})-1$$
$$= \{139,796/(64,720+[1-0.3000]$$
$$[1-0.3000]\{51,294[1-0.1900$$
$$-0.5800]-48,851*0.0100$$
$$[1+0.0500]\})-1$$
$$= 99.00\%$$

Corporate IFRS-GAAP (B/S-I/S), ISBN 13: **978-1983566332**, ISBN 10: **1983566330**

Law-5423:

If (f= F/S= Fixed Portion= 58.00%), (f= L/E= Leverage or Gearing Ratio= 99.00%), (S'= Sales of Past Year= 48,851 millions USD), (S= Sales or Revenues= 51,294 millions USD), (t= T/B= Tax Rate= 30.00%), (s= [S/S']-1= Sales Growth= 5.00%), (d= D/A= Dividend Portion or Payout= 30.00%), (U= Utilized or Starting Capital= 64,720 millions USD), (v= V/S= Variable Portion= 19.00%), and (I= Interest Expense= 513 millions USD), then it's (W= Wealth or Total Assets planned), is:

$$W = [1+I](U+[1-d][1-t]\{S-Sv-S'f[1+s]-I\})$$
$$= [1+I](U+[1-d][1-t]\{S[1-v]$$
$$-S'f[1+s]-I\})$$
$$= [1+0.9900](64,720+[1-0.3000]$$
$$[1-0.3000]\{51,294[1-0.1900]$$
$$-48,851*0.5800[1+0.0500]$$
$$-513\})$$
$$= 139,796 \text{ millions USD}$$

Corporate IFRS-GAAP (B/S-I/S), ISBN 13: **978-1983566332**, ISBN 10: **1983566330**

Law-5424:

If (f= F/S= Fixed Portion= 58.00%), (l= L/E= Leverage or Gearing Ratio= 99.00%), (S'= Sales of Past Year= 48,851 millions USD), (W= Wealth or Total Assets= 139,796 millions USD), (t= T/B= Tax Rate= 30.00%), (s= [S/S']-1= Sales Growth= 5.00%), (d= D/A= Dividend Portion or Payout= 30.00%), (U= Utilized or Starting Capital= 64,720 millions USD), (v= V/S= Variable Portion= 19.00%), and (l= Interest Expense= 513 millions USD), then it's (S= Sales or Revenues planned), is:

$$S = (S'f[1+s]+l+\{W/[1+l]-U\}/\{[1-d][1-t]\})$$
$$/[1-v]$$
$$= (48,851*0.5800[1+0.0500]+513$$
$$+\{139,796/[1+0.9900]$$
$$-64,720\}/\{[1-0.3000]$$
$$[1-0.3000]\})/[1-0.1900]$$
$$= 51,294 \text{ millions USD}$$

Corporate IFRS-GAAP (B/S-I/S), ISBN 13: **978-1983566332**, ISBN 10: **1983566330**

<u>Law-5425</u>:

If (**f**= F/S= Fixed Portion= <u>58.00%</u>), (**l**= L/E= Leverage or Gearing Ratio= <u>99.00%</u>), (**$'**= Sales of Past Year= <u>48,851</u> millions USD), (**W**= Wealth or Total Assets= <u>139,796</u> millions USD), (**t**= T/B= Tax Rate= <u>30.00%</u>), (**s**= [S/S']-1= Sales Growth= <u>5.00%</u>), (**d**= D/A= Dividend Portion or Payout= <u>30.00%</u>), (**U**= Utilized or Starting Capital= <u>64,720</u> millions USD), (**$**= Sales or Revenues= <u>51,294</u> millions USD), and (**l**= Interest Expense= <u>513</u> millions USD), then it's (**v**= Variable Portion planned), is:

$$\mathbf{v}= 1-(\mathbf{\$'f}[1+\mathbf{s}]+\mathbf{l}+\{\mathbf{W}/[1+\mathbf{l}]-\mathbf{U}\}$$
$$/\{[1-\mathbf{d}][1-\mathbf{t}]\})/\mathbf{\$}$$
$$= 1-(48,851*0.5800[1+0.0500]$$
$$+513+\{139,796/[1+0.9900]$$
$$-64,720\}/\{[1-0.3000]$$
$$[1-0.3000]\})/51,294$$
$$= \underline{19.00\%}$$

Corporate IFRS-GAAP (B/S-I/S), ISBN 13: **978-1983566332**, ISBN 10: **1983566330**

Law-5426:

If (**f**= F/S=Fixed Portion= 58.00%), (**l**= L/E=
Leverage or Gearing Ratio= 99.00%), (**v**= V/S=
Variable Portion= 19.00%), (**W**= Wealth or Total
Assets= 139,796 millions USD), (**t**= T/B= Tax Rate=
30.00%), (**s**= [S/S']-1= Sales Growth= 5.00%), (**d**=
D/A= Dividend Portion or Payout= 30.00%), (**U**=
Utilized or Starting Capital= 64,720 millions USD),
(**S**= Sales or Revenues= 51,294 millions USD), and
(**I**= Interest Expense= 513 millions USD), then it's
(**S'**= Sales Past), must be:

$$S' = (S[1-v]-I-\{W/[1+l]-U\}/\{[1-d][1-t]\})$$
$$/\{f[1+s]\}$$
$$= (51,294[1-0.1900]-513$$
$$-\{139,796/[1+0.9900]$$
$$-64,720\}/\{[1-0.3000]$$
$$[1-0.3000]\})/\{0.5800$$
$$[1+0.0500]\}$$
$$= 48,851 \text{ millions USD}$$

Corporate IFRS-GAAP (B/S-I/S), ISBN 13: **978-1983566332**, ISBN 10: **1983566330**

Law-5427:

If (**I**= L/E= Leverage or Gearing Ratio= 99.00%), (**v**= V/S= Variable Portion= 19.00%), (**W**= Wealth or Total Assets= 139,796 millions USD), (**t**= T/B= Tax Rate= 30.00%),(**\$'**= Sales of Past Year= 48,851 millions USD), (**s**= [S/S']-1= Sales Growth= 5.00%), (**d**= D/A= Dividend Portion or Payout= 30.00%), (**U**= Utilized or Starting Capital= 64,720 millions USD), (**\$**= Sales or Revenues= 51,294 millions USD), and (**I**= Interest Expense= 513 millions USD), then it's (**f**= Fixed Portion planned), is:

$$f = (\$[1\text{-}v]\text{-}I\text{-}\{W/[1+I]\text{-}U\}/\{[1\text{-}d][1\text{-}t]\})$$
$$/\{\$'[1+s]\}$$
$$= (51,294[1\text{-}0.1900]\text{-}513$$
$$-\{139,796/[1+0.9900]$$
$$-64,720\}/\{[1\text{-}0.3000]$$
$$[1\text{-}0.3000]\})/\{48,851$$
$$[1+0.0500]\}$$
$$= 58.00\%$$

Corporate IFRS-GAAP (B/S-I/S), ISBN 13: 978-1983566332, ISBN 10: 1983566330

Law-5428:

If (**I**= L/E= Leverage or Gearing Ratio= 99.00%),
(**v**= V/S= Variable Portion= 19.00%), (**W**= Wealth or
Total Assets= 139,796 millions USD), (**t**= T/B= Tax
Rate= 30.00%),(**$'**= Sales of Past Year= 48,851
millions USD), (**f**= F/S= Fixed Portion= 58.00%),
(**d**= D/A= Dividend Portion or Payout= 30.00%), (**U**=
Utilized or Starting Capital= 64,720 millions USD),
(**$**= Sales or Revenues= 51,294 millions USD), and
(**I**= Interest Expense= 513 millions USD), then it's
(**s**= Sales Growth planned), is:

$$s= (\$[1\text{-}v]\text{-}I\text{-}\{W/[1+I]\text{-}U\}/\{[1\text{-}d][1\text{-}t]\})$$
$$/[\$'f]\text{-}1$$
$$= (51,294[1\text{-}0.1900]\text{-}513$$
$$-\{139,796/\ [1+0.9900]$$
$$-64,720\}/\{[1\text{-}0.3000]$$
$$[1\text{-}0.3000]\})/[48,851$$
$$*0.5800]\text{-}1$$

$$= \underline{5.00\%}$$

Corporate IFRS-GAAP (B/S-I/S), ISBN 13: **978-1983566332**, ISBN 10: **1983566330**

Law-5429:

If (l= L/E= Leverage or Gearing Ratio= 99.00%),
(v= V/S= Variable Portion= 19.00%), (W= Wealth or
Total Assets= 139,796 millions USD), (t= T/B= Tax
Rate= 30.00%),(S'= Sales of Past Year= 48,851
millions USD), (f= F/S= Fixed Portion= 58.00%),
(d= D/A= Dividend Portion or Payout= 30.00%), (U=
Utilized or Starting Capital= 64,720 millions USD),
(S= Sales or Revenues= 51,294 millions USD), and
(s= [S/S']-1= Sales Growth= 5.00%), then it's (I=
Interest Portion planned), is:

$$I= S[1-v]-S'f[1+s]-\{W/[1+l]-U\}/\{[1-d][1-t]\}$$
$$= 51,294[1-0.1900]- 48,851*0.5800$$
$$[1+0.0500]-\{139,796$$
$$/[1+0.9900]-64,720\}$$
$$/\{[1-0.3000][1-0.3000]\}$$
$$= \underline{513} \text{ millions USD}$$

Corporate IFRS-GAAP (B/S-I/S), ISBN 13: **978-1983566332**, ISBN 10: **1983566330**

Law-5430:

If (I= L/E= Leverage or Gearing Ratio= <u>99.00%</u>), (v= V/S= Variable Portion= <u>19.00%</u>), (W= Wealth or Total Assets= <u>139,796</u> millions USD), (I= Interest Expense= <u>513</u> millions USD), (S'= Sales of Past Year= <u>48,851</u> millions USD), (f= F/S= Fixed Portion= <u>58.00%</u>), (d= D/A= Dividend Portion or Payout= <u>30.00%</u>), (U= Utilized or Starting Capital= <u>64,720</u> millions USD), (S= Sales or Revenues= <u>51,294</u> millions USD), and (s= [S/S']-1= Sales Growth= <u>5.00%</u>), then it's (t= Tax Rate planned), is:

$$t = 1 - \{W/[1+I]-U\}/([1-d]\{S[1-v]-S'f[1+s]-I\})$$
$$= 1 - \{139,796/[1+0.9900]-64,720\}$$
$$/([1-0.3000]\{51,294$$
$$[1-0.1900]-48,851*0.5800$$
$$[1+0.0500]-513\})$$
$$= \underline{30.00\%}$$

Corporate IFRS-GAAP (B/S-I/S), ISBN 13: **978-1983566332**, ISBN 10: **1983566330**

Law-5431:

If (**f**= L/E= Leverage or Gearing Ratio= 99.00%), (**v**= V/S= Variable Portion= 19.00%), (**W**= Wealth or Total Assets= 139,796 millions USD), (**I**= Interest Expense= 513 millions USD), (**$'**= Sales of Past Year= 48,851 millions USD), (**f**= F/S= Fixed Portion= 58.00%), (**d**= D/A= Dividend Portion or Payout= 30.00%), (**U**= Utilized or Starting Capital= 64,720 millions USD), (**$**= Sales or Revenues= 51,294 millions USD), and (**s**= [S/S']-1= Sales Growth= 5.00%), then it's (**d**= Dividend Portion or Payout planned), is:

$$d= 1-\{W/[1+I]-U\}$$
$$/([1-t]\{\$[1-v]-\$'f[1+s]-I\})$$
$$= 1-\{139,796/[1+0.9900]-64,720\}$$
$$/([1-0.3000]\{51,294$$
$$[1-0.1900]- 48,851*0.5800$$
$$[1+0.0500]-513\})$$
$$= 30.00\%$$

Corporate IFRS-GAAP (B/S-I/S), ISBN 13: **978-1983566332**, ISBN 10: **1983566330**

<u>Law-5432</u>:

If (**I**= L/E= Leverage or Gearing Ratio= <u>99.00%</u>), (**v**= V/S= Variable Portion= <u>19.00%</u>), (**W**= Wealth or Total Assets= <u>139,796</u> millions USD), (**I**= Interest Expense= <u>513</u> millions USD), (**S'**= Sales of Past Year= <u>48,851</u> millions USD), (**f**= F/S= Fixed Portion= <u>58.00%</u>), (**d**= D/A= Dividend Portion or Payout= <u>30.00%</u>), (**t**= T/B= Tax Rate= <u>30.00%</u>), (**S**= Sales or Revenues= <u>51,294</u> millions USD), and (**s**= [S/S']-1= Sales Growth= <u>5.00%</u>), then it's (**U**= Utilized or Starting Capital planned), is:

$$U= W/[1+I]-[1-d][1-t]\{S[1-v]-S'f[1+s]-I\}$$
$$= 139,796/[1+0.9900]-[1-0.3000]$$
$$[1-0.3000]\{51,294$$
$$[1-0.1900]-48,851$$
$$*0.5800[1+0.0500]-513\}$$
$$= \underline{64,720} \text{ millions USD}$$

Corporate IFRS-GAAP (B/S-I/S), ISBN 13: **978-1983566332**, ISBN 10: **1983566330**

Law-5433:

If (**U**= Utilized or Starting Capital= 64,720 millions USD), (**v**= V/S= Variable Portion= 19.00%), (**W**= Wealth or Total Assets= 139,796 millions USD), (**I**= Interest Expense= 513 millions USD), (**S'**= Sales of Past Year= 48,851 millions USD), (**f**= F/S= Fixed Portion= 58.00%), (**d**= D/A= Dividend Portion or Payout= 30.00%), (**t**= T/B= Tax Rate= 30.00%), (**S**= Sales or Revenues= 51,294 millions USD), and (**s**= [S/S']-1= Sales Growth= 5.00%), then it's (**l**= Leverage or Gearing Ratio planned), is:

$$l = W/(U+[1-d][1-t]\{S[1-v]-S'f[1+s]-I\})-1$$
$$= 139,796/(64,720+[1-0.3000]$$
$$[1-0.3000]\{51,294$$
$$[1-0.1900]-48,851*0.5800$$
$$[1+0.0500]-513\})-1$$
$$= 99.00\%$$

Corporate IFRS-GAAP (B/S-I/S), ISBN 13: **978-1983566332**, ISBN 10: **1983566330**

<u>Law-5434</u>:

If (**U**= Utilized or Starting Capital= <u>64,720</u> millions USD), (**v**= V/S= Variable Portion= <u>19.00%</u>), (**l**= L/E= Leverage or Gearing Ratio= <u>99.00%</u>), (**i**= I/S= Interest Portion= <u>1.00%</u>), (**$'**= Sales of Past Year= <u>48,851</u> millions USD), (**f**= F/S= Fixed Portion= <u>58.00%</u>), (**d**= D/A= Dividend Portion or Payout= <u>30.00%</u>), (**t**= T/B= Tax Rate= <u>30.00%</u>), (**$**= Sales or Revenues= <u>51,294</u> millions USD), and (**s**= [S/S']-1= Sales Growth= <u>5.00%</u>), then it's (**W** = Wealth or Total Assets planned), is:

$$W= [1+l](U+[1-d][1-t]\{\$-\$v]$$
$$-\$'f[1+s]-\$i\})$$
$$=[1+l](U+[1-d][1-t]\{\$[1-v-i]$$
$$-\$'f[1+s]\})$$
$$= [1+0.9900](64,720+[1-0.3000]$$
$$[1-0.3000]\{51,294$$
$$[1-0.1900-0.0100]- 48,851$$
$$*0.5800[1+0.0500]\})$$
$$= \underline{139,796} \text{ millions USD}$$

Corporate IFRS-GAAP (B/S-I/S), ISBN 13: **978-1983566332**, ISBN 10: **1983566330**

<u>Law-5435</u>:

If (U= Utilized or Starting Capital= <u>64,720</u> millions USD), (v= V/S= Variable Portion= <u>19.00%</u>), (I= L/E= Leverage or Gearing Ratio= <u>99.00%</u>), (W= Wealth or Total Assets= <u>139,796</u> millions USD), (i= I/S= Interest Portion= <u>1.00%</u>), (S'= Sales of Past Year= <u>48,851</u> millions USD), (f= F/S= Fixed Portion= <u>58.00%</u>), (d= D/A= Dividend Portion or Payout= <u>30.00%</u>), (t= T/B= Tax Rate= <u>30.00%</u>), and (s= [S/S']-1= Sales Growth= <u>5.00%</u>), then it's (S= Sales or Revenues planned), is:

$$S = (S'f[1+s]+\{W/[1+I]-U\}/\{[1-d][1-t]\})/[1-v-i]$$

$$= (48{,}851*0.5800[1+0.0500]$$
$$+\{139{,}796/[1+0.9900]$$
$$-64{,}720\}/\{[1-0.3000]$$
$$[1-0.3000]\})$$
$$/[1-0.1900-0.0100]$$

$$= \underline{51{,}294} \text{ millions USD}$$

Corporate IFRS-GAAP (B/S-I/S), ISBN 13: **978-1983566332**, ISBN 10: **1983566330**

Law-5436:

If (**U**= Utilized or Starting Capital= <u>64,720</u> millions USD), (**S**= Sales or Revenues= <u>51,294</u> millions USD), (**l**= L/E= Leverage or Gearing Ratio= <u>99.00%</u>), (**W**= Wealth or Total Assets= <u>139,796</u> millions USD), (**i**= I/S= Interest Portion= <u>1.00%</u>), (**S'**= Sales of Past Year= <u>48,851</u> millions USD), (**f**= F/S= Fixed Portion= <u>58.00%</u>), (**d**= D/A= Dividend Portion or Payout= <u>30.00%</u>), (**t**= T/B= Tax Rate= <u>30.00%</u>), and (**s**= [S/S']-1= Sales Growth= <u>5.00%</u>), then it's (**v**= Variable Portion planned), is:

$$v= 1\text{-}\mathbf{i}\text{-}(\mathbf{S'f}[1+\mathbf{s}]+\{\mathbf{W}/[1+\mathbf{l}]\text{-}\mathbf{U}\}$$
$$/\{[1\text{-}\mathbf{d}][1\text{-}\mathbf{t}]\})/\mathbf{S}$$
$$= 1\text{-}0.0100\text{-}(48,851*0.5800$$
$$[1+0.0500]+\{139,796$$
$$/[1+0.9900]\text{-}64,720\}$$
$$/\{[1\text{-}0.3000][1\text{-}0.3000]\})$$
$$/51,294$$

$$= \underline{19.00\%}$$

Corporate IFRS-GAAP (B/S-I/S), ISBN 13: **978-1983566332**, ISBN 10: **1983566330**

Law-5437:

If (**U**= Utilized or Starting Capital= 64,720 millions USD), (**S**= Sales or Revenues= 51,294 millions USD), (**I**= L/E= Leverage or Gearing Ratio= 99.00%), (**W**= Wealth or Total Assets= 139,796 millions USD), (**v**= V/S= Variable Portion= 19.00%), (**S'**= Sales of Past Year= 48,851 millions USD), (**f**= F/S= Fixed Portion= 58.00%), (**d**= D/A= Dividend Portion or Payout= 30.00%), (**t**= T/B= Tax Rate= 30.00%), and (**s**= [S/S']-1= Sales Growth= 5.00%), then it's (**i**= Interest Portion planned), is:

$$i = 1 - v - (S'f[1+s] + \{W/[1+I] - U\}$$
$$/\{[1-d][1-t]\})/S$$
$$= 1 - 0.1900 - (48,851*0.5800$$
$$[1+0.0500] + \{139,796$$
$$/[1+0.9900] - 64,720\}$$
$$/\{[1-0.3000][1-0.3000]\})$$
$$/51,294$$
$$= 1.00\%$$

Corporate IFRS-GAAP (B/S-I/S), ISBN 13: **978-1983566332**, ISBN 10: **1983566330**

Law-5438:

If (**U**= Utilized or Starting Capital= 64,720 millions USD), (**$**= Sales or Revenues= 51,294 millions USD), (**l**= L/E= Leverage or Gearing Ratio= 99.00%), (**W**= Wealth or Total Assets= 139,796 millions USD), (**v**= V/S= Variable Portion= 19.00%), (**i**= I/S= Interest Portion= 1.00%), (**f**= F/S= Fixed Portion= 58.00%), (**d**= D/A= Dividend Portion or Payout= 30.00%), (**t**= T/B= Tax Rate= 30.00%), and (**s**= [S/S']-1= Sales Growth= 5.00%), then it's (**$'**= Sales Past), must be:

$$\$' = (\$[1\text{-}v\text{-}i]\text{-}\{W/[1+l]\text{-}U\}/\{[1\text{-}d][1\text{-}t]\})/\{f[1+s]\}$$

$$= (51,294[1\text{-}0.1900\text{-}0.0100]$$
$$-\{139,796/[1+0.9900]$$
$$-64,720\}/\{[1\text{-}0.3000]$$
$$[1\text{-}0.3000]\})$$
$$/\{0.5800[1+0.0500]\}$$

$$= 48,851 \text{ millions USD}$$

Corporate IFRS-GAAP (B/S-I/S), ISBN 13: **978-1983566332**, ISBN 10: **1983566330**

Law-5439:

If (**U**= Utilized or Starting Capital= 64,720 millions USD), (**S**= Sales or Revenues= 51,294 millions USD), (**I**= L/E= Leverage or Gearing Ratio= 99.00%), (**W**= Wealth or Total Assets= 139,796 millions USD), (**v**= V/S= Variable Portion= 19.00%), (**i**= I/S= Interest Portion= 1.00%), (**S'**= Sales of Past Year= 48,851 millions USD), (**d**= D/A= Dividend Portion or Payout= 30.00%), (**t**= T/B= Tax Rate= 30.00%), and (**s**= [S/S']-1= Sales Growth= 5.00%), then it's (**f**= Fixed Portion planned), is:

$$\mathbf{f} = (\mathbf{S}[1\text{-}\mathbf{v}\text{-}\mathbf{i}]\text{-}\{\mathbf{W}/[1+\mathbf{I}]\text{-}\mathbf{U}\}/\{[1\text{-}\mathbf{d}][1\text{-}\mathbf{t}]\})$$
$$/\{\mathbf{S'}[1+\mathbf{s}]\}$$
$$= (51,294[1\text{-}0.1900\text{-}0.0100]$$
$$-\{139,796/[1+0.9900]$$
$$-64,720\}/\{[1\text{-}0.3000]$$
$$[1\text{-}0.3000]\})$$
$$/\{48,851[1+0.0500]\}$$
$$= \underline{58.00\%}$$

Corporate IFRS-GAAP (B/S-I/S), ISBN 13: **978-1983566332**, ISBN 10: **1983566330**

Law-5440:

If (**U**= Utilized or Starting Capital= 64,720 millions USD), (**$**= Sales or Revenues= 51,294 millions USD), (**I**= L/E= Leverage or Gearing Ratio= 99.00%), (**W**= Wealth or Total Assets= 139,796 millions USD), (**v**= V/S= Variable Portion= 19.00%), (**i**= I/S= Interest Portion= 1.00%), (**$'**= Sales of Past Year= 48,851 millions USD), (**d**= D/A= Dividend Portion or Payout= 30.00%), (**t**= T/B= Tax Rate= 30.00%), and (**f**= F/S= Fixed Portion= 58.00%), then it's (**s**= Sales Growth planned), is:

$$s = (\$[1\text{-}v\text{-}i]\text{-}\{W/[1+I]\text{-}U\}/\{[1\text{-}d][1\text{-}t]\})$$
$$/[\$'f]\text{-}1$$
$$= (51{,}294[1\text{-}0.1900\text{-}0.0100]$$
$$-\{139{,}796/[1+0.9900]$$
$$-64{,}720\}/\{[1\text{-}0.3000]$$
$$[1\text{-}0.3000]\})/[48{,}851$$
$$*0.5800]\text{-}1$$
$$= 5.00\%$$

Corporate IFRS-GAAP (B/S-I/S), ISBN 13: **978-1983566332**, ISBN 10: **1983566330**

Law-5441:

If (U= Utilized or Starting Capital= 64,720 millions USD), (S= Sales or Revenues= 51,294 millions USD), (I= L/E= Leverage or Gearing Ratio= 99.00%), (W= Wealth or Total Assets= 139,796 millions USD), (v= V/S= Variable Portion= 19.00%), (i= I/S= Interest Portion= 1.00%), (S'= Sales of Past Year= 48,851 millions USD), (d= D/A= Dividend Portion or Payout= 30.00%), (s= [S/S']-1= Sales Growth= 5.00%), and (f= F/S= Fixed Portion= 58.00%), then it's (t= Tax Rate planned), is:

$$t = 1-\{W/[1+I]-U\}$$
$$/([1-d]\{S[1-v-i]-S'f[1+s]\})$$
$$= 1-\{139,796/[1+0.9900]-64,720\}$$
$$/\{[1-0.3000]\{51,294$$
$$[1-0.1900-0.0100]-48,851$$
$$*0.5800[1+0.0500]\})$$

$$= 30.00\%$$

Corporate IFRS-GAAP (B/S-I/S), ISBN 13: **978-1983566332**, ISBN 10: **1983566330**

<u>Law-5442</u>:

If (**U**= Utilized or Starting Capital= <u>64,720</u> millions USD), (**$**= Sales or Revenues= <u>51,294</u> millions USD), (**⌿**= L/E= Leverage or Gearing Ratio= <u>99.00%</u>), (**W**= Wealth or Total Assets= <u>139,796</u> millions USD), (**v**= V/S= Variable Portion= <u>19.00%</u>), (**i**= I/S= Interest Portion= <u>1.00%</u>), (**$'**= Sales of Past Year= <u>48,851</u> millions USD), (**t**= T/B= Tax Rate= <u>30.00%</u>), (**s**= [S/S']-1= Sales Growth= <u>5.00%</u>), and (**f**= F/S= Fixed Portion= <u>58.00%</u>), then it's (**d**= Dividend Portion or Payout planned), is:

$$\mathbf{d}= 1-\{\mathbf{W}/[1+\mathbf{⌿}]-\mathbf{U}\}/([1-\mathbf{t}]\{\mathbf{\$}[1-\mathbf{v}-\mathbf{i}]-\mathbf{\$'f}[1+\mathbf{s}]\})$$

$$= 1-\{139{,}796/[1+0.9900]-64{,}720\}$$
$$/\{[1-0.3000]\{51{,}294$$
$$[1-0.1900-0.0100]-48{,}851$$
$$*0.5800[1+0.0500]\})$$

$$= \underline{30.00\%}$$

Corporate IFRS-GAAP (B/S-I/S), ISBN 13: **978-1983566332**, ISBN 10: **1983566330**

Law-5443:

If (**d**= D/A= Dividend Portion or Payout= 30.00%),
(**$**= Sales or Revenues= 51,294 millions USD), (**l**=
L/E= Leverage or Gearing Ratio= 99.00%), (**W**=
Wealth or Total Assets= 139,796 millions USD), (**v**=
V/S= Variable Portion= 19.00%), (**i**= I/S= Interest
Portion= 1.00%), (**$'**= Sales of Past Year= 48,851
millions USD), (**t**= T/B= Tax Rate= 30.00%), (**s**=
[S/S']-1= Sales Growth= 5.00%), and (**f**= F/S= Fixed
Portion= 58.00%), then it's (**U**= Utilized or Starting
Capital planned), is:

$$U= W/[1+l]-[1-d][1-t]\{\$[1-v-i]-\$'f[1+s]\}$$
$$= 139,796/[1+0.9900]-[1-0.3000]$$
$$[1-0.3000]\{51,294[1-0.1900$$
$$-0.0100]-48,851*0.5800$$
$$[1+0.0500]\}$$
$$= 64,720 \text{ millions USD}$$

Corporate IFRS-GAAP (B/S-I/S), ISBN 13: **978-1983566332**, ISBN 10: **1983566330**

<u>Law-5444</u>:

If (**d**= D/A= Dividend Portion or Payout= <u>30.00%</u>),
(**$**= Sales or Revenues= <u>51,294</u> millions USD), (**U**=
Utilized or Starting Capital= <u>64,720</u> millions USD),
(**W**= Wealth or Total Assets= <u>139,796</u> millions
USD), (**v**= V/S= Variable Portion= <u>19.00%</u>), (**i**= I/S=
Interest Portion= <u>1.00%</u>), (**$'**= Sales of Past Year=
<u>48,851</u> millions USD), (**t**= T/B= Tax Rate= <u>30.00%</u>),
(**s**= [S/S']-1= Sales Growth= <u>5.00%</u>), and (**f**= F/S=
Fixed Portion= <u>58.00%</u>), then it's (***I*** = Leverage or
Gearing Ratio planned), is:

$$I = W/(U+[1-d][1-t]\{\$[1-v-i]-\$'f[1+s]\})-1$$
$$= 139,796/(64,720+[1-0.3000]$$
$$[1-0.3000]\{51,294$$
$$[1-0.1900-0.0100]$$
$$-48,851*0.5800$$
$$[1+0.0500]\})-1$$
$$= \underline{99.00\%}$$

Corporate IFRS-GAAP (B/S-I/S), ISBN 13: **978-1983566332**, ISBN 10: **1983566330**

<u>Law-5445</u>:

If (**d**= D/A= Dividend Portion or Payout= <u>30.00%</u>), (**$**= Sales or Revenues= <u>51,294</u> millions USD), (**U**= Utilized or Starting Capital= <u>64,720</u> millions USD), (**⌐** L/E= Leverage or Gearing Ratio= <u>99.00%</u>), (**υ**= V/S= Variable Portion= <u>19.00%</u>), (**i**= I/S= Interest Portion= <u>1.00%</u>), (**$'**= Sales of Past Year= <u>48,851</u> millions USD), (**t**= T/B= Tax Rate= <u>30.00%</u>), (**s**= [S/S']-1= Sales Growth= <u>5.00%</u>), and (**f**= F/S= Fixed Portion= <u>58.00%</u>), then it's (**W**= Wealth or Total Assets planned), is:

$$\mathbf{W}= [1+\mathbf{⌐}](\mathbf{U}+[1-\mathbf{d}][1-\mathbf{t}]$$
$$\{\mathbf{\$}-\mathbf{\$}\mathbf{υ}-\mathbf{\$'f}[1+\mathbf{s}]-\mathbf{\$'i}[1+\mathbf{s}]\})$$
$$= [1+\mathbf{⌐}](\mathbf{U}+[1-\mathbf{d}][1-\mathbf{t}]\{\mathbf{\$}[1-\mathbf{υ}]$$
$$-\mathbf{\$'}[1+\mathbf{s}][\mathbf{f}+\mathbf{i}]\})$$
$$= [1+0.9900](64,720+[1-0.3000]$$
$$[1-0.3000]\{51,294$$
$$[1-0.1900]-48,851$$
$$[1+0.0500]$$
$$[0.5800+0.0100]\})$$
$$= \underline{139,796} \text{ millions USD}$$

Corporate IFRS-GAAP (B/S-I/S), ISBN 13: **978-1983566332**, ISBN 10: **1983566330**

Law-5446:

If (**d**= D/A= Dividend Portion or Payout= 30.00%), (**W**= Wealth or Total Assets= 139,796 millions USD), (**U**= Utilized or Starting Capital= 64,720 millions USD), (**l**= L/E= Leverage or Gearing Ratio= 99.00%), (**v**= V/S= Variable Portion= 19.00%), (**i**= I/S= Interest Portion= 1.00%), (**$'**= Sales of Past Year= 48,851 millions USD), (**t**= T/B= Tax Rate= 30.00%), (**s**= [S/S']-1= Sales Growth= 5.00%), and (**f**= F/S= Fixed Portion= 58.00%), then it's (**$** = Sales or Revenues planned), is:

$$\$= (\$'[1+s][f+i]+\{W/[1+l]-U\} /\{[1-d][1-t]\})/[1-v]$$
$$= (48,851[1+0.0500][0.5800+0.0100] +\{139,796/[1+0.9900] -64,720\}/\{[1-0.3000] [1-0.3000]\})/[1-0.1900]$$
$$= 51,294 \text{ millions USD}$$

Corporate IFRS-GAAP (B/S-I/S), ISBN 13: **978-1983566332**, ISBN 10: **1983566330**

Law-5447:

If (d= D/A= Dividend Portion or Payout= 30.00%), (W= Wealth or Total Assets= 139,796 millions USD), (U= Utilized or Starting Capital= 64,720 millions USD), (l= L/E= Leverage or Gearing Ratio= 99.00%), (S= Sales or Revenues= 51,294 millions USD), (i= I/S= Interest Portion= 1.00%), (S'= Sales of Past Year= 48,851 millions USD), (t= T/B= Tax Rate= 30.00%), (s= [S/S']-1= Sales Growth= 5.00%), and (f= F/S= Fixed Portion= 58.00%), then it's (v= Variable Portion planned), is:

$$v= 1-(S'[1+s][f+i]+\{W/[1+l]-U\}$$
$$/\{[1-d][1-t]\})/S$$
$$= 1-(48,851[1+0.0500][0.5800$$
$$+0.0100]+\{139,796$$
$$/[1+0.9900]-64,720\}$$
$$/\{[1-0.3000][1-0.3000]\})$$
$$/51.294$$

$$= 19.00\%$$

Law-5448:

If (**d**= D/A= Dividend Portion or Payout= 30.00%),
(**W**= Wealth or Total Assets= 139,796 millions
USD), (**U**= Utilized or Starting Capital= 64,720
millions USD), (**l**= L/E= Leverage or Gearing Ratio=
99.00%), (**$**= Sales or Revenues= 51,294 millions
USD), (**i**= I/S= Interest Portion= 1.00%), (**v**= V/S=
Variable Portion= 19.00%), (**t**= T/B= Tax Rate=
30.00%), (**s**= [S/S']-1= Sales Growth= 5.00%), and
(**f**= F/S= Fixed Portion= 58.00%), then it's (**$'**= Sales
Past), must be:

$$\$' = (\$[1-\text{v}]-\$'[1+\text{s}][\text{f}+\text{i}]-\{\text{W}/[1+\textit{l}\,]-\text{U}\}$$
$$/\{[1-\text{d}][1-\text{t}]\})/\{[1+\text{s}][\text{f}+\text{i}]\}$$
$$= (51,294[1-0.1900]-\{139,796$$
$$/[1+0.9900]-64,720\}$$
$$/\{[1-0.3000][1-0.3000]\})$$
$$/\{[1+0.0500][0.5800$$
$$+0.0100]\}$$
$$= 48,851 \text{ millions USD}$$

Corporate IFRS-GAAP (B/S-I/S), ISBN 13: **978-1983566332**, ISBN 10: **1983566330**

<u>Law-5449</u>:

If (**d**= D/A= Dividend Portion or Payout= <u>30.00%</u>), (**W**= Wealth or Total Assets= <u>139,796</u> millions USD), (**U**= Utilized or Starting Capital= <u>64,720</u> millions USD), (**I**= L/E= Leverage or Gearing Ratio= <u>99.00%</u>), (**S**= Sales or Revenues= <u>51,294</u> millions USD), (**i**= I/S= Interest Portion= <u>1.00%</u>), (**v**= V/S= Variable Portion= <u>19.00%</u>), (**t**= T/B= Tax Rate= <u>30.00%</u>), (**S'**= Sales of Past Year= <u>48,851</u> millions USD), and (**f**= F/S= Fixed Portion= <u>58.00%</u>), then it's (**s**= Sales Growth planned), is:

$$s = (S[1-v]-S'[1+s][f+i]-\{W/[1+I]-U\}$$
$$/\{[1-d][1-t]\})/\{S'[f+i]\}-1$$
$$= (51,294[1-0.1900]-\{139,796$$
$$/[1+0.9900]-64,720\}$$
$$/\{[1-0.3000][1-0.3000]\})$$
$$/\{48,851[0.5800$$
$$+0.0100]\}-1$$
$$= \underline{5.00\%}$$

Corporate IFRS-GAAP (B/S-I/S), ISBN 13: **978-1983566332**, ISBN 10: **1983566330**

Law-5450:

If (**d**= D/A= Dividend Portion or Payout= 30.00%), (**W**= Wealth or Total Assets= 139,796 millions USD), (**U**= Utilized or Starting Capital= 64,720 millions USD), (**I**= L/E= Leverage or Gearing Ratio= 99.00%), (**S**= Sales or Revenues= 51,294 millions USD), (**i**= I/S= Interest Portion= 1.00%), (**v**= V/S= Variable Portion= 19.00%), (**t**= T/B= Tax Rate= 30.00%), (**S'**= Sales of Past Year= 48,851 millions USD), and (**s**= [S/S']-1= Sales Growth= 5.00%), then it's (**f**= Fixed Portion planned), is:

$$f = (S[1-v]-S'[1+s][f+i]-\{W/[1+I]-U\}$$
$$/\{[1-d][1-t]\})/\{S'[1+s]\}-I$$
$$= (51,294[1-0.1900]-\{139,796$$
$$/[1+0.9900]-64,720\}$$
$$/\{[1-0.3000][1-0.3000]\})$$
$$/\{48,851[1+0.0500]\}$$
$$-0.0100$$
$$= 58.00\%$$

Corporate IFRS-GAAP (B/S-I/S), ISBN 13: **978-1983566332**, ISBN 10: **1983566330**

Law-5451:

If (**d**= D/A= Dividend Portion or Payout= 30.00%), (**W**= Wealth or Total Assets= 139,796 millions USD), (**U**= Utilized or Starting Capital= 64,720 millions USD), (**l**= L/E= Leverage or Gearing Ratio= 99.00%), (**S**= Sales or Revenues= 51,294 millions USD), (**f**= F/S= Fixed Portion= 58.00%), (**v**= V/S= Variable Portion= 19.00%), (**t**= T/B= Tax Rate= 30.00%), (**S'**= Sales of Past Year= 48,851 millions USD), and (**s**= [S/S']-1= Sales Growth= 5.00%), then it's (**i**= Interest Portion planned), is:

$$i = (S[1-v]-S'[1+s][f+i]-\{W/[1+l]-U\}$$
$$/\{[1-d][1-t]\})/\{S'[1+s]\}-f$$
$$= (51,294[1-0.1900]-\{139,796$$
$$/[1+0.9900]-64,720\}$$
$$/\{[1-0.3000][1-0.3000]\})$$
$$/\{48,851[1+0.0500]\}$$
$$-0.0100$$
$$= \underline{1.00\%}$$

Corporate IFRS-GAAP (B/S-I/S), ISBN 13: **978-1983566332**, ISBN 10: **1983566330**

Law-5452:

If (d= D/A=Dividend Portion or Payout= 30.00%), (W= Wealth or Total Assets= 139,796 millions USD), (U= Utilized or Starting Capital= 64,720 millions USD), (I= L/E= Leverage or Gearing Ratio= 99.00%), (S= Sales or Revenues= 51,294 millions USD), (f= F/S=Fixed Portion= 58.00%), (v= V/S= Variable Portion= 19.00%), (i= I/S= Interest Portion= 1.00%), (S'= Sales of Past Year= 48,851 millions USD), and (s= [S/S']-1= Sales Growth= 5.00%), then it's (t= Tax Rate planned), is:

$$t= 1-\{W/[1+I]-U\}/([1-d]\{S[1-v]$$
$$-S'[1+s][f+i]\})$$
$$= 1-\{139.796/[1+0.9900]-64,720\}$$
$$/([1-0.3000]\{51,294$$
$$[1-0.1900]-48,851$$
$$[1+0.0500]$$
$$[0.5800+0.0100]\})$$

$$= 30.00\%$$

Corporate IFRS-GAAP (B/S-I/S), ISBN 13: **978-1983566332**, ISBN 10: **1983566330**

<u>Law-5453</u>:

If (**t**= T/B= Tax Rate= <u>30.00%</u>), (**W**= Wealth or Total Assets= <u>139,796</u> millions USD), (**U**= Utilized or Starting Capital= <u>64,720</u> millions USD), (**f**= L/E= Leverage or Gearing Ratio= <u>99.00%</u>), (**$**= Sales or Revenues= <u>51,294</u> millions USD), (**f**= F/S= Fixed Portion= <u>58.00%</u>), (**v**= V/S= Variable Portion= <u>19.00%</u>), (**i**= I/S= Interest Portion= <u>1.00%</u>), (**$'**= Sales of Past Year= <u>48,851</u> millions USD), and (**s**= [S/S']-1= Sales Growth= <u>5.00%</u>), then it's (**d**= Dividend Portion or Payout planned), is:

$$
\begin{aligned}
\mathbf{d} &= 1-\{\mathbf{W}/[1+\mathbf{f}]-\mathbf{U}\}/([1-\mathbf{t}]\{\mathbf{\$}[1-\mathbf{v}] \\
&\quad -\mathbf{\$'}[1+\mathbf{s}][\mathbf{f}+\mathbf{i}]\}) \\
&= 1-\{139,796/[1+0.9900]-64,720\} \\
&\quad /([1-0.3000]\{51,294 \\
&\quad [1-0.1900]-48,851 \\
&\quad [1+0.0500] \\
&\quad [0.5800+0.0100]\}) \\
&= \underline{30.00\%}
\end{aligned}
$$

Corporate IFRS-GAAP (B/S-I/S), ISBN 13: **978-1983566332**, ISBN 10: **1983566330**

Law-5454:

If (**t**= T/B= Tax Rate= 30.00%), (**W**= Wealth or Total Assets= 139,796 millions USD), (**d**= D/A= Dividend Portion or Payout= 30.00%), (**l**= L/E= Leverage or Gearing Ratio= 99.00%), (**$**= Sales or Revenues= 51,294 millions USD), (**f**= F/S= Fixed Portion= 58.00%), (**v**= V/S= Variable Portion= 19.00%), (**i**= I/S= Interest Portion= 1.00%), (**$'**= Sales of Past Year= 48,851 millions USD), and (**s**= [S/S']-1= Sales Growth= 5.00%), then it's (**U**= Utilized or Starting Capital planned), is:

$$U= W/[1+l]-[1-d][1-t]\{\$[1-v]$$
$$-\$'[1+s][f+i]\}$$
$$= 139{,}796/[1+0.9900]-[1-0.3000]$$
$$[1-0.3000]\{51{,}294$$
$$[1-0.1900]-48{,}851$$
$$[1+0.0500]$$
$$[0.5800+0.0100]\}$$
$$= 64{,}720 \text{ millions USD}$$

Corporate IFRS-GAAP (B/S-I/S), ISBN 13: **978-1983566332**, ISBN 10: **1983566330**

Law-5455:

If (**t**= T/B= Tax Rate= 30.00%), (**W**= Wealth or Total Assets= 139,796 millions USD), (**d**= D/A= Dividend Portion or Payout= 30.00%), (**U**= Utilized or Starting Capital= 64,720 millions USD), (**$**= Sales or Revenues= 51,294 millions USD), (**f**= F/S= Fixed Portion= 58.00%), (**v**= V/S= Variable Portion= 19.00%), (**i**= I/S= Interest Portion= 1.00%), (**$'**= Sales of Past Year= 48,851 millions USD), and (**s**= [S/S']-1= Sales Growth= 5.00%), then it's (**l**= Leverage or Gearing Ratio planned), is:

$$l = W/(U+[1\text{-}d][1\text{-}t]\{\$[1\text{-}v]$$
$$-\$'[1+s][f+i]\})\text{-}1$$
$$= 139,796/(64,720+[1\text{-}0.3000]$$
$$[1\text{-}0.3000]\{51,294$$
$$[1\text{-}0.1900]\text{-}48,851$$
$$[1+0.0500]$$
$$[0.5800+0.0100]\})\text{-}1$$
$$= 99.00\%$$

Corporate IFRS-GAAP (B/S-I/S), ISBN 13: **978-1983566332**, ISBN 10: **1983566330**

Law-5456:

If (**t**= T/B= Tax Rate= 30.00%), (**I**= L/E= Leverage
or Gearing Ratio= 99.00%), (**d**= D/A= Dividend
Portion or Payout= 30.00%), (**U**= Utilized or Starting
Capital= 64,720 millions USD), (**S**= Sales or
Revenues= 51,294 millions USD), (**F**= Fixed Cost=
29,740 millions USD), (**v**= V/S= Variable Portion=
19.00%), (**I**= Interest Expense= 513 millions USD),
(**S'**= Sales of Past Year= 48,851 millions USD), and
(**s**= [S/S']-1= Sales Growth= 5.00%), then it's (**W**=
Wealth or Total Assets planned), is:

$$W= [1+I](U+[1-d][1-t]\{S-F-I-S'v[1+s]\})$$
$$= [1+0.9900]\{64,720+[1-0.3000]$$
$$[1-0.3000]\{51,294-29,750$$
$$-513-48,851*0.1900$$
$$[1+0.0500]\})$$
$$= 139,796 \text{ millions USD}$$

Corporate IFRS-GAAP (B/S-I/S), ISBN 13: **978-1983566332**, ISBN 10: **1983566330**

Law-5457:

If (**t**= T/B= Tax Rate= 30.00%), (**l**= L/E= Leverage or Gearing Ratio= 99.00%), (**d**= D/A= Dividend Portion or Payout= 30.00%), (**U**= Utilized or Starting Capital= 64,720 millions USD), (**F**= Fixed Cost= 29,740 millions USD), (**W**= Wealth or Total Assets= 139,796 millions USD), (**v**= V/S= Variable Portion= 19.00%), (**I**= Interest Expense= 513 millions USD), (**S'**= Sales of Past Year= 48,851 millions USD), and (**s**= [S/S']-1= Sales Growth= 5.00%), then it's (**S**= Sales or Revenues planned), is:

$$S= S'v[1+s]+F+I+\{W/[1+l]-U\}$$
$$/\{[1-d][1-t]\}$$
$$= 48,851*0.1900\ [1+0.0500]$$
$$+29,750+513+\{139,796$$
$$/[1+0.9900]-64,720$$
$$+[1-0.3000][1-0.3000]\}$$
$$= 51,294 \text{ millions USD}$$

Corporate IFRS-GAAP (B/S-I/S), ISBN 13: **978-1983566332**, ISBN 10: **1983566330**

Law-5458:

If (**t**= T/B= Tax Rate= 30.00%), (**l**= L/E= Leverage or Gearing Ratio= 99.00%), (**d**= D/A= Dividend Portion or Payout= 30.00%), (**U**= Utilized or Starting Capital= 64,720 millions USD), (**F**= Fixed Cost= 29,740 millions USD), (**W**= Wealth or Total Assets= 139,796 millions USD), (**v**= V/S= Variable Portion= 19.00%), (**I**= Interest Expense= 513 millions USD), (**$**= Sales or Revenues= 51,294 millions USD), and (**s**= [S/S']-1= Sales Growth= 5.00%), then it's (**$'**= Sales Past), must be:

$$\$' = (\$\text{-}F\text{-}I\text{-}\{W/[1+I]\text{-}U\}/\{[1\text{-}d][1\text{-}t]\})$$
$$/\{v[1+s]\}$$
$$= (51,294\text{-}29,750\text{-}513\text{-}\{139,796$$
$$/[1+0.9900]\text{-}64,720\}$$
$$/\{[1\text{-}0.3000][1\text{-}0.3000]\})$$
$$/\{0.1900[1+0.0500]\}$$
$$= \underline{48,851} \text{ millions USD}$$

Corporate IFRS-GAAP (B/S-I/S), ISBN 13: **978-1983566332**, ISBN 10: **1983566330**

Law-5459:

If (**t**= T/B= Tax Rate= 30.00%), (**I**= L/E= Leverage or Gearing Ratio= 99.00%), (**d**= D/A= Dividend Portion or Payout= 30.00%), (**U**= Utilized or Starting Capital= 64,720 millions USD), (**F**= Fixed Cost= 29,740 millions USD), (**W**= Wealth or Total Assets= 139,796 millions USD), (**S'**= Sales of Past Year= 48,851 millions USD), (**I**= Interest Expense= 513 millions USD), (**S**= Sales or Revenues= 51,294 millions USD), and (**s**= [S/S']-1= Sales Growth= 5.00%), then it's (**v**= Variable Portion planned), is:

$$v = (S\text{-}F\text{-}I\text{-}\{W/[1+I]\text{-}U\}/\{[1\text{-}d][1\text{-}t]\})$$
$$/\{S'[1+s]\}$$
$$= (51,294\text{-}29,750\text{-}513\text{-}\{139,796$$
$$/[1+0.9900]\text{-}64,720\}$$
$$/\{[1\text{-}0.3000][1\text{-}0.3000]\})$$
$$/\{48,851[1+0.0500]\}$$
$$= 19.00\%$$

Corporate IFRS-GAAP (B/S-I/S), ISBN 13: **978-1983566332**, ISBN 10: **1983566330**

Law-5460:

If (t= T/B= Tax Rate= 30.00%), (l= L/E= Leverage or Gearing Ratio= 99.00%), (d= D/A= Dividend Portion or Payout= 30.00%), (U= Utilized or Starting Capital= 64,720 millions USD), (F= Fixed Cost= 29,740 millions USD), (W= Wealth or Total Assets= 139,796 millions USD), (S'= Sales of Past Year= 48,851 millions USD), (l= Interest Expense= 513 millions USD), (S= Sales or Revenues= 51,294 millions USD), and (v= V/S= Variable Portion= 19.00%), then it's (s= Sales Growth planned), is:

$$s = (S\text{-}F\text{-}l\text{-}\{W/[1+l\,]\text{-}U\}/\{[1\text{-}d][1\text{-}t]\})$$
$$/[S'v]\text{-}1$$
$$= (51,294\text{-}29,750\text{-}513\text{-}\{139,796$$
$$/[1+0.9900]\text{-}64,720\}$$
$$/\{[1\text{-}0.3000][1\text{-}0.3000]\})$$
$$/[48,851*0.1900]\text{-}1$$
$$= \underline{5.00\%}$$

Corporate IFRS-GAAP (B/S-I/S), ISBN 13: **978-1983566332**, ISBN 10: **1983566330**

Law-5461:

If (**t**= T/B= Tax Rate= 30.00%), (**I**= L/E= Leverage or Gearing Ratio= 99.00%), (**d**= D/A= Dividend Portion or Payout= 30.00%), (**U**= Utilized or Starting Capital= 64,720 millions USD), (**s**= [S/S']-1= Sales Growth= 5.00%), (**W**= Wealth or Total Assets= 139,796 millions USD), (**S'**= Sales of Past Year= 48,851 millions USD), (**I**= Interest Expense= 513 millions USD), (**S**= Sales or Revenues= 51,294 millions USD), and (**v**= V/S= Variable Portion= 19.00%), then it's (**F**= Fixed Cost planned), is:

$$\mathbf{F}= \mathbf{S}\text{-}\mathbf{S'v}[1+\mathbf{s}]\text{-}\mathbf{I}\text{-}\{\mathbf{W}/[1+\mathbf{I}]\text{-}\mathbf{U}\}$$
$$/\{[1\text{-}\mathbf{d}][1\text{-}\mathbf{t}]\}$$
$$= 51,294\text{-}48,851*0.1900$$
$$[1+0.0500]\text{-}513\text{-}\{139,796$$
$$/[1+0.9900]\text{-}64,720\}$$
$$/\{[1\text{-}0.3000][1\text{-}0.3000]\}$$
$$= 29,750 \text{ millions USD}$$

Corporate IFRS-GAAP (B/S-I/S), ISBN 13: **978-1983566332**, ISBN 10: **1983566330**

<u>Law-5462</u>:

If (**t**= T/B= Tax Rate= <u>30.00%</u>), (**I**= L/E= Leverage or Gearing Ratio= <u>99.00%</u>), (**d**= D/A= Dividend Portion or Payout= <u>30.00%</u>), (**U**= Utilized or Starting Capital= <u>64,720</u> millions USD), (**s**= [S/S']-1= Sales Growth= <u>5.00%</u>), (**W**= Wealth or Total Assets= <u>139,796</u> millions USD), (**S'**= Sales of Past Year= <u>48,851</u> millions USD), (**F**= Fixed Cost= <u>29,740</u> millions USD), (**S**= Sales or Revenues= <u>51,294</u> millions USD), and (**v**= V/S= Variable Portion= <u>19.00%</u>), then it's (**I**= Interest Expense planned), is:

$$I= S-S'v[1+s]-F-\{W/[1+I]-U\}/\{[1-d][1-t]\}$$
$$= 51,294-48,851*0.1900[1+0.0500]$$
$$-29,750-\{139,796$$
$$/[1+0.9900]-64,720\}$$
$$/\{[1-0.3000][1-0.3000]\}$$
$$= \underline{513} \text{ millions USD}$$

Corporate IFRS-GAAP (B/S-I/S), ISBN 13: **978-1983566332**, ISBN 10: **1983566330**

Law-5463:

If (I= Interest Expense= 513 millions USD), (I= L/E= Leverage or Gearing Ratio= 99.00%), (d= D/A= Dividend Portion or Payout= 30.00%), (U= Utilized or Starting Capital= 64,720 millions USD), (s= [S/S']-1= Sales Growth= 5.00%), (W= Wealth or Total Assets= 139,796 millions USD), (S'= Sales of Past Year= 48,851 millions USD), (F= Fixed Cost= 29,740 millions USD), (S= Sales or Revenues= 51,294 millions USD), and (v= V/S= Variable Portion= 19.00%), then it's (t= Tax Rate planned), is:

t= 1-{W/[1+I]-U}/([1-d]{S-$S'v$[1+s]-F-I})

= 1-{139,796/[1+0.9900]-64,720}
/([1-0.3000]{51,294
-48,851*0.1900
[1+0.0500]-29,750-513})

= 30.00%

Corporate IFRS-GAAP (B/S-I/S), ISBN 13: **978-1983566332**, ISBN 10: **1983566330**

Law-5464:

If (**I**= Interest Expense= 513 millions USD), (**I**= L/E= Leverage or Gearing Ratio= 99.00%), (**t**= T/B= Tax Rate= 30.00%), (**U**= Utilized or Starting Capital= 64,720 millions USD), (**s**= [S/S']-1= Sales Growth= 5.00%), (**W**= Wealth or Total Assets= 139,796 millions USD), (**S'**= Sales of Past Year= 48,851 millions USD), (**F**= Fixed Cost= 29,740 millions USD), (**S**= Sales or Revenues= 51,294 millions USD), and (**v**= V/S= Variable Portion= 19.00%), then it's (**d**= Dividend Portion or Payout planned), is:

$$d= 1-\{W/[1+I]-U\}/([1-t]\{S-S'v[1+s]-F-I\})$$
$$= 1-\{139,796/[1+0.9900]-64,720\}$$
$$/([1-0.3000]\{51,294$$
$$-48,851*0.1900[1+0.0500]$$
$$-29,750-513\})$$

$$= 30.00\%$$

Corporate IFRS-GAAP (B/S-I/S), ISBN 13: **978-1983566332**, ISBN 10: **1983566330**

Law-5465:

If (**I**= Interest Expense= 513 millions USD), (**l**= L/E= Leverage or Gearing Ratio= 99.00%), (**t**= T/B= Tax Rate= 30.00%), (**d**= D/A= Dividend Portion or Payout= 30.00%), (**s**= [S/S']-1= Sales Growth= 5.00%), (**W**= Wealth or Total Assets= 139,796 millions USD), (**S'**= Sales of Past Year= 48,851 millions USD), (**F**= Fixed Cost= 29,740 millions USD), (**S**= Sales or Revenues= 51,294 millions USD), and (**v**= V/S= Variable Portion= 19.00%), then it's (**U**= Utilized or Starting Capital planned), is:

$$U = W/[1+l]-[1-d][1-t]\{S-S'v[1+s]-F-I\}$$
$$= 139,796/[1+0.9900]-[1-0.3000]$$
$$[1-0.3000]\{51,294-48,851$$
$$*0.1900[1+0.0500]-29,750$$
$$-513\}$$
$$= 64,720 \text{ millions USD}$$

Corporate IFRS-GAAP (B/S-I/S), ISBN 13: **978-1983566332**, ISBN 10: **1983566330**

Law-5466:

If (**I**= Interest Expense= 513 millions USD), (**U**= Utilized or Starting Capital= 64,720 millions USD), (**t**= T/B= Tax Rate= 30.00%), (**d**= D/A= Dividend Portion or Payout= 30.00%), (**s**= [S/S']-1= Sales Growth= 5.00%), (**W**= Wealth or Total Assets= 139,796 millions USD), (**S'**= Sales of Past Year= 48,851 millions USD), (**F**= Fixed Cost= 29,740 millions USD), (**S**= Sales or Revenues= 51,294 millions USD), and (**v**= V/S= Variable Portion= 19.00%), then it's (**I** = Leverage or gearing Ratio planned), is:

$$I = W/(U+[1-d][1-t]\{S-S'v[1+s]-F-I\})-1$$
$$= 139,796/(64,720+[1-0.3000]$$
$$[1-0.3000]\{51,294-48,851$$
$$*0.1900[1+0.0500]$$
$$-29,750-513\})-1$$
$$= 99.00\%$$

<u>Law-5467</u>:

If (**i**= I/S= Interest Portion= <u>1.00%</u>), (**U**= Utilized or Starting Capital= <u>64,720</u> millions USD), (**t**= T/B= Tax Rate= <u>30.00%</u>), (**d**= D/A= Dividend Portion or Payout= <u>30.00%</u>), (**s**= [S/S']-1= Sales Growth= <u>5.00%</u>), (**l**= L/E= Leverage or Gearing Ratio= <u>99.00%</u>), (**S'**= Sales of Past Year= <u>48,851</u> millions USD), (**F**= Fixed Cost= <u>29,740</u> millions USD), (**S**= Sales or Revenues= <u>51,294</u> millions USD), and (**v**= V/S= Variable Portion= <u>19.00%</u>), then it's (**W**= Wealth or Total Assets planned), is:

$$W= [1+l](U+[1-d][1-t]\{S-S'v[1+s]-F-Si\})$$
$$=[1+l](U+[1-d][1-t]\{S[1-i]$$
$$-S'v[1+s]-F\})$$
$$= [1+0.9900](64,720+[1-0.3000]$$
$$[1-0.3000]\{51,294$$
$$[1-0.0100]-48,851$$
$$*0.1900[1+0.0500]$$
$$-29,750\})$$
$$= \underline{139,796} \text{ millions USD}$$

Corporate IFRS-GAAP (B/S-I/S), ISBN 13: **978-1983566332**, ISBN 10: **1983566330**

<u>Law-5468</u>:

If (**i**= I/S= Interest Portion= <u>1.00%</u>), (**U**= Utilized or Starting Capital= <u>64,720</u> millions USD), (**t**= T/B= Tax Rate= <u>30.00%</u>), (**d**= D/A= Dividend Portion or Payout= <u>30.00%</u>), (**s**= [S/S']-1= Sales Growth= <u>5.00%</u>), (**l**= L/E= Leverage or Gearing Ratio= <u>99.00%</u>), (**S'**= Sales of Past Year= <u>48,851</u> millions USD), (**F**= Fixed Cost= <u>29,740</u> millions USD), (**W**= Wealth or Total Assets= <u>139,796</u> millions USD), and (**v**= V/S= Variable Portion= <u>19.00%</u>), then it's (**S**= Sales or Revenues planned), is:

$$S= (S'v[1+s]+F+\{W/[1+l]-U\}/\{[1-d][1-t]\})/[1-i]$$

$$= (48,851*0.1900[1+0.0500]$$
$$+29,750 +\{139,796$$
$$/[1+0.9900]-64,720\}$$
$$/\{[1-0.3000][1-0.3000]\})$$
$$/[1-0.0100]$$

$$= \underline{51,294} \text{ millions USD}$$

Corporate IFRS-GAAP (B/S-I/S), ISBN 13: **978-1983566332**, ISBN 10: **1983566330**

Law-5469:

If ($= Sales or Revenues= <u>51,294</u> millions USD), (**U**= Utilized or Starting Capital= <u>64,720</u> millions USD), (**t**= T/B= Tax Rate= <u>30.00%</u>), (**d**= D/A= Dividend Portion or Payout= <u>30.00%</u>), (**s**= [S/S']-1= Sales Growth= <u>5.00%</u>), (**ſ**= L/E= Leverage or Gearing Ratio= <u>99.00%</u>), (**$'**= Sales of Past Year= <u>48,851</u> millions USD), (**F**= Fixed Cost= <u>29,740</u> millions USD), (**W**= Wealth or Total Assets= <u>139,796</u> millions USD), and (**v**= V/S= Variable Portion= <u>19.00%</u>), then it's (**i**= Interest Portion planned), is:

$$i= 1-(\$'v[1+s]+F+\{W/[1+ſ]-U\}$$
$$/\{[1-d][1-t]\})/\$$$
$$= 1-(48,851*0.1900[1+0.0500]$$
$$+29,750 +\{139,796$$
$$/[1+0.9900]-64,720\}$$
$$/\{[1-0.3000][1-0.3000]\})$$
$$/51,294$$

$$= \underline{1.00\%}$$

Corporate IFRS-GAAP (B/S-I/S), ISBN 13: **978-1983566332**, ISBN 10: **1983566330**

<u>Law-5470</u>:

If (**$**= Sales or Revenues= <u>51,294</u> millions USD), (**U**= Utilized or Starting Capital= <u>64,720</u> millions USD), (**t**= T/B= Tax Rate= <u>30.00%</u>), (**d**= D/A= Dividend Portion or Payout= <u>30.00%</u>), (**s**= [S/S']-1= Sales Growth= <u>5.00%</u>), (**l**= L/E= Leverage or Gearing Ratio= <u>99.00%</u>), (**i**= I/S= Interest Portion= <u>1.00%</u>), (**F**= Fixed Cost= <u>29,740</u> millions USD), (**W**= Wealth or Total Assets= <u>139,796</u> millions USD), and (**v**= V/S= Variable Portion= <u>19.00%</u>), then it's (**$'**= Sales Past), must be:

$$\$' = (\$[1-i]-F-\{W/[1+l]-U\}/\{[1-d][1-t]\})$$
$$/\{v[1+s]\}$$
$$= (51,294[1-0.0100]-29,750$$
$$-\{139,796/[1+0.9900]$$
$$-64,720\}/\{[1-0.3000]$$
$$[1-0.3000]\})$$
$$/\{0.1900[1+0.0500]\}$$
$$= \underline{48,851} \text{ millions USD}$$

Corporate IFRS-GAAP (B/S-I/S), ISBN 13: **978-1983566332**, ISBN 10: **1983566330**

Law-5471:

If ($\textbf{S}$= Sales or Revenues= 51,294 millions USD), ($\textbf{U}$= Utilized or Starting Capital= 64,720 millions USD), ($\textbf{t}$= T/B= Tax Rate= 30.00%), ($\textbf{d}$= D/A= Dividend Portion or Payout= 30.00%), ($\textbf{s}$= [S/S']-1= Sales Growth= 5.00%), ($\textbf{f}$= L/E= Leverage or Gearing Ratio= 99.00%), ($\textbf{i}$= I/S= Interest Portion= 1.00%), ($\textbf{F}$= Fixed Cost= 29,740 millions USD), ($\textbf{W}$= Wealth or Total Assets= 139,796 millions USD), and ($\textbf{S'}$= Sales of Past Year= 48,851 millions USD), then it's ($\textbf{v}$= Variable Portion planned), is:

$$\textbf{v}= (\textbf{S}[1\text{-}\textbf{i}]\text{-}\textbf{F}\text{-}\{\textbf{W}/[1+\textbf{f}]\text{-}\textbf{U}\}/\{[1\text{-}\textbf{d}][1\text{-}\textbf{t}]\})/\{\textbf{S'}[1+\textbf{s}]\}$$

$$= (51,294[1\text{-}0.0100]\text{-}29,750$$
$$-\{139,796/[1+0.9900]$$
$$-64,720\}/\{[1\text{-}0.3000]$$
$$[1\text{-}0.3000]\})$$
$$/\{48,851[1+0.0500]\}$$

$$= \underline{19.00\%}$$

Corporate IFRS-GAAP (B/S-I/S), ISBN 13: **978-1983566332**, ISBN 10: **1983566330**

Law-5472:

If ($= Sales or Revenues= 51,294 millions USD), (**U**= Utilized or Starting Capital= 64,720 millions USD), (**t**= T/B= Tax Rate= 30.00%), (**d**= D/A= Dividend Portion or Payout= 30.00%), (**v**= V/S= Variable Portion= 19.00%), (**l**= L/E= Leverage or Gearing Ratio= 99.00%), (**i**= I/S= Interest Portion= 1.00%), (**F**= Fixed Cost= 29,740 millions USD), (**W**= Wealth or Total Assets= 139,796 millions USD), and ($'= Sales of Past Year= 48,851 millions USD), then it's (**s**= Sales Growth planned), is:

$$s = (\$[1-\mathbf{i}]-\mathbf{F}-\{\mathbf{W}/[1+\mathbf{l}]-\mathbf{U}\}/\{[1-\mathbf{d}][1-\mathbf{t}]\})/[\$'\mathbf{v}]-1$$

$$= (51,294[1-0.0100]-29,750$$
$$-\{139,796/[1+0.9900]$$
$$-64,720\}/\{[1-0.3000]$$
$$[1-0.3000]\})/[48,851$$
$$*0.1900]-1$$

$$= 5.00\%$$

Corporate IFRS-GAAP (B/S-I/S), ISBN 13: **978-1983566332**, ISBN 10: **1983566330**

Law-5473:

If (**$**= Sales or Revenues= 51,294 millions USD), (**U**= Utilized or Starting Capital= 64,720 millions USD), (**t**= T/B= Tax Rate= 30.00%), (**d**= D/A= Dividend Portion or Payout= 30.00%), (**v**= V/S= Variable Portion= 19.00%), (**I**= L/E= Leverage or Gearing Ratio= 99.00%), (**i**= I/S= Interest Portion= 1.00%), (**s**= [S/S']-1= Sales Growth= 5.00%), (**W**= Wealth or Total Assets= 139,796 millions USD), and (**$'**= Sales of Past Year= 48,851 millions USD), then it's (**F**= Fixed Cost planned), is:

$$F= \$[1\text{-}i]\text{-}\$'v][1+s]\text{-}\{W/[1+I]\text{-}U\}$$
$$/\{[1\text{-}d][1\text{-}t]\}$$
$$= (51,294[1\text{-}0.0100]\text{-} 48,851$$
$$*0.1900][1+0.0500]$$
$$-\{139,796/[1+0.9900]$$
$$-64,720\}/\{[1\text{-}0.3000]$$
$$[1\text{-}0.3000]\}$$
$$= 29,750 \text{ millions USD}$$

Corporate IFRS-GAAP (B/S-I/S), ISBN 13: **978-1983566332**, ISBN 10: **1983566330**

<u>Law-5474</u>:

If (**$**= Sales or Revenues= <u>51,294</u> millions USD), (**U**= Utilized or Starting Capital= <u>64,720</u> millions USD), (**F**= Fixed Cost= <u>29,740</u> millions USD), (**d**= D/A= Dividend Portion or Payout= <u>30.00%</u>), (**v**= V/S= Variable Portion= <u>19.00%</u>), (**ℓ**= L/E= Leverage or Gearing Ratio= <u>99.00%</u>), (**i**= I/S= Interest Portion= <u>1.00%</u>), (**s**= [S/S']-1= Sales Growth= <u>5.00%</u>), (**W**= Wealth or Total Assets= <u>139,796</u> millions USD), and (**$'**= Sales of Past Year= <u>48,851</u> millions USD), then it's (**t**= Tax Rate planned), is:

$$t = 1 - \{W/[1+\ell]-U\}/([1-d]\{\$[1-i]-\$'v]$$
$$[1+s]-F\})$$
$$= 1 - \{139{,}796/[1+0.9900]-64{,}720\}$$
$$/([1-0.3000]\{51{,}294$$
$$[1-0.0100]-48{,}851*0.1900]$$
$$[1+0.0500]-29{,}750\})$$

$$= \underline{30.00\%}$$

Corporate IFRS-GAAP (B/S-I/S), ISBN 13: **978-1983566332**, ISBN 10: **1983566330**

Law-5475:

If (**$**= Sales or Revenues= 51,294 millions USD), (**U**= Utilized or Starting Capital= 64,720 millions USD), (**F**= Fixed Cost= 29,740 millions USD), (**t**= T/B= Tax Rate= 30.00%), (**v**= V/S= Variable Portion= 19.00%), (**l**= L/E= Leverage or Gearing Ratio= 99.00%), (**i**= I/S= Interest Portion= 1.00%), (**s**= [S/S']-1= Sales Growth= 5.00%), (**W**= Wealth or Total Assets= 139,796 millions USD), and (**$'**= Sales of Past Year= 48,851 millions USD), then it's (**d**= Dividend Portion or Payout planned), is:

$$d= 1-\{W/[1+l]-U\}/([1-t]\{\$[1-i]-\$'v][1+s]-F\})$$
$$= 1-\{139,796/[1+0.9900]-64,720\}/([1-0.3000]\{51,294[1-0.0100]-48,851*0.1900][1+0.0500]-29,750\})$$
$$= 30.00\%$$

Corporate IFRS-GAAP (B/S-I/S), ISBN 13: **978-1983566332**, ISBN 10: **1983566330**

Law-5476:

If ($\$$= Sales or Revenues= 51,294 millions USD), (**d**= D/A= Dividend Portion or Payout= 30.00%), (**F**= Fixed Cost= 29,740 millions USD), (**t**= T/B= Tax Rate= 30.00%), (**v**= V/S= Variable Portion= 19.00%), (**I**= L/E= Leverage or Gearing Ratio= 99.00%), (**i**= I/S= Interest Portion= 1.00%), (**s**= [S/S']-1= Sales Growth= 5.00%), (**W**= Wealth or Total Assets= 139,796 millions USD), and ($\$'$= Sales of Past Year= 48,851 millions USD), then it's (**U**= Utilized or Starting Capital planned), is:

$$\mathbf{U}= \mathbf{W}/[1+\mathbf{I}]-[1-\mathbf{d}][1-\mathbf{t}]\{\$[1-\mathbf{i}]-\$'\mathbf{v}][1+\mathbf{s}]-\mathbf{F}\}$$

$$= 139,796/[1+0.9900]-[1-0.3000]$$
$$[1-0.3000]\{51,294$$
$$[1-0.0100]-48.851*0.1900]$$
$$[1+0.0500]-29,750\}$$

$$= \underline{64,720} \text{ millions USD}$$

Corporate IFRS-GAAP (B/S-I/S), ISBN 13: **978-1983566332**, ISBN 10: **1983566330**

Law-5477:

If ($ = Sales or Revenues= 51,294 millions USD), ($d=$ D/A= Dividend Portion or Payout= 30.00%), ($F=$ Fixed Cost= 29,740 millions USD), ($t=$ T/B= Tax Rate= 30.00%), ($v=$ V/S= Variable Portion= 19.00%), ($U=$ Utilized or Starting Capital= 64,720 millions USD), ($i=$ I/S= Interest Portion= 1.00%), ($s=$ [S/S']-1= Sales Growth= 5.00%), ($W=$ Wealth or Total Assets= 139,796 millions USD), and ($'=$ Sales of Past Year= 48,851 millions USD), then it's ($I=$ Leverage or Gearing Ratio planned), is:

$$I = W/(U+[1-d][1-t]\{\$[1-i]-\$'v][1+s]-F\})-1$$
$$= 139,796/(64,720+[1-0.3000]$$
$$[1-0.3000]\{51,294[1-0.0100]$$
$$-48,851*0.1900][1+0.0500]$$
$$-29,750\})-1$$
$$= 99.00\%$$

Corporate IFRS-GAAP (B/S-I/S), ISBN 13: **978-1983566332**, ISBN 10: **1983566330**

Law-5478:

If ($= Sales or Revenues= 51,294 millions USD), (**d**= D/A= Dividend Portion or Payout= 30.00%), (**F**= Fixed Cost= 29,740 millions USD), (**t**= T/B= Tax Rate= 30.00%), (**v**= V/S= Variable Portion= 19.00%), (**U**= Utilized or Starting Capital= 64,720 millions USD), (**i**= I/S= Interest Portion= 1.00%), (**s**= [S/S']-1= Sales Growth= 5.00%), (**l**= L/E= Leverage or Gearing Ratio= 99.00%), and (**$'**= Sales of Past Year= 48,851 millions USD), then it's (**W**= Wealth or Total Assets planned), is:

$$\mathbf{W}= [1+\mathbf{\mathit{l}}](\mathbf{U}+[1-\mathbf{d}][1-\mathbf{t}]\{\$-\$'\mathbf{v}][1+\mathbf{s}]-\mathbf{F}$$
$$-\$'\mathbf{i}[1+\mathbf{s}]\})$$

$$= [1+\mathbf{\mathit{l}}](\mathbf{U}+[1-\mathbf{d}][1-\mathbf{t}]\{\$-\mathbf{F}-\$'[1+\mathbf{s}]$$
$$[\mathbf{v}+\mathbf{i}]\})$$

$$= [1+0.9900](64,720+[1-0.3000]$$
$$[1-0.3000]\{51,294-29,750$$
$$-48,851[1+0.0500]$$
$$[0.1900+0.0100]\})$$

$$= 139,796 \text{ millions USD}$$

Corporate IFRS-GAAP (B/S-I/S), ISBN 13: **978-1983566332**, ISBN 10: **1983566330**

<u>Law-5479</u>:

If (**W**= Wealth or Total Assets= <u>139,796</u> millions USD), (**d**= D/A= Dividend Portion or Payout= <u>30.00%</u>), (**F**= Fixed Cost= <u>29,740</u> millions USD), (**t**= T/B= Tax Rate= <u>30.00%</u>), (**v**= V/S= Variable Portion= <u>19.00%</u>), (**U**= Utilized or Starting Capital= <u>64,720</u> millions USD), (**i**= I/S= Interest Portion= <u>1.00%</u>), (**s**= [S/S']-1= Sales Growth= <u>5.00%</u>), (**l**= L/E= Leverage or Gearing Ratio= <u>99.00%</u>), and (**S'**= Sales of Past Year= <u>48,851</u> millions USD), then it's (**S**= Sales or Revenues planned), is:

$$S = F + S'[1+s][v+i] + \{W/[1+l] - U\}$$
$$/\{[1-d][1-t]\}$$
$$= 29,750 + 48,851[1+0.0500]$$
$$[0.1900+0.0100] + \{139,796$$
$$/[1+0.9900] - 64,720\}$$
$$/\{[1-0.3000][1-0.3000]\}$$
$$= \underline{51,294} \text{ millions USD}$$

Corporate IFRS-GAAP (B/S-I/S), ISBN 13: **978-1983566332**, ISBN 10: **1983566330**

Law-5480:

If (**W**= Wealth or Total Assets= 139,796 millions USD), (**d**= D/A= Dividend Portion or Payout= 30.00%), (**F**= Fixed Cost= 29,740 millions USD), (**t**= T/B= Tax Rate= 30.00%), (**v**= V/S= Variable Portion= 19.00%), (**U**= Utilized or Starting Capital= 64,720 millions USD), (**i**= I/S= Interest Portion= 1.00%), (**s**= [S/S']-1= Sales Growth= 5.00%), (**l**= L/E= Leverage or Gearing Ratio= 99.00%), and (**S**= Sales or Revenues= 51,294 millions USD), then it's (**S'**= Sales Past), must be:

$$S' = (S-F-W/[1+l]-U\}/\{[1-d][1-t]\})$$
$$/\{[1+s][v+i]\}$$
$$= (51,294-29,750-\{139,796$$
$$/[1+0.9900]-64,720\}$$
$$/\{[1-0.3000][1-0.3000]\})$$
$$/\{[1+0.0500]$$
$$[0.1900+0.0100]\}$$
$$= \underline{48,851} \text{ millions USD}$$

Corporate IFRS-GAAP (B/S-I/S), ISBN 13: **978-1983566332**, ISBN 10: **1983566330**

Law-5481:

If (W= Wealth or Total Assets= 139,796 millions USD), (d= D/A= Dividend Portion or Payout= 30.00%), (F= Fixed Cost= 29,740 millions USD), (t= T/B= Tax Rate= 30.00%), (S'= Sales of Past Year= 48,851 millions USD), (U= Utilized or Starting Capital= 64,720 millions USD), (i= I/S= Interest Portion= 1.00%), (s= [S/S']-1= Sales Growth= 5.00%), (l= L/E= Leverage or Gearing Ratio= 99.00%), and (S= Sales or Revenues= 51,294 millions USD), then it's (v= Variable Portion planned), is:

$$v = (S\text{-}F\text{-}W/[1+l]\text{-}U\}/\{[1\text{-}d][1\text{-}t]\})/\{S'[1+s]\}\text{-}i$$
$$= (51,294\text{-}29,750\text{-}\{139,796$$
$$/[1+0.9900]\text{-}64,720\}$$
$$/\{[1\text{-}0.3000][1\text{-}0.3000]\})$$
$$/\{48,851[1+0.0500]\}\text{-}0.0100$$
$$= 19.00\%$$

Corporate IFRS-GAAP (B/S-I/S), ISBN 13: **978-1983566332**, ISBN 10: **1983566330**

Law-5482:

If (**W**= Wealth or Total Assets= 139,796 millions USD), (**d**= D/A= Dividend Portion or Payout= 30.00%), (**F**= Fixed Cost= 29,740 millions USD), (**t**= T/B= Tax Rate= 30.00%), (**S'**= Sales of Past Year= 48,851 millions USD), (**U**= Utilized or Starting Capital= 64,720 millions USD), (**v**= V/S= Variable Portion= 19.00%), (**s**= [S/S']-1= Sales Growth= 5.00%), (**l**= L/E= Leverage or Gearing Ratio= 99.00%), and (**S**= Sales or Revenues= 51,294 millions USD), then it's (**i**= Interest Portion planned), is:

$$i = (S\text{-}F\text{-}W/[1+l]\text{-}U\}/\{[1\text{-}d][1\text{-}t]\})$$
$$/\{S'[1+s]\}\text{-}v$$
$$= (51,294\text{-}29,750\text{-}\{139,796$$
$$/[1+0.9900]\text{-}64,720\}$$
$$/\{[1\text{-}0.3000][1\text{-}0.3000]\})$$
$$/\{48,851[1+0.0500]\}\text{-}0.1900$$
$$= \underline{1.00}\%$$

Corporate IFRS-GAAP (B/S-I/S), ISBN 13: **978-1983566332**, ISBN 10: **1983566330**

<u>Law-5483</u>:

If (**W**= Wealth or Total Assets= <u>139,796</u> millions USD), (**d**= D/A= Dividend Portion or Payout= <u>30.00%</u>), (**F**= Fixed Cost= <u>29,740</u> millions USD), (**t**= T/B= Tax Rate= <u>30.00%</u>), (**S'**= Sales of Past Year= <u>48,851</u> millions USD), (**U**= Utilized or Starting Capital= <u>64,720</u> millions USD), (**v**= V/S= Variable Portion= <u>19.00%</u>), (**i**= I/S= Interest Portion= <u>1.00%</u>), (**l**= L/E= Leverage or Gearing Ratio= <u>99.00%</u>), and (**S**= Sales or Revenues= <u>51,294</u> millions USD), then it's (**s**= Sales Growth planned), is:

$$s = (\textbf{S-F-W}/[1+\textbf{\textit{l}}]-\textbf{U}\}/\{[1-\textbf{d}][1-\textbf{t}]\})$$
$$/\{\textbf{S'}[\textbf{v+i}]\}-1$$
$$= (51,294-29,750-\{139,796$$
$$/[1+0.9900]-64,720\}$$
$$/\{[1-0.3000][1-0.3000]\})$$
$$/\{48,851[0.1900+0.0100]\}-1$$
$$= \underline{5.00\%}$$

Corporate IFRS-GAAP (B/S-I/S), ISBN 13: **978-1983566332**, ISBN 10: **1983566330**

Law-5484:

If (**W**= Wealth or Total Assets= 139,796 millions USD), (**d**= D/A= Dividend Portion or Payout= 30.00%), (**s**= [S/S']-1= Sales Growth= 5.00%), (**t**= T/B= Tax Rate= 30.00%), (**S'**= Sales of Past Year= 48,851 millions USD), (**U**= Utilized or Starting Capital= 64,720 millions USD), (**v**= V/S= Variable Portion= 19.00%), (**i**= I/S= Interest Portion= 1.00%), (**I**= L/E= Leverage or Gearing Ratio= 99.00%), and (**S**= Sales or Revenues= 51,294 millions USD), then it's (**F**= Fixed Cost planned), is:

$$F= S-S'[1+s][v+i]-W/[1+I]-U\}/\{[1-d][1-t]\}$$
$$= 51,294-48,851[0.1900+0.0100]$$
$$-\{139,796/[1+0.9900]$$
$$-64,720\}/\{[1-0.3000]$$
$$[1-0.3000]\}$$
$$= 29,750 \text{ millions USD}$$

Corporate IFRS-GAAP (B/S-I/S), ISBN 13: **978-1983566332**, ISBN 10: **1983566330**

<u>Law-5485</u>:

If (**W**= Wealth or Total Assets= <u>139,796</u> millions USD), (**d**= D/A= Dividend Portion or Payout= <u>30.00%</u>), (**s**= [S/S']-1= Sales Growth= <u>5.00%</u>), (**F**= Fixed Cost= <u>29,740</u> millions USD), (**S'**= Sales of Past Year= <u>48,851</u> millions USD), (**U**= Utilized or Starting Capital= <u>64,720</u> millions USD), (**v**= V/S= Variable Portion= <u>19.00%</u>), (**i**= I/S= Interest Portion= <u>1.00%</u>), (**l**= L/E= Leverage or Gearing Ratio= <u>99.00%</u>), and (**S**= Sales or Revenues= <u>51,294</u> millions USD), then it's (**t**= Tax Rate planned), is:

$$t= 1-\{W/[1+l]-U\}/([1-d]\{S-S'[1+s][v+i]-F\})$$
$$= 1-\{139,796/[1+0.9900]-64,720\}$$
$$/([1-0.3000]\{51,294-48,851$$
$$[0.1900+0.0100]-29,750\})$$
$$= \underline{30.00\%}$$

Corporate IFRS-GAAP (B/S-I/S), ISBN 13: **978-1983566332**, ISBN 10: **1983566330**

Law-5486:

If (**W**= Wealth or Total Assets= 139,796 millions USD), (**t**= T/B= Tax Rate= 30.00%), (**s**= [S/S']-1= Sales Growth= 5.00%), (**F**= Fixed Cost= 29,740 millions USD), (**$'**= Sales of Past Year= 48,851 millions USD), (**U**= Utilized or Starting Capital= 64,720 millions USD), (**v**= V/S= Variable Portion= 19.00%), (**i**= I/S= Interest Portion= 1.00%), (**l**= L/E= Leverage or Gearing Ratio= 99.00%), and (**$**= Sales or Revenues= 51,294 millions USD), then it's (**d**= Dividend Portion or Payout planned), is:

$$\mathbf{d} = 1 - \{\mathbf{W}/[1+\mathbf{l}] - \mathbf{U}\}/([1-\mathbf{t}]\{\mathbf{\$}-\mathbf{\$'}[1+\mathbf{s}][\mathbf{v}+\mathbf{i}]-\mathbf{F}\})$$

$$= 1 - \{139{,}796/[1+0.9900] - 64{,}720\}$$
$$/([1-0.3000]\{51{,}294-48{,}851$$
$$[0.1900+0.0100]-29{,}750\})$$

$$= 30.00\%$$

Corporate IFRS-GAAP (B/S-I/S), ISBN 13: **978-1983566332**, ISBN 10: **1983566330**

<u>Law-5487</u>:

If (**W**= Wealth or Total Assets= <u>139,796</u> millions USD), (**t**= T/B= Tax Rate= <u>30.00%</u>), (**s**= [S/S']-1= Sales Growth= <u>5.00%</u>), (**F**= Fixed Cost= <u>29,740</u> millions USD), (**S'**= Sales of Past Year= <u>48,851</u> millions USD), (**d**= D/A= Dividend Portion or Payout= <u>30.00%</u>), (**v**= V/S= Variable Portion= <u>19.00%</u>), (**i**= I/S= Interest Portion= <u>1.00%</u>), (**l**= L/E= Leverage or Gearing Ratio= <u>99.00%</u>), and (**S**= Sales or Revenues= <u>51,294</u> millions USD), then it's (**U**= Utilized or Starting Capital planned), is:

$$\mathbf{U}= \mathbf{W}/[1+\mathbf{\mathit{l}}\,]-[1-\mathbf{d}][1-\mathbf{t}]\{\mathbf{S}-\mathbf{S'}[1+\mathbf{s}][\mathbf{v}+\mathbf{i}]-\mathbf{F}\}$$

$$= 139{,}796/[1+0.9900]-[1-0.3000]$$
$$[1-0.3000]\{51{,}294-48{,}851$$
$$[0.1900+0.0100]-29{,}750\}$$

$$= \underline{64{,}720} \text{ millions USD}$$

Corporate IFRS-GAAP (B/S-I/S), ISBN 13: **978-1983566332**, ISBN 10: **1983566330**

<u>Law-5488</u>:

If (**W**= Wealth or Total Assets= <u>139,796</u> millions USD), (**t**= T/B= Tax Rate= <u>30.00%</u>), (**s**= [S/S']-1= Sales Growth= <u>5.00%</u>), (**F**= Fixed Cost= <u>29,740</u> millions USD), (**\$'**= Sales of Past Year= <u>48,851</u> millions USD), (**d**= D/A= Dividend Portion or Payout= <u>30.00%</u>), (**v**= V/S= Variable Portion= <u>19.00%</u>), (**i**= I/S= Interest Portion= <u>1.00%</u>), (**U**= Utilized or Starting Capital= <u>64,720</u> millions USD), and (**\$**= Sales or Revenues= <u>51,294</u> millions USD), then it's (**I** = Leverage or Gearing Ratio planned), is:

$$I = W/(U+[1-d][1-t]\{S-S'[1+s][v+i]-F\})-1$$
$$= 139,796/(64,720+[1-0.3000]$$
$$[1-0.3000]\{51,294-48,851$$
$$[0.1900+0.0100]-29,750\})-1$$
$$= \underline{99.00\%}$$

Corporate IFRS-GAAP (B/S-I/S), ISBN 13: **978-1983566332**, ISBN 10: **1983566330**

Law-5489:

If ($\textit{\textbf{f}}$= L/E= Leverage or Gearing Ratio= 99.00%), ($\textbf{t}$= T/B= Tax Rate= 30.00%), ($\textbf{s}$= [S/S']-1= Sales Growth= 5.00%), ($\textbf{f}$= F/S= Fixed Portion= 58.00%), ($\textbf{S'}$= Sales of Past Year= 48,851 millions USD), ($\textbf{d}$= D/A= Dividend Portion or Payout= 30.00%), ($\textbf{v}$= V/S= Variable Portion= 19.00%), ($\textbf{I}$= Interest Expense= 513 millions USD), ($\textbf{U}$= Utilized or Starting Capital= 64,720 millions USD), and ($\textbf{S}$= Sales or Revenues= 51,294 millions USD), then it's ($\textbf{W}$= Wealth or Total Assets planned), is:

$$\textbf{W} = [1+\textit{\textbf{f}}](\textbf{U}+[1-\textbf{d}][1-\textbf{t}]\{\textbf{S}-\textbf{S'v}[1+\textbf{s}]-\textbf{Sf}-\textbf{I}\})$$

$$=[1+\textit{\textbf{f}}](\textbf{U}+[1-\textbf{d}][1-\textbf{t}]\{\textbf{S}[1-\textbf{f}]$$
$$-\textbf{S'v}[1+\textbf{s}]-\textbf{I}\})$$

$$= [1+0.9900](64{,}720+[1-0.3000]$$
$$[1-0.3000]\{51{,}294[1-0.5800]$$
$$-48{,}851*0.1900\,[1+0.0500]$$
$$-513\})$$

$$= 139{,}796 \text{ millions USD}$$

Corporate IFRS-GAAP (B/S-I/S), ISBN 13: **978-1983566332**, ISBN 10: **1983566330**

<u>Law-5490</u>:

If (I= L/E= Leverage or Gearing Ratio= <u>99.00%</u>), (t= T/B= Tax Rate= <u>30.00%</u>), (s= [S/S']-1= Sales Growth= <u>5.00%</u>), (f= F/S= Fixed Portion= <u>58.00%</u>), (S'= Sales of Past Year= <u>48,851</u> millions USD), (d= D/A= Dividend Portion or Payout= <u>30.00%</u>), (v= V/S= Variable Portion= <u>19.00%</u>), (I= Interest Expense= <u>513</u> millions USD),(U= Utilized or Starting Capital= <u>64,720</u> millions USD), and (W= Wealth or Total Assets= <u>139,796</u> millions USD), then it's (S= Sales or Revenues planned), is:

$$S= (S'v[1+s]+I+\{W/[1+I]-U\}/\{[1-d][1-t]\})$$
$$/[1-f]$$
$$= (48,851*0.1900][1+0.0500]+513$$
$$+\{139,796/[1+0.9900]$$
$$-64,720\}/\{[1-0.3000]$$
$$[1-0.3000]\})/[1-0.5800]$$
$$= \underline{51,294} \text{ millions USD}$$

Corporate IFRS-GAAP (B/S-I/S), ISBN 13: **978-1983566332**, ISBN 10: **1983566330**

Law-5491:

If ($\textbf{\textit{l}}$= L/E= Leverage or Gearing Ratio= 99.00%), ($\textbf{t}$= T/B= Tax Rate= 30.00%), ($\textbf{s}$= [S/S']-1= Sales Growth= 5.00%), ($\textbf{\textit{S}}$= Sales or Revenues= 51,294 millions USD), ($\textbf{\textit{S}'}$= Sales of Past Year= 48,851 millions USD), ($\textbf{d}$= D/A= Dividend Portion or Payout= 30.00%), ($\textbf{v}$= V/S= Variable Portion= 19.00%), ($\textbf{\textit{l}}$= Interest Expense= 513 millions USD),($\textbf{U}$= Utilized or Starting Capital= 64,720 millions USD), and ($\textbf{W}$= Wealth or Total Assets= 139,796 millions USD), then it's ($\textbf{f}$= Fixed Portion planned), is:

$$\textbf{f}= 1-(\textbf{\textit{S}'}\textbf{v}[1+\textbf{s}]+\textbf{\textit{l}}+\{\textbf{W}/[1+\textbf{\textit{l}}]-\textbf{U}\}$$
$$/\{[1-\textbf{d}][1-\textbf{t}]\})/\textbf{\textit{S}}$$
$$= 1-(48,851*0.1900[1+0.0500]+513$$
$$+\{139,796/[1+0.9900]$$
$$-64,720\}/\{[1-0.3000]$$
$$[1-0.3000]\})/51,294$$
$$= 58.00\%$$

Corporate IFRS-GAAP (B/S-I/S), ISBN 13: **978-1983566332**, ISBN 10: **1983566330**

<u>Law-5492</u>:

If (I= L/E= Leverage or Gearing Ratio= <u>99.00%</u>), (t= T/B= Tax Rate= <u>30.00%</u>), (s= [S/S']-1= Sales Growth= <u>5.00%</u>), (S= Sales or Revenues= <u>51,294</u> millions USD), (f= F/S=Fixed Portion= <u>58.00%</u>), (d= D/A= Dividend Portion or Payout= <u>30.00%</u>), (v= V/S= Variable Portion= <u>19.00%</u>), (I= Interest Expense= <u>513</u> millions USD),(U= Utilized or Starting Capital= <u>64,720</u> millions USD), and (W= Wealth or Total Assets= <u>139,796</u> millions USD), then it's (S'= Sales Past), must be:

$$S' = (S[1\text{-}f]\text{-}I\text{-}\{W/[1+I]\text{-}U\}/\{[1\text{-}d][1\text{-}t]\})$$
$$/\{v[1+s]\}$$
$$= (51,294[1\text{-}0.5800]\text{-}513\text{-}\{139,796$$
$$/[1+0.9900]\text{-}64,720\}$$
$$/\{[1\text{-}0.3000][1\text{-}0.3000]\})$$
$$/\{0.1900][1+0.0500]\}$$
$$= \underline{48,851} \text{ millions USD}$$

Corporate IFRS-GAAP (B/S-I/S), ISBN 13: **978-1983566332**, ISBN 10: **1983566330**

Law-5493:

If (I= L/E= Leverage or Gearing Ratio= 99.00%), (t= T/B= Tax Rate= 30.00%), (s= [S/S']-1= Sales Growth= 5.00%), (S= Sales or Revenues= 51,294 millions USD), (f= F/S= Fixed Portion= 58.00%), (d= D/A= Dividend Portion or Payout= 30.00%), (S'= Sales of Past Year= 48,851 millions USD), (I= Interest Expense= 513 millions USD),(U= Utilized or Starting Capital= 64,720 millions USD), and (W= Wealth or Total Assets= 139,796 millions USD), then it's (v= Variable Portion planned), is:

$$v= (S[1\text{-}f]\text{-}I\text{-}\{W/[1+I]\text{-}U\}/\{[1\text{-}d][1\text{-}t]\})$$
$$/\{S'[1+s]\}$$
$$= (51,294[1\text{-}0.5800]\text{-}513\text{-}\{139,796$$
$$/[1+0.9900]\text{-}64,720\}$$
$$/\{[1\text{-}0.3000][1\text{-}0.3000]\})$$
$$/\{48,851[1+0.0500]\}$$
$$= 19.00\%$$

Corporate IFRS-GAAP (B/S-I/S), ISBN 13: **978-1983566332**, ISBN 10: **1983566330**

Law-5494:

If (**f**= L/E= Leverage or Gearing Ratio= 99.00%), (**t**= T/B= Tax Rate= 30.00%), (**v**= V/S= Variable Portion= 19.00%), (**$**= Sales or Revenues= 51,294 millions USD), (**f**= F/S= Fixed Portion= 58.00%), (**d**= D/A= Dividend Portion or Payout= 30.00%), (**$'**= Sales of Past Year= 48,851 millions USD), (**I**= Interest Expense= 513 millions USD),(**U**= Utilized or Starting Capital= 64,720 millions USD), and (**W**= Wealth or Total Assets= 139,796 millions USD), then it's (**s**= Sales Growth planned), is:

$$s= (\$[1-f]-I-\{W/[1+I]-U\}/\{[1-d][1-t]\})/[\$'v]-1$$
$$= (51,294[1-0.5800]-513-\{139,796$$
$$/[1+0.9900]-64,720\}$$
$$/\{[1-0.3000][1-0.3000]\})$$
$$/[48,851*0.1900]-1$$
$$= 5.00\%$$

Corporate IFRS-GAAP (B/S-I/S), ISBN 13: **978-1983566332**, ISBN 10: **1983566330**

<u>Law-5495</u>:

If (**I**= L/E= Leverage or Gearing Ratio= <u>99.00%</u>), (**t**= T/B= Tax Rate= <u>30.00%</u>), (**v**= V/S= Variable Portion= <u>19.00%</u>), (**S**= Sales or Revenues= <u>51,294</u> millions USD), (**f**= F/S= Fixed Portion= <u>58.00%</u>), (**d**= D/A= Dividend Portion or Payout= <u>30.00%</u>), (**S'**= Sales of Past Year= <u>48,851</u> millions USD), (**s**= [S/S']-1= Sales Growth= <u>5.00%</u>), (**U**= Utilized or Starting Capital= <u>64,720</u> millions USD), and (**W**= Wealth or Total Assets= <u>139,796</u> millions USD), then it's (**I**= interest Expense planned), is:

$$I = S[1-f] - S'v[1+s] - \{W/[1+I] - U\}$$
$$/\{[1-d][1-t]\} - 1$$
$$= (51,294[1-0.5800] - 48,851 * 0.1900$$
$$[1+0.0500] - \{139,796$$
$$/[1+0.9900] - 64,720\}$$
$$/\{[1-0.3000][1-0.3000]\})$$
$$= \underline{513} \text{ millions USD}$$

Corporate IFRS-GAAP (B/S-I/S), ISBN 13: **978-1983566332**, ISBN 10: **1983566330**

Law-5496:

If (l= L/E= Leverage or Gearing Ratio= 99.00%), (I= Interest Expense= 513 millions USD), (v= V/S= Variable Portion= 19.00%), (S= Sales or Revenues= 51,294 millions USD), (f= F/S= Fixed Portion= 58.00%), (d= D/A= Dividend Portion or Payout= 30.00%), (S'= Sales of Past Year= 48,851 millions USD), (s= [S/S']-1= Sales Growth= 5.00%), (U= Utilized or Starting Capital= 64,720 millions USD), and (W= Wealth or Total Assets= 139,796 millions USD), then it's (t= Tax Rate planned), is:

$$t= 1-\{W/[1+l]-U\}/([1-d]\{S[1-f]-S'v[1+s]-I\})$$
$$= 1-\{139,796/[1+0.9900]-64,720\}$$
$$/([1-0.3000]\{51,294$$
$$[1-0.5800]-48,851*0.1900$$
$$[1+0.0500]-513\})$$
$$= 30.00\%$$

Corporate IFRS-GAAP (B/S-I/S), ISBN 13: **978-1983566332**, ISBN 10: **1983566330**

<u>Law-5497</u>:

If (**I**= L/E= Leverage or Gearing Ratio= <u>99.00%</u>), (**I**= Interest Expense= <u>513</u> millions USD), (**v**= V/S= Variable Portion= <u>19.00%</u>), (**S**= Sales or Revenues= <u>51,294</u> millions USD), (**f**= F/S= Fixed Portion= <u>58.00%</u>), (**t**= T/B= Tax Rate= <u>30.00%</u>), (**S'**= Sales of Past Year= <u>48,851</u> millions USD), (**s**= [S/S']-1= Sales Growth= <u>5.00%</u>), (**U**= Utilized or Starting Capital= <u>64,720</u> millions USD), and (**W**= Wealth or Total Assets= <u>139,796</u> millions USD), then it's (**d**= Dividend Portion or Payout planned), is:

$$d = 1 - \{W/[1+I]-U\}/([1-t]\{S[1-f]-S'v[1+s]-I\})$$
$$= 1-\{139,796/[1+0.9900]-64,720\}$$
$$/([1-0.3000]\{51,294$$
$$[1-0.5800]-48,851*0.1900$$
$$[1+0.0500]-513\})$$

$$= \underline{30.00\%}$$

Corporate IFRS-GAAP (B/S-I/S), ISBN 13: **978-1983566332**, ISBN 10: **1983566330**

Law-5498:

If (I= L/E= Leverage or Gearing Ratio= 99.00%), (I= Interest Expense= 513 millions USD), (v= V/S= Variable Portion= 19.00%), (S= Sales or Revenues= 51,294 millions USD), (f= F/S= Fixed Portion= 58.00%), (t= T/B= Tax Rate= 30.00%), (S'= Sales of Past Year= 48,851 millions USD), (s= [S/S']-1= Sales Growth= 5.00%), (d= D/A= Dividend Portion or Payout= 30.00%), and (W= Wealth or Total Assets= 139,796 millions USD), then it's (U= Utilized or Starting Capital planned), is:

$$U = W/[1+I]-[1-d][1-t]\{S[1-f]-S'v[1+s]-I\}$$
$$= 139,796/[1+0.9900]-[1-0.3000]$$
$$[1-0.3000]\{51,294$$
$$[1-0.5800]-48,851*0.1900$$
$$[1+0.0500]-513\}$$
$$= 64,720 \text{ millions USD}$$

Corporate IFRS-GAAP (B/S-I/S), ISBN 13: **978-1983566332**, ISBN 10: **1983566330**

Law-5499:

If (**U**= Utilized or Starting Capital= 64,720 millions USD), (**I**= Interest Expense= 513 millions USD), (**v**= V/S= Variable Portion= 19.00%), (**$**= Sales or Revenues= 51,294 millions USD), (**f**= F/S= Fixed Portion= 58.00%), (**t**= T/B= Tax Rate= 30.00%), (**$'**= Sales of Past Year= 48,851 millions USD), (**s**= [S/S']-1= Sales Growth= 5.00%), (**d**= D/A= Dividend Portion or Payout= 30.00%), and (**W**= Wealth or Total Assets= 139,796 millions USD), then it's (**I** = Leverage or Gearing Ratio planned), is:

$$I = W/(U+[1-d][1-t]\{\$[1-f]-\$'v[1+s]-I\})-1$$
$$= 139,796/(64,720+[1-0.3000]$$
$$[1-0.3000]\{51,294$$
$$[1-0.5800]-48,851*0.1900$$
$$[1+0.0500]-513\})-1$$
$$= 99.00\%$$

Corporate IFRS-GAAP (B/S-I/S), ISBN 13: **978-1983566332**, ISBN 10: **1983566330**

Law-5500:

If (**U**= Utilized or Starting Capital= 64,720 millions USD), (**i**= I/S= Interest Portion= 1.00%), (**v**= V/S= Variable Portion= 19.00%), (**S**= Sales or Revenues= 51,294 millions USD), (**f**= F/S= Fixed Portion= 58.00%), (**t**= T/B= Tax Rate= 30.00%), (**S'**= Sales of Past Year= 48,851 millions USD), (**s**= [S/S']-1= Sales Growth= 5.00%), (**d**= D/A= Dividend Portion or Payout= 30.00%), and (**l**= L/E= Leverage or Gearing Ratio= 99.00%), then it's (**W**= Wealth or Total Assests planned), is:

$$\begin{aligned}
\textbf{W} &= [1+\textbf{\textit{l}}\,](\textbf{U}+[1-\textbf{d}][1-\textbf{t}] \\
&\qquad \{\textbf{S}-\textbf{S'v}[1+\textbf{s}]-\textbf{Sf}-\textbf{Si}\}) \\
&= [1+\textbf{\textit{l}}\,](\textbf{U}+[1-\textbf{d}][1-\textbf{t}] \\
&\qquad \{\textbf{S}[1-\textbf{f}-\textbf{i}]-\textbf{S'v}[1+\textbf{s}]\}) \\
&= [1+0.9900]-64,720\}/\{[1-0.3000] \\
&\qquad [1-0.3000]\{51,294 \\
&\qquad [1-0.5800-0.0100]-48,851 \\
&\qquad *0.1900[1+0.0500]\}) \\
&= \underline{139,796} \text{ millions USD}
\end{aligned}$$

Corporate IFRS-GAAP (B/S-I/S), ISBN 13: **978-1983566332**, ISBN 10: **1983566330**

Finance Construction-6, *Tim Asikin, Steve Asikin, Indra Senihardja*

The Rest will be Continued in
The Series of Book

Finance Construction 7

Corporate IFRS-GAAP (B/S-I/S) Engineering
Technologies No. 5,501-6,000 of 111,111 Laws

At
http: //www.Amazon.Com

Corporate IFRS-GAAP (B/S-I/S), ISBN 13: **978-1983566332**, ISBN 10: **1983566330**

REFERENCES:

A. Books:

1) **Amin, A.R. & Asikin, S.**(2011a- 110 pages), *CEO Time Matrix: Productivity Management Skill that Increase CEO Effectiveness by 12400%*, Seattle (US): Amazon Createspace. http:/www.amazon.com/CEO-Time-Matrix-Productivity-Effectiveness/dp/1461066069/ref=sr_1_1?s=books&ie=UTF8&qid=1387091880&sr=1-1&keywords=ceo+time+matrix. ISBN-13: 978-1461066064, ISBN-10: 1461066069

2) **Amin, A.R. & Asikin, S.** (2014- 76 pages), *No Urgency CEO: A High-Tech Book in Modern Management*, Seattle (US): Amazon Createspace. http:/www.amazon.com/No-Urgency-CEO-Hi-Tech-Management/dp/1489509356/ref=sr_1_1?s=books&ie=UTF8&qid=1396440688&sr=1-1&keywords=no+urgency. ISBN-13: 978-1489509352, ISBN-10: 1489509356

3} **Amin, A.R. & Asikin, S.** (2011b- 292 pages), *The Celestial Management: Spiritual Wisdom for All Human Beings*, Seattle (US): Amazon Createspace. http:/www.amazon.com/Celestial-Management-Islamic-Wisdom-Beings/dp/1456450700/ref=sr_1_1?s=books&ie=UTF8&qid=1387092036&sr=1-1&keywords=celestial+management. ISBN-13: 978-1456450700, ISBN-10: 1456450700

4) **Asikin, S.** (2009- 548 pages), *Arigato, Obrigado... (Thanks): Portuguese-Japan 1550-1647 Historic Novel*, Seattle (US): Amazon Createspace. http:/www.amazon.com/Arigato-Obrigado-Thanks-Portuguese-Japan-1550-1647/dp/1453786368/ref=sr_1_1?s=books&ie=UTF8&qid=1396442162&sr=1-1&keywords=arigato. ISBN-13: 978-1453786369, ISBN-10: 1453786368

5) **Asikin, S.** (2013a- 824 pages), *Econometric Rex: 888 Marketing Promo 6666 Six Sigma Parametric Theories (of 10 Data)*, Seattle (US): Amazon Createspace. http:/www.amazon.co.uk/Econometric-Rex-Marketing-Parametric-Theories/dp/1482745259, ISBN-13: 978-1482745252, ISBN-10: 1482745259

6) **Asikin, S.** (2010b- 224 pages), *FINANCE Architecture: VISUAL art, for Corporate Finance's Beauty, Safety and Simplicity*, Seattle (US): Amazon Createspace. http:/www.amazon.co.uk/FINANCE-Architecture-Corporate-Finances-Simplicity/dp/1453829547/ref=sr_1_14?s=books&ie=UTF8&qid=1385371842&sr=1-14&keywords=finance+architecture. ISBN-13: 978-1453829547, ISBN-10: 1453829547

7) **Asikin, S.** (2010c- 150 pages), *Geometric FINANCE: Useful concepts that work better than Economic Noble Laureates'* Seattle (US): Amazon Createspace. http:/www.amazon.co.uk/Magic-Geometric-FINANCE-concepts-Laureates/dp/1453790284/ref=tmm_pap_title_0, ISBN-13: 978-1453790281, ISBN-10: 1453790284

8) **Asikin, S.** (2010a- 290 pages), *Investment ALGEBRA: Strategic Profitable Comprehensive Business Finance Engineering Technology*, Seattle (US): Amazon Createspace. http:/www.amazon.co.uk/Investment-ALGEBRA-Profitable-Comprehensive-Engineering/dp/1453856811/ref=tmm_pap_title_0?ie=UTF8&qid=1385371309&sr=1-1, ISBN-13: 978-1453856819, ISBN-10: 1453856811

9) **Asikin, S.** (2011a- 352 pages), *State Economics: Comprehensive Macro-Micro Economics' Simple Fiscal-Monetary Export-Import Accouting, Integrated Supply-Demand Managerial ... Mathematical Engineering Visual* . Seattle (US): Amazon Createspace. http:/www.amazon.co.uk/STATE-ECONOMICS-Steve-Asikin-ebook/dp/B00G89Q5SS, http:/www.amazon.co.uk State-Economics-Steve-Asikin-ebook/dp/B00B8PLQ0I, ISBN-13: 978-1461066095, ISBN-10: 1461066093

10) **Asikin, S.** (2013b- 260 pages), *Technoeconomistat: Economic &Statistical Technology for Financial Planning*, Seattle (US): Amazon Createspace. http:/www.amazon.co.uk/Technoeconomistat-Economic-Statistical-Technology-Financial/dp/1482304244, ISBN-13: 978-1482304244, ISBN-10: 1482304244

11) **Asikin, S.** (2011b- 540 pages), *Terima Kasih: Novel Internasional Sejarah Jepang-Portugis 1550-1647*, Seattle (US): Amazon Createspace. http:/www.amazon.com/Terima-Kasih-Internasional-Jepang-Portugis-1550-1647/dp/146351106X/ref=sr_1_1?s=books&ie=UTF8&qid=1396442361&sr=1-1&keywords=terima, ISBN-13: 978-1463511067, ISBN-10: 146351106X

12) **Asikin, S.** (2010d- 294 pages), *TREASURY Planology: Strategic Profitable Comprehensive Business Finance Engineering Technology*, Seattle (US): Amazon Createspace. http:/www.amazon.co.uk/TREASURY-PLANOLOGY-Steve-Asikin-ebook/dp/B00G88MTFW/ref=sr_1_1?s=books&ie=UTF8&qid=1385403690&sr=1-1&keywords=Treasury+Planology. ISBN-13: 978-1453891476, ISBN-10: 1453891471

13) **Asikin, S.** (2014a- 190 pages), *Calculus Accountancy: Leibnitz Newton Pacioli's Polynomial Quantitative Finance Risk Modelling Optimization*, Seattle (US): Amazon Createspace. http:/www.amazon.com/Calculus-Accountancy-Polynomial-Quantitative-Optimization/dp/1495463028/ref=sr_1_1?s=books&ie=UTF8&qid=1396441338&sr=1-1&keywords=calculus+accountancy, ISBN-13: 978-1495463020, ISBN-10: 1495463028

14) **Asikin, S.**(2014b- 318 pages), *No Parametric: No Parametric: Hyperbolic Octahedron Central 3-Dimensional Orthogonal Risk Distribution Topology*, Seattle (US): Amazon Createspace. http:/www.amazon.com/Parametric-Hyperbolic-Octahedron-3-Dimensional-Distribution/dp/1496089731/ref=sr_1_1?s=books&ie=UTF8&qid=1396441543&sr=1-1&keywords=No+Parametric, ISBN-13: 978-1496089731, ISBN-10: 1496089731

Corporate IFRS-GAAP (B/S-I/S), ISBN 13: **978-1983566332**, ISBN 10: **1983566330**

15) **Asikin, S.** (2014c- 138 pages), *Quick Help: Religion vs Real-Legion Philosophy*, Seattle (US): Amazon Createspace.
http://www.amazon.com/Quick-Help-Religion-Real-Legion-Philosophy/dp/1497466520/ref=sr_1_2?s=books&ie=UTF8&qid=1396441740&sr=1-2&keywords=quick+help, ISBN-13: 978-1497466524, ISBN-10: 1497466520

16) **Asikin, S.** (2014d- 824 pages), *Anne of Denmark: King James I the Bible, Historic Plays Drama*, Seattle (US): Amazon Createspace.
http://www.amazon.com/Anne-Denmark-James-Bible-Drama/dp/149299734X/ref=sr_1_1?s=books&ie=UTF8&qid=1396441857&sr=1-1&keywords=anne+asikin, ISBN-13: 978-1492997344, ISBN-10: 149299734X

17) **Asikin, S.** (2014e- 184 pages), *Constructive Spells: Positive Spiritual Mentality Personal Development*: Seattle (US): Amazon Createspace.
http://www.amazon.com/Constructive-Spells-Spiritual-Mentality-Development/dp/1497498899/ref=sr_1_1?s=books&ie=UTF8&qid=1397664512&sr=1-1&keywords=constructive+spells; ISBN-13: 978-1497498891, ISBN-10: 1497498899

18) **Asikin, S.** (2014f- 466 pages), *Elizabeth of Russia: Another Virgin Queen Gloariana: A Romanov Plays Drama*, Seattle (US): Amazon Createspace.
http://www.amazon.com/Elizabeth-RUSSIA-Another-Gloriana-Romanov/dp/1505542480/ref=sr_1_4?s=books&ie=UTF8&qid=1423888482&sr=1-4&keywords=elizabeth+russia; ISBN-13: 978-1505542486, ISBN-10: 1505542480

19) **Asikin, S.** (2014f- 708 pages), *Shinmen Munisai:Miyamoto Musashi's Father, A Sword Samurai Plays Drama*, Seattle (US): Amazon Createspace. http://www.amazon.com/Shinmen-Munisai-Miyamoto-Musashis-Samurai/dp/1507857799/ref=sr_1_1?s=books&ie=UTF8&qid=1428094280&sr=1-1&keywords=munisai; ISBN-13: 978-1507857793, ISBN-10: 1507857799

20) **Asikin, S.** (2015a-767 pages), *Math Fin Law-1: Mathematical Financial Law for Public Listed Firms Rule 1-4441*, Seattle (US): Amazon Createspace.
http://www.amazon.com/Math-Fin-Law-Mathematical-Financial-ebook/dp/B00WLNX3Y4/ref=asap_bc?ie=UTF8; ISBN-13: 978-1511792219, ISBN-10: 1511792213

21) **Asikin, S.** (2015b-813 pages), *Math Fin Law-2: Mathematical Financial Law for Public Listed Firms Rule 4442-8712*, Seattle (US): Amazon Createspace. http://www.amazon.com/Math-Fin-Law-Mathematical-No-4442-8712/dp/1512285080/ref=sr_1_1?s=books&ie=UTF8&qid=1432732199&sr=1-1&keywords=math+fin+law+2; ISBN-13: 978-1512285086, ISBN-10: 1512285080

22) **Asikin, S.** (2015c-813 pages), *Math Fin Law-3: Mathematical Financial Law for Public Listed Firms Rule 8713-12575*, Seattle (US): Amazon Createspace. http://www.amazon.com/Math-Fin-Law-Mathematical-8712-12575/dp/1514276763/ref=sr_1_1?s=books&ie=UTF8&qid=1433982149&sr=1-1&keywords=math+fin+law+3, ISBN-13: 978-1514276761. ISBN-10: 1514276763

Corporate IFRS-GAAP (B/S-I/S), ISBN 13: **978-1983566332**, ISBN 10: **1983566330**

23) **Asikin, S**. (2015d-813 pages), *Math Fin Law-4: Mathematical Financial Law for Public Listed Firms Rule 12576-16333,* Seattle (US): Amazon Createspace. http://www.amazon.com/Math-Fin-Law-Mathematical-12576-16333/dp/1514685132/ref=sr_1_1?s=books&ie=UTF8&qid=1435393628&sr=1-1&keywords=math+fin+law+4, ISBN-13: 978-1514685136, ISBN-10: 1514685132

24) **Asikin, S**. (2015d-816 pages), *Math Fin Law-5: Mathematical Financial Law for Public Listed Firms Rule 16334-19904,* Seattle (US): Amazon Createspace. http://www.amazon.com/Math-Fin-Law-Mathematical-No-16334-19904/dp/1515073009/ref=sr_1_1?s=books&ie=UTF8&qid=1437224693&sr=1-1&keywords=math+fin+law+5. ISBN-13: 978-1515073000, ISBN-10: 1515073009

25) **Asikin, S**. (2015d-728 pages), *Math Fin Law-6: Mathematical Financial Law for Public Listed Firms Rule 19905-23237,* Seattle (US): Amazon Createspace. https://www.amazon.com/Math-Fin-Law-Mathematical-No-19905-23238-ebook/dp/B0129ILVNK/ref=sr_1_2?ie=UTF8&qid=1478017624&sr=8-2&keywords=Math+Fin+Law+Asikin. ISBN-13: 978-1515158691, ISBN-10: 1515158691

26) **Asikin, S**. (2016a-728 pages), *Math Fin Law-7: Mathematical Financial Law for Public Listed Firms Rule 23238-27115,* Seattle (US): Amazon Createspace. https://www.amazon.com/Math-Fin-Law-Mathematical-No-23238-27115/dp/1508518521/ref=la_B00428V6JU_1_27?s=books. ISBN-13: 978-1508518525, ISBN-10: 1508518521

27) **Asikin, S**. (2016b-728 pages), *Math Fin Law-8: Mathematical Financial Law for Public Listed Firms Rule 27116-30330,* Seattle (US): Amazon Createspace. https://www.amazon.com/dp/1541091884, https://www.amazon.com/Math-Fin-Law-Mathematical-No-27116-30330/dp/1541091884/ref=la_B00428V6JU_1_27?s=books&ie=UTF8&qid=1479398538&sr=1-27&refinements=p_82%3AB00428V6JU, ISBN-13: 978-1508518525, ISBN-10: 1508518521

28) **Asikin, S**. (2017a-828 pages), *Math Fin Law-9: Mathematical Financial Law for Public Listed Firms Rule 30330-33373,* Seattle (US): Amazon Createspace. https://www.amazon.com/Math-Fin-Law-Mathematical-30331-33373-ebook/dp/B0IMRA8HZQ/ref=la_B00428V6JU_1_4?s=books&ie=UTF8&qid=1485921040&sr=1-4. ISBN-13: 978-1541092433, ISBN-10: 1541092430

29) **Asikin, S**. (2017b-823 pages), *Math Fin Law-10: Mathematical Financial Law for Public Listed Firms Rule 33374-36242,* Seattle (US): Amazon Createspace. https://www.amazon.com/s/ref=nb_sb_noss?url=search-alias%3Dstripbooks&field-keywords=math+fin+law+10, https://www.amazon.com/s/ref=nb_sb_noss?url=search-alias%3Dstripbooks&field-keywords=math+fin+law+10, ISBN-13: 978-1541092761, ISBN-10: 1541092767

Corporate IFRS-GAAP (B/S-I/S), ISBN 13: **978-1983566332**, ISBN 10: **1983566330**

30) **Asikin, S**. (2017c-823 pages), *Math Finance Law-11: Mathematical Financial Law for Public Listed Firms Rule 36243-39158,* Seattle (US): Amazon Createspace. https://www.amazon.com/s/ref=nb_sb_noss?url=search-alias%3Dstripbooks&field-keywords=math+finance+law+11, https://www.amazon.com/dp/1541215265/ref=sr_1_1?s=books&ie=UTF8&qid=1488029978& sr=1-1&keywords=math+finance+law, ISBN-13: 978-1541215269. ISBN-10: 1541215265

31) **Asikin, S**. (2017d-823 pages), *Math Finance Law-12: Mathematical Financial Law for Public Listed Firms Rule 30150-42152,* Seattle (US): Amazon Createspace. https://www.amazon.com/s/ref=nb_sb_noss?url=search-alias%3Dstripbooks&field-keywords=math+finance+law+12. https://www.amazon.com/Math-Finance-Law-Mathematical-39159-42152/dp/1541215516/ref=sr_1_2?s=books&ie=UTF8&qid=1490990342&sr=1-2&keywords=math+finance+law+12. ISBN-13: 978-1541215511, ISBN-10: 1541215516

32) **Asikin, S**. (2017e-824 pages), *Math Finance Law-13: Mathematical Financial Law for Public Listed Firms Rule 42153-44938,* Seattle (US): Amazon Createspace. https://www.amazon.com/dp/1542675219/ref=sr_1_1?s=books&ie=UTF8&qid=1491146227&s r=1-1&keywords=math+finance+law+13 . ISBN-13: 978-1542675215, ISBN-10: 1542675219

33) **Asikin, S**. (2017f-824 pages), *Math Finance Law-14: Mathematical Financial Law for Public Listed Firms Rule 44939-47745,* Seattle (US): Amazon Createspace. https://www.amazon.com/dp/1542675219/ref=sr_1_1?s=books&ie=UTF8&qid=1491146227&s r=1-1&keywords=math+finance+law+14 . ISBN-13: 978-1544628967, ISBN-10: 154462896X

34) **Asikin, S**. (2017g-810 pages), *Math Finance Law-15: Mathematical Financial Law for Public Listed Firms Rule 47746-50465,* Seattle (US): Amazon Createspace. https://www.amazon.com/dp/1545118647/ref=sr_1_1?s=books&ie=UTF8&qid=1493626425&s r=1-1&keywords=math+finance+law+15, . ISBN-13: 978-1545118641, ISBN-10: 1545118647

35) **Asikin, S**. (2017h-816 pages), *Math Finance Law 16: Mathematical Financial Law Public Listed Firm Rule No.50,466-53,320,* Seattle (US), Amazon CreateSpace, https://www.amazon.com/dp/1545320179/ref=sr_1_1?s=books&ie=UTF8&qid=1495692695& sr=1-1&keywords=math+finance+law+16. ISBN-13: 978-1545320174, ISBN-10: 1545320179

Corporate IFRS-GAAP (B/S-I/S), ISBN 13: **978-1983566332**, ISBN 10: **1983566330**

36) **Asikin, S**. (2017i-809 pages), *Math Finance Law 17*: *Mathematical Financial Law Public Listed Firm Rule No.53,231-56,124*, Seattle (US), Amazon CreateSpace, https://www.amazon.com/dp/1545320179/ref=sr_1_1?s=books&ie=UTF8&qid=1495692695& sr=1-1&keywords=math+finance+law+17. ISBN-13: 978-1545486474, ISBN-10: 1545486476

37) **Asikin, S**. (2017j-815 pages), *Math Finance Law 18*: *Mathematical Financial Law Public Listed Firm Rule No.56,125-58,942*, Seattle (US), Amazon CreateSpace, https://www.amazon.com/dp/1545320179/ref=sr_1_1?s=books&ie=UTF8&qid=1495692695& sr=1-1&keywords=math+finance+law+18. ISBN-13: 978-1546912361, ISBN-10: 1546912363

38) **Asikin, S.** (2014-47 pages), *Doing Business in Comprehensive Interactive Interdisciplinary Micro-Macro Eco nomic International Accounting Architecture of Serial Well Known General Great Theories*, Proceedings in Finance and Risk Perspective 13, ACRN Oxford Publishing House, UK, p.197-243, https://books.google.com/books?id=4e4KAwAAQBAJ&pg=PA219&lpg=PA219&dq=steve+asiki n+oxford&source=bl&ots=qVq9KyAmem&sig=ABdBliwuzUwRigIsuLimuJ5ZdW4&hl=en&sa=X&s qi=2&ved=0ahUKEwjZkMrF7KjTAhUDSiYKHREDB-qQ6AEILjAC#v=onepage&q=steve%20asikin%20oxford&f=false. ISBN 978-3-9503518-1-1.

39) **Asikin, T.& Asikin, S**. (2013- 828 pages), *448 Poems: KARAOKE Song Book: 24 Hours Story of 30 Languages*, Seattle (US): Amazon Createspace. http:/www.amazon.com/448-Poems-KARAOKE-Song-Book/dp/1480034363/ref=sr_1_1?s=books&ie=UTF8&qid=1396442034&sr=1-1&keywords=448+poems, ISBN-13: 978-1480034365, ISBN-10: 1480034363

40) **Asikin, T, Asikin, S, Senihardja, I.** (2017a- 508 pages), *Finance Construction 1*: *Corporate IFRS-GAAP (B/S-I/S) Engineering Technologies No. 1-1,000 of 111,111 Laws*, Seattle (US): Amazon Createspace. https://www.amazon.com/s/ref=nb_sb_noss?url=search-alias%3Dstripbooks&field-keywords=finance+construction+1+asikin. ISBN-13: 978-1544059525, ISBN-10: 1544059523

41) **Asikin, T, Asikin, S, Senihardja, I.** (2017b- 526 pages), *Finance Construction 2*: *Corporate IFRS-GAAP (B/S-I/S) Engineering Technologies No. 1,001-2,000 of 111,111 Laws*, Seattle (US): Amazon Createspace. https://www.amazon.com/s/ref=nb_sb_noss?url=search-alias%3Dstripbooks&field-keywords=finance+construction+2+asikin. ISBN-13: 978-1542402415, ISBN-10: 1542402417

Corporate IFRS-GAAP (B/S-I/S), ISBN 13: **978-1983566332**, ISBN 10: **1983566330**

42) **Asikin, T, Asikin, S, Senihardja, I.** (2017c- 738 pages), *Finance Construction 3: Corporate IFRS-GAAP (B/S-I/S) Engineering Technologies No. 2,001-3,000 of 111,111 Laws*, Seattle (US): Amazon Createspace. https://www.amazon.com/s/ref=nb_sb_noss?url=search-alias%3Dstripbooks&field-keywords=finance+construction+3+asikin. ISBN-13: 978-1975999292, ISBN-10: 1975999290

43) **Asikin, T, Asikin, S, Senihardja, I.** (2018a- 656 pages), *Finance Construction 4: Corporate IFRS-GAAP (B/S-I/S) Engineering Technologies No. 3,001-4,000 of 111,111 Laws*, Seattle (US): Amazon Createspace. https://www.amazon.com/s/ref=nb_sb_noss?url=search-alias%3Dstripbooks&field-keywords=finance+construction+4+asikin. ISBN-13: 978-1977659187, ISBN-10: 1977659187

44) **Asikin, T, Asikin, S, Senihardja, I.** (2018b- 656 pages), *Finance Construction 5: Corporate IFRS-GAAP (B/S-I/S) Engineering Technologies No. 4,001-5,000 of 111,111 Laws*, Seattle (US): Amazon Createspace. https://www.amazon.com/s/ref=nb_sb_noss?url=search-alias%3Dstripbooks&field-keywords=finance+construction+5+asikin. ISBN-978-1983566141, ISBN-10: 1983566144

45) **Brigham, EF & Gapenski, LC,** . (1988- Textbook), *Financial Management: Theory and Practice (Study Guide)*, NY (US): Dryden Press. http:/www.amazon.com/Financial-Management-Theory-Practice-Study/dp/0030125391/ref=sr_1_1?s=books&ie=UTF8&qid=1429380835&sr=1-1 ;

46) **Chandonet, RE,** . (2011-32 pages), *Balancesheet Management*, Trends & Strategies, NY (US): Sandler O'Neil. https://www.ncbankers.org/uploads/File/Meetings/Presentations/2011CONV-Chandonnet.pdf;

47) **Cornuejols G & Tutuncu, R,** . (2006-358 pages), *Optimization Methods in Finance*, Cambridge (UK): Cambridge University Press. http://www.math.ku.dk/~rolf/CT_FinOpt.pdf;

48) **Crook, M.** (2015-6 pages), *Optimization Methods in Finance*, UBS FS (Swiss): CIO WM Research. https://www.ubs.com/content/dam/WealthManagementAmericas/documents/investment-strategy-insights-2015-01-09-balance-sheet-optimization.pdf;

49) **Fee, Greg.** (2012-127 pages), *Catalan's Constant: (Ramanujan's Formula}: Catalan Constant to 300,000 digits*, Seattle (US): Amazon Kindle. http://www.Amazon.com/gp/aw/d/B00827ZIEW/ref=mp_sa_1_4?ie=UTF8%Qqid=1 4843678186sr=8-4&pi=AC_SX236_SY340_QL65&keywords=ramanujan&dpPl=1&dpID=51FqIwDZpRL&ref=plSrch

50) **Gardenfors, Peter.** (2014) *The Geometry of Meaning: Semantics Based on Conceptual Spaces*. Cambridge, Mass.: MIT Press. ISBN 0-262-02678-3.

Corporate IFRS-GAAP (B/S-I/S), ISBN 13: **978-1983566332**, ISBN 10: **1983566330**

51) **Gardenfors, Peter.** (2000), *Conceptual Spaces: The Geometry of Thought*. Cambridge, Mass.: MIT Press. ISBN 0-262-07199-1.

52) **Kanigel, Robert.** (2013-465 pages), *The Man Who Knew Infinity:A Life of the Genius Ramanujan*, Washington (US): Washington Square Press&Amazon Kindle.
http://www.Amazon.com/gp/aw/d/B008BW4VEGM/ref=pd_aw_sim35l_1?ie=UTF8&psc=1&refRID=VYN345GVDBJ6ATDQ35W2&dpID=9lZgpc4ETdL

53) **Limlingan, Victor S.** (1993-1 page), *The Limlingan Financial Model*, Manila: Asian Manager. http://limlingan.com/images/article.gif

54) **Newton, Isaac.** (2015-631 pages), *Principia:The Mathematical Principles of Natural Philosophy (Active Content)*, London (UK): RSM&Amazon Kindle.
http://www.Amazon.com/gp/aw/sitb/B0IISFJSJ0?ref=sib_dp__aw_kb_udp

55) **Nikisa, Evets.** (2014g- 828 pages), *Wonderful Live Spells: 144-Steps Recreative Nice Children Story, Succesful Development:* Seattle (US): Amazon Createspace. http://www.amazon.com/WONDERFUL-Live-Spells-Recreative-
Development/dp/1499248423/ref=sr_1_1?s=books&ie=UTF8&qid=1418698379&sr=1-1&keywords=wonderful+asikin, ISBN-13: 978-1499248425, ISBN-10: 1499248423

56) **Nielson, SS.** (2007- 430 pages), *Practical Financial Optimization*: *Decision Making for Financial Engineers:* NY (US): Wiley.
https://books.google.com/books/about/Practical_Financial_Optimization.html?id=XkwLAAAACAAJ.

57) **Niswonger CR, Fess PE & Warren CE.** (1993-Textbook), *Accounting Principles:* Seattle (US): South Western Publishing Co.
http:/www.amazon.com/Accounting-Principles-Chapters-C-Rollin-Niswonger/dp/0538818603

58) **Omerod, P.** (1993-Textbook), (1994), *The Death of Ecomics*, St Martin's Press, London, . http:/www.amazon.com/Accounting-Principles-Chapters-C-Rollin-Niswonger/dp/0538818603, ISBN-13: 978-0312134648, ISBN-10: 0312134649

59) **Paccioli, L. (1989-Textbook),** *Summa de Arithmetica Geometria Proportioni et Proportionalita*, Venice 1494.
http://www.amazon.com/Arithmetica-Geometria-Proportioni-Proportionalita-Venice/dp/B00T4HXTIW/ref=sr_1_fkmr0_2?s=books&ie=UTF8&qid=1429597966&sr=1-2-fkmr0&keywords=summa+mhematica+arithmetica

60) **Persson, Torsten & Tabellini, Guido** (2000), *Political Economics – Explaining Economic Policy*, MIT Press.
https://mitpress.mit.edu/authors/torsten-persson

61) **Persson, Torsten & Tabellini, Guido** (2003), *The Economic Effects of Constitutions*, MIT Press. https://mitpress.mit.edu/authors/torsten-persson

62) **Persson, Torsten & Tabellini, Guido** (2000), *Monetary and Fiscal Policy –1 Credibility*, MIT Press. https://mitpress.mit.edu/authors/torsten-persson

Corporate IFRS-GAAP (B/S-I/S), ISBN 13: **978-1983566332**, ISBN 10: **1983566330**

63) **Persson, Torsten & Tabellini, Guido** (2003), *Monetary and Fiscal Policy –2 Politics*, MIT Press. https://mitpress.mit.edu/authors/torsten-persson
64) **Puts, J** (2012-102 pages), *Balancesheet Optimization Under Basel III*, Amsterdam (Netherlads): Vrije Universiteit. http://www.math.vu.nl/~sbhulai/papers/thesis-puts.pdf
65) **Timmermans, K** (2012-30 pages), *Balancesheet Optimization Under Basel III*, Amsterdam (Netherlads): ING Investor Day. https://www.ing.com/web/file?uuid=09c803d9-79a1-47ec-bfe5...owner...
66) **Weston JF, Copeland TE & Shastri, K.** (2004-Textbook), *Financial Theory and Corporate Policy:* NY (US): Addison Wesley. http:/www.amazon.com/Financial-Corporate-Copeland-Kuldeep-Paperback/dp/B008YSUE4M/ref=sr_1_1?s=books&ie=UTF8&qid=1429381700&sr=1-1&keywords=weston+copeland
67) **Zenios SA.** (2002-350 pages), *Financial Optimization:* Cambridge (UK): Cambridge University Press. https://books.google.com/books/about/Financial_Optimization.html?id=-FJdgjZ0qSAC, .

506
Corporate IFRS-GAAP (B/S-I/S), ISBN 13: **978-1983566332**, ISBN 10: **1983566330**

B. Intellectual Properties:

1)<u>Indonesian Banker Certification 1993</u>, *Sertifikasi Bankir Indonesia 1993*;
2)<u>MATHFINPLAN</u> (Mathematical Financial Planning);
   Perencanaan Keuangan Matematis
3)<u>BANK MATHPLAN</u> (Bank's Mathematical Planning);
   Perencanaan Matematis Bank
4)<u>TECHNO-ECONOMISTAT</u> (Technology for Economical Statistics);
   Teknologi Statistik untuk Ekonomi
5)<u>FINANCIAL GEOMETRY</u> (IT Program); *Ilmu Geometri Keuangan*
6)<u>BAK DOBEL</u> (Car Gasoline Tank: Left and Right Side Fuel Fill-in).
   Tangki Mobil yang bisa diisi dari Kiri atau Kanan
7)<u>Mister GO-CHENG</u> (Cheap Food Franchising). *Waralaba Pangan Murah*

507

Corporate IFRS-GAAP (B/S-I/S), ISBN 13: **978-1983566332**, ISBN 10: **1983566330**

C. SCIENTIFIC PAPERS:

1) **Eichhornia Crassipes**, as Anti Polutant and Fertile Sediment, *Pemanfaatan Eceng Gondok untuk Menetralkan Polusi dan Pengurukan Subur pada Genangan Air*, LKIR, Lomba Karya Ilmiah Remaja, LIPI (1979)

2) **Archimedes Catesian Buoyancy Law**, *Penyimpangan Hukum Archimedes pada Model Pelampung Cartesian*, Lomba Karya Ilmiah Remaja, Depdikbud-RI (1980)

3) **SURAMADU**, Surabaya-Madura Inter Island Bridge, Design Supervisor, *Sumarsono Sumali Engineering Contractor*, Tambak-V, Tandes, Surabaya (1982),

4) **GAZOLINE Taxi-Pool Depot**, Jemur Handayani, Surabaya, *Sumarsono Sumali Engineering Contractor*, Tambak-V, Tandes, Surabaya (1982),

5) **SOEKARNO-HATTA**, Jakarta International Airport, Garuda Maintenance Facilities, *Junior Architect*, As Bulit Drawings, Coinstruction Management, Aerospatiale-Encona Engineering (1983-1985),

6) **ADISUCIPTO**, Yogya International Airport, *Junior Architect*, Preliminary Technical Drawings, Budiono Surasno-Encona Engineering (1986),

7) **BLOK-M PLAZA**, "New Garden Hall", Kebayoran Baru Supermall, *Preliminary Design Architect*, Danisworo-James Ferry- Jasa Ferry-Encona (1986),

8) **NORTH JAKARTA PUBLIC LIBRARY**, Semper, Tanjung Priok, *Chief Architect*, Agung Darma Sarana-Asep Hermawan (1987),

9) **The Banker's Trainer**: A Manual for The Training Head in Banking. *Profesi Bankir Pelatih, Tuntunan bagi Kepala Diklat Perbankan*, (1200 Pages, 1993),

10) **Indonesian Bankers Certification**, a Strategic Move for the National Banking Improvement. *Sertifikasi Bankir Indonesia, Langkah Strategis Pembenahan Perbankan Nasional*, (KOMPAS, 20 Juli 1993).

11) **Indonesian Rupiah's Exchange Rate**, on a Quiet but Steady Movement. *Nilai Tukar Rupiah, Diam-diam Meningkat Pesat*, (MANAJEMEN, Jan-Feb 1994).

12) **The Bank's Credit**, for the Best Practices. *Kredit Praktis* (200 pages, 1994),

13) **Down to Earth Credit Meeting Credo**, *Kiat Rapat Pemberian Kredit yang Realistik*. (MANAJEMEN, Sep-Oct 1994).

14) **SCHIPHOL**, Amsterdam International Airport, 1st Anthony Fokker Weg, Sloten, *Airport Engineer*, Samsung- Fokker-Beurlin, BV (1996),

15) **Ultrapeneurship: The Secret of Starting Business Without Capital**. Ultrapreneur: *Kiat Usaha Tanpa Modal*, (MANAJEMEN, Nov-Dec 1996),

16) **Recipe from the Corporate Raider's Kitchen**, *Menguak Dapur Corporate Raiders*, (MANAJEMEN, Jan-Feb 1997).

17) **The 10-Tips and Tricks of Engineered Conglomeration**. *Sepuluh Rekayasa Keuangan Konglomerat*, (MANAJEMEN, Mar-Apr 1997),

18) **The Management of Succession in Five Parts of the World**. *Manajemen Suksesi di Lima bagian Dunia*, (MANAJEMEN, Mei-Jun 1997),

Corporate IFRS-GAAP (B/S-I/S), ISBN 13: **978-1983566332**, ISBN 10: **1983566330**

19)**Auditing**: The Global Language for the 2003's Human Resources. *Audit : Bahasa Global untuk SDM 2003*, (MANAJEMEN, Jul-Aug 1997),

20)**The Elected Parliament 1997-2002**, and the Hope of Professionals. *DPR 1997– 2002 dan Harapan Profesional*, (MANAJEMEN, Sep-Okt 1997),

21)**Disappearance of World's Soft Currencies**, and Proposition for Global Economic Thinking. *Lenyapnya Mata Uang Lunak Dunia dan Usulan Pemikiran Ekonomi Global*, (MANAJEMEN, Nov-Dec 1997),

22)**The Human Resources and Its Accountability**, in Critical Situation. *Akuntabilitas Sumberdaya Manusia di Masa Krisis*, (MANAJEMEN, Jan-Feb 1998), '

23)**The Investment Economic Model**, and the Failure of Growth Economic Theories. *Model Ekonomi Investasi dan Kegagalan Teori Ekonomi Pertumbuhan*, (MANAJEMEN, Nov 1998),

24)**The Exact Equilibrium of Capital Budgeting**, *Persamaan-persamaan akurat dalam Penganggaran Modal Perusahaan.* (MANAGEMENT EXPOSE, November 1998),

25)**The Corruptive Economic Model**, and Its Solution for Long Term Debt's Problem. *Model Ekonomi Korupsi dan Penyelesaian Utang Jangka Panjang*, (MANAJEMEN, Dec 1998),

26)**The Self Equity Budgeting**, and Operations Research in Investments. *Penganggaran Modal Sendiri dan Riset Operasi dalam Investasi*, (MANAJEMEN, Jan 1999),

27)**Actual Impact of Inflation**, and the Need of Geometric Approach in Statistic. *Pengaruh Riil Inflasi dan Perlunya Statistika Geometri*, (MANAJEMEN, Feb 1999),

28)**Multiplication of Money**, to Overcome the Crisis. *Melipat Gandakan Uang untuk Mengusir Krisis*, (MANAJEMEN, Apr 1999),

29)**The Interactive Mathematical Model**, for Multi Constraints Financial Budgeting, *Model Matematika Ekonomi Interaktif bagi Penganggaran Keuangan Multi Kendala.* (SCRIPTA ECONOMICA, Vol. 2, April 1999)

30)**Application of Ethics**, for the Management Accounting Practices. *Aplikasi Etika dalam Praktek Akuntansi Manajemen*, (MANAJEMEN, May 1999),

31)**The Curriculum of Magistrate of Management in Finance**: Its Needs and Practices. *Kurikulum Magister Manajemen Keuangan 1988-1998: Antara Kebutuhan dan Praktek*, (MANAJEMEN, Jun 1999),

32)**The Attractiveness**, Personal Power and Management. *Pesona Pribadi: "Personal Power" dan Manajemen*, (MANAJEMEN, Jul 1999),

33)**The Market Research**, for Magistrate Degree in Management. *Soal Jawab Riset Pasar untuk Pascasarjana Manajemen* (126 Pages, 1999),

34)**The Thesis Manual**, Question and Answer for Magistrate Degree in Management. *Soal Jawab Tugas Akhir untuk Pascasarjana Manajemen* (126 Pages, 1999),

35)**The Risk Management**, Question and Answers for Magistrate Degree in Management. *Soal Jawab Manajemen Resiko untuk Pascasarjana Manajemen* (126 Pages, 1999),

Corporate IFRS-GAAP (B/S-I/S), ISBN 13: **978-1983566332**, ISBN 10: **1983566330**

36) **The Credit Manual**, Question and Answers for Magistrate Degree in Management. *Soal Jawab Manajemen Kredit untuk Pascasarjana Manajemen* (126 Pages, 1999),

37) **Enhancing the Morality Structure**. *Solidkan Struktur Moralitas*, (MANAJEMEN, Oct 1999)

38) **Performance of PT. Astra Otoparts**, Plc, in its Press Releases and Public Financial Reports. *Kinerja PT. Astra Otoparts dalam Siaran Pers dan Laporan Keuangan Publik*, (MANAJEMEN, Oct 1999)

39) **The Management of Conflicts**, for the New Indonesian's "Rainbow Colored" Cabinet. *Manajemen Konflik dan Kabinet Pelangi*, (MANAJEMEN, Dec 1999),

40) **Technoeconometric**: An Engineering Elaboration to Matrix Multi Variate Econometrics, *Tekno Ekonometrik sebagai sumbangan teknologi untuk Ekonometrika Matrik Multi Variat.* (MANAGEMENT EXPOSE, November 1999),

41) **Technoeconometric** : A Mathematical Shortcut for Non Linear Bivariate Econometrics, *Tekno Ekonometrik sebagai Usulan Jalan Pintas Matematik untuk Ekonometrika non Linear Bivariat.* (SCRIPTA ECONOMICA, Vol. 2, No. 3, Desember 1999),

42) **Developing "New Image"**, for the New Indonesia. *Membangun Citra Indonesia Baru*, (MANAJEMEN, Jan 2000),

43) **Segregations of Duties**, the Needs of Both Indonesian Governments and Private Sectors. *Pemerintah dan Swasta Perlu Berbagi Tugas*, MANAJEMEN, Mar 2000),

44) **The 812.500 Financial Models**, for the Autonomous Country Governments. *Tentang 812.500 Model Keuangan Otonomi Daerah*, (MANAJEMEN, Mar 2000),

45) **The Forecast of 10 Most Attractive Share Values**, in the Jakarta Stock Exchange, April 2000. *Prakiraan Harga Sepuluh Saham Unggulan di Bursa Efek Jakarta, April 2000*, (MANAJEMEN, Mar 2000),

46) **Let Us Merchandise the Intelectual Properties**!. *Jual "Intelectual Property" Saja!* (MANAJEMEN, Apr 2000),

47) **The Forecast of ten Most Attractive Share Values**, in the Jakarta Stock Exchange May 2000. *Prakiraan Harga Sepuluh Saham Unggulan di Bursa Efek Jakarta Mei 2000*, (MANAJEMEN, Apr 2000),

48) **HEATHROW**, London International Airport, Second Runway, *Q-Basic Algorithmic Engineer*, Beasley-Krishnamoorthy, Imperial College, Cambridge, Post Doctoral Project (2000),

49) **Why the Banking Sector Should be Healthy**, Before Others?. *Mengapa Perbankan Harus Sehat?* (MANAJEMEN, Mei 2000),

50) **Recycling of Wastes as a Negative Costing**. *Memanfaatkan Sampah atau Negative Costing*, (MANAJEMEN, Agt 2000),

51) **The Guidance for Indonesia Corporate Restructuring**. *Kiat Restrukturisasi Perusahaan Indonesia*, (MANAJEMEN, Agt 2000),

Corporate IFRS-GAAP (B/S-I/S), ISBN 13: **978-1983566332**, ISBN 10: **1983566330**

52)<u>Redefinition of the Ethical Comprehension</u>. *Pemaknaan Kembali atas Pemahaman Etika,* (MANAJEMEN, Sep 2000),

53)<u>The Critical View the 1999's Indonesians Central Bank Financial Statement</u>. *Analisa Laporan Keuangan Bank Indonesia 1999,* (MANAJEMEN, Oct 2000),

54)<u>Becoming an "Achiever"</u>, not Just Common "Worker". *Bukan Sekadar "Kerja", Tapi Menghasilkan Karya!* (MANAJEMEN, Feb 2001),

55)<u>Ten Steps to Real Wealth</u>! *Sepuluh Langkah Menjadi Kaya,* (MANAJEMEN, Jun 2001),

56)<u>Creative Financial Breakthrough</u>, *Seminar Sehari tentang Terobosan Keuangan yang Kreatif,* (Gramedia-PPM dan MANAJEMEN, Jun 2001),

57)<u>The Marketing Practices of Financial Softwares</u>, for Publics Listed Corporation. *Praktek Pemasaran Program Perangkat Lunak untuk Keuangan Perusahaan Publik* (Univ Satyagama, Jakarta Juni 24, 2001).

58)<u>The Mathematical Financial Planning</u>, *Perancangan Keuangan Secara Eksak dengan Matematik.* (MANAJEMEN, Jul 2001),

59)<u>The Real Research and Development, not Rework and Destruction</u>. *Litbang yang Bukan "Sulit Berkembang",* (MANAJEMEN, Jul 2001),

60)<u>The Treatise on the Advantages and Disadvantages</u>, of Revolutionary Learning Methods. *Baik Buruknya: Revolusi Cara Belajar,* (MANAJEMEN, Jul 2001),

61)<u>Better Make an Indonesian "Corruption Prevention" Body</u>, not Just a "Corruption Watch". *Buat Indonesian Corruption Prevention (ICP), Bukan Indonesian Corruption Watch, ICW!* (MANAJEMEN, Okt 2001),

62)<u>Preparation of the Worlds Business Leaders</u>, for our Century and the Century After. *Mempersiapkan Manusia-manusia yang akan menjadi Khafilah Bisnis Dunia Abad ini dan Abad Mendatang.* (Univ Satyagama – Gregorio Araneta Univ, Jakarta, Oktober 27, 2001).

63)<u>The Good Corporate Governance</u>, is not just a Morale Condition. *Kepemerintahan Perusahaan yang Baik, Bukan Sekedar Moralitas,* (MANAJEMEN, Nov 2001),

64)<u>Corporate Healthiness and Cleanliness as a Challenge</u>. *Bersih, Sehat Perusahaan, sebuah Tantangan* (MANAJEMEN, Jan 2002),

65)<u>Developing the Cultural Base for Future Competitions</u>. *Budaya untuk Bersaing di Masa Depan* (MANAJEMEN, Jan 2002),

66)<u>"Catapult", "Bullet" or "Ballistic Missiles" Approach</u>, for Corporate Healthiness and Cleanliness. *Pola "Ketapel", "Peluru Biasa" atau "Rudal" untuk Keuangan Perusahaan yang Bersih Sehat.* (MANAJEMEN, Feb 2002),

67)<u>The Mathematical Invesment Solution Model</u>, for Indonesian Foreign Debt and Its Corruption, *Solusi Matematika Investasi bagi Penyelesaian Masalah Utang Luar Negeri dan Korupsi di Indonesia.* (MANAGEMENT EXPOSE, Mar 2002),

Corporate IFRS-GAAP (B/S-I/S), ISBN 13: **978-1983566332**, ISBN 10: **1983566330**

68)<u>The Attitude, Skills and Knowledge (ASK)</u>, in Educational System Approach. *Sikap, Keterampilan dan Pengetahuan Menurut Pendekatan Sistemik Pendidikan,* (MANAJEMEN, Apr 2002),

69)<u>Strategic Treatise on Regional Autonomy</u>, in Futurization, Digitalization, Securitization and Globalization Era in Accordance to the Awakening of the Mercantile Nations. *Kajian Strategik Otonomi Daerah : Era Futurisasi, Digitalisasi, Strukturisasi, dan Globalisasi dalam Kebangkitan Negara-Negara Dagang.* (Training for Provincial Government Regional Management, Indonesian Association of Provincial Governments, *Pelatihan Pengembangan Manajemen Pemerintah Daerah bagi Perangkat Pemerintah Propinsi, Angkatan II,* (APPSI), June 27, 2002.

70)<u>The Suggestion for the Doctoral Course</u>, for Islamic Financial Management System. *Usulan-usulan Perbaikan dalam Kurikulum Doktor Manajemen Keuangan Islam.* (December, 17-19, 2002, Canadian International Development Agency - CIDA, ADSGM, Universiti Utara Malaysia, Kedah, Da'arul Sintok).

71)<u>The Defensive Strategy for the Market Leader</u>, a Nokia Cellular Telephone's Case. *Strategi Bertahan Sang Pemimpin Pasar dengan Menggunakan Multi Merk, Kasus Telepon Seluler Nokia.* (March, 12, 2004, Univ Tarumanegara, Jakarta).

Corporate IFRS-GAAP (B/S-I/S), ISBN 13: **978-1983566332**, ISBN 10: **1983566330**

Finance Construction-6, *Tim Asikin, Steve Asikin, Indra Senihardja*

Corporate IFRS-GAAP (B/S-I/S), ISBN 13: **978-1983566332**, ISBN 10: **1983566330**

Finance Construction-6, *Tim Asikin, Steve Asikin, Indra Senihardja*

Corporate IFRS-GAAP (B/S-I/S), ISBN 13: **978-1983566332**, ISBN 10: **1983566330**

Finance Construction-6, *Tim Asikin, Steve Asikin, Indra Senihardja*

Corporate IFRS-GAAP (B/S-I/S), ISBN 13: **978-1983566332**, ISBN 10: **1983566330**

Finance Construction-6, *Tim Asikin, Steve Asikin, Indra Senihardja*

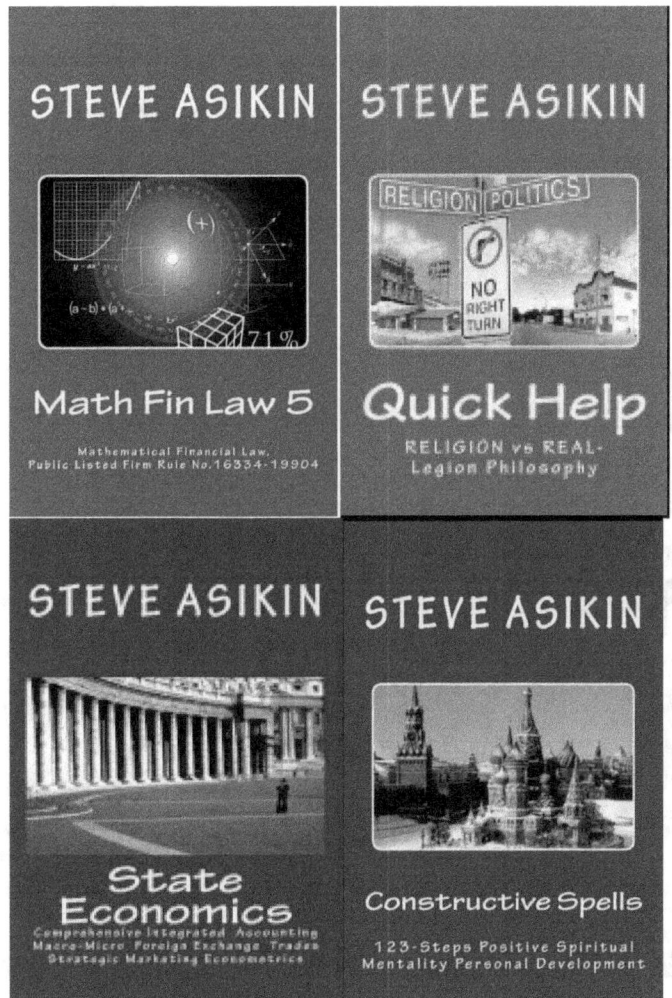

Corporate IFRS-GAAP (B/S-I/S), ISBN 13: **978-1983566332**, ISBN 10: **1983566330**

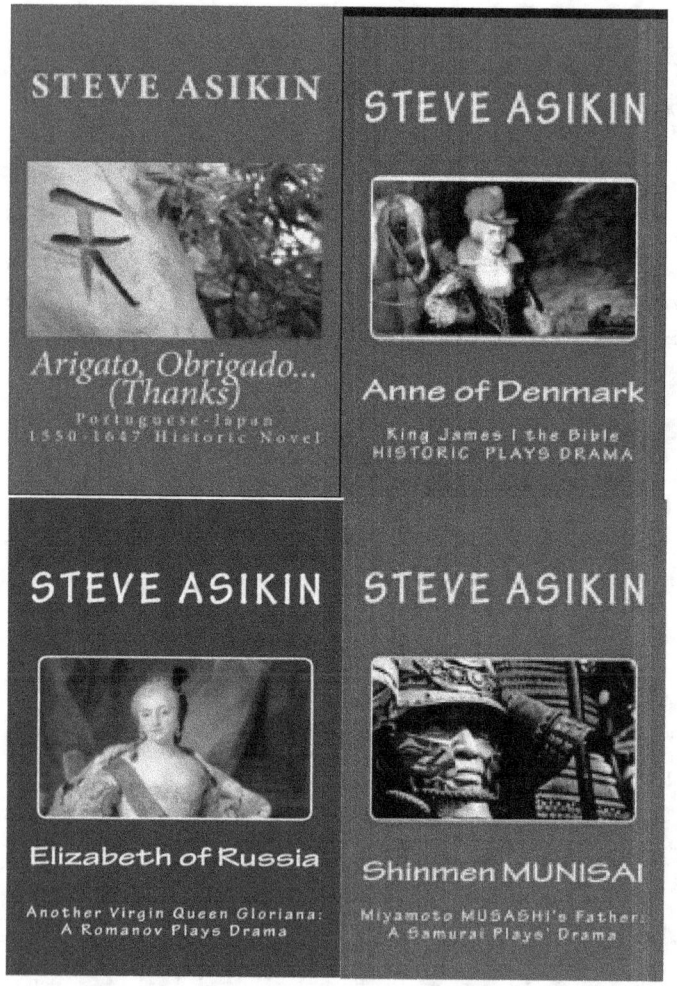

Corporate IFRS-GAAP (B/S-I/S), ISBN 13: **978-1983566332**, ISBN 10: **1983566330**

Corporate IFRS-GAAP (B/S-I/S), ISBN 13: **978-1983566332**, ISBN 10: **1983566330**

Finance Construction-6, *Tim Asikin, Steve Asikin, Indra Senihardja*

Corporate IFRS-GAAP (B/S-I/S), ISBN 13: **978-1983566332**, ISBN 10: **1983566330**

Finance Construction-6, *Tim Asikin, Steve Asikin, Indra Senihardja*

Corporate IFRS-GAAP (B/S-I/S), ISBN 13: **978-1983566332**, ISBN 10: **1983566330**

Finance Construction-6, *Tim Asikin, Steve Asikin, Indra Senihardja*

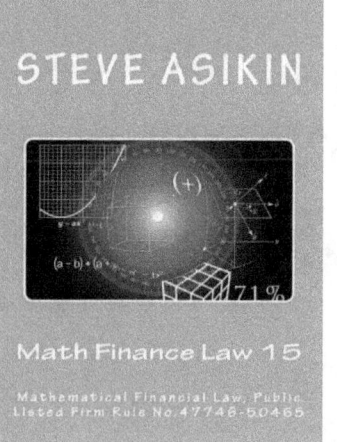

Corporate IFRS-GAAP (B/S-I/S), ISBN 13: **978-1983566332**, ISBN 10: **1983566330**

Finance Construction-6, *Tim Asikin, Steve Asikin, Indra Senihardja*

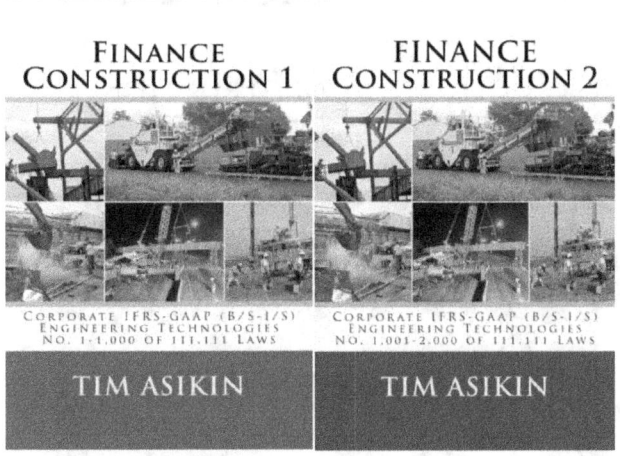

522
Corporate IFRS-GAAP (B/S-I/S), ISBN 13: **978-1983566332**, ISBN 10: **1983566330**

Finance Construction-6, *Tim Asikin, Steve Asikin, Indra Senihardja*

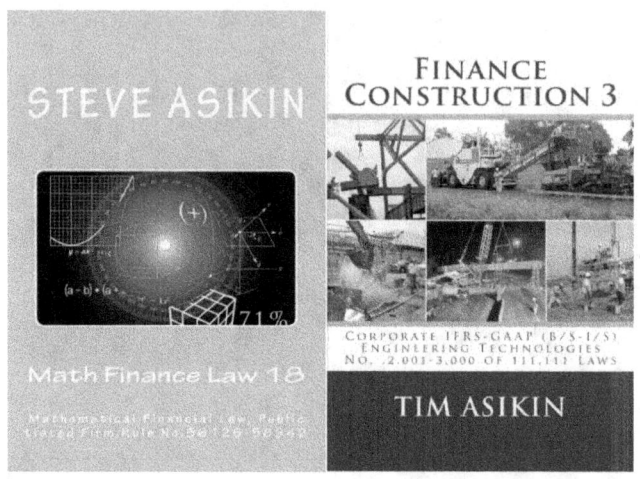

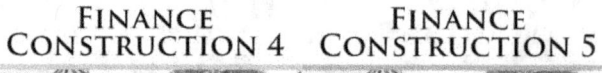

Corporate IFRS-GAAP (B/S-I/S), ISBN 13: **978-1983566332**, ISBN 10: **1983566330**

www.ingramcontent.com/pod-product-compliance
Lightning Source LLC
Chambersburg PA
CBHW071246220526
45468CB00001B/11